STUDENT SOLUTIONS

SEARS & ZEMANSKY'S

COLLEGE PHYSICS

9TH EDITION

HUGH D. YOUNG

Forrest Newman
Sacramento City College

PEARSON

Boston Columbus Indianapolis New York San Francisco Upper Saddle River
Amsterdam Cape Town Dubai London Madrid Milan Munich Paris Montreal Toronto
Delhi Mexico City São Paulo Sydney Hong Kong Seoul Singapore Taipei Tokyo

Executive Editor:	Nancy Whilton
Senior Project Editor:	Katie Conley
Editorial Manager:	Laura Kenney
Managing Editor:	Corinne Benson
Production Project Manager:	Beth Collins
Production Management and Compositor:	PreMediaGlobal
Manufacturing Buyer:	Jeffrey Sargent
Senior Marketing Manager:	Kerry Chapman
Cover Design:	Derek Bacchus and Seventeenth Street Design
Cover Photo Credits:	Mike Kemp/Rubberball/Corbis
Cover and Text Printer:	Bind-Rite Graphics

ISBN 10: 0-321-74769-0
ISBN 13: 978-0-321-74769-3

1 2 3 4 5 6 7 8 9 10—BR—14 13 12 11

CONTENTS

Electricity and Magnetism

Light and Optics

Modern Physics

PREFACE

This Student Solutions Manual contains detailed solutions for select odd-numbered end-of-chapter problems. The problems included were carefully chosen to include at least one representative example of each problem type. The remaining problems, for which solutions are not given here, constitute an ample set of problems for students to tackle on their own. All solutions are written in the Set Up/Solve/Reflect framework used in the textbook. In most cases, rounding was done in intermediate steps, so you may obtain slightly different results if you handle the rounding differently. This manual greatly expands the set of worked-out examples that accompany the presentation of physics laws and concepts in the text. It was written to provide students with models to follow in working physics problems. The problems are worked out in the manner and style in which students should work out their own problem solutions.

We have made every effort to be accurate and correct in the solutions, but if you find errors or ambiguities it would be very helpful if you would point these out to the publisher.

MATHEMATICS REVIEW

Problems 3, 7, 11, 13, 17, 21, 23, 25, 29, 31

*0.3. Set Up and Solve: $\left(\dfrac{4x^2}{2y^3}\right)^2 = \dfrac{4^2(x^2)^2}{2^2(y^3)^2} = \dfrac{16x^4}{4y^6} = 4x^4y^{-6}.$

Reflect: The answer could also be written as $\dfrac{4x^4}{y^6}$.

*0.7. Set Up and Solve: The decimal point must be moved 2 places to the left to change 123 into a number between 1 and 10. Thus, we have $123 \times 10^{-6} = 1.23 \times 10^2 \times 10^{-6} = 1.23 \times 10^{-4}$.

Reflect: Alternatively, we could have written our original number in decimal form by moving the decimal point 6 places to the left to obtain $123 \times 10^{-6} = 0.000123$. Finally we could convert the result into scientific notation by moving the decimal point 4 places to the right to obtain $0.000123 = 1.23 \times 10^{-4}$.

*0.11. Set Up and Solve: Add $-3x^2 - 6$ to both sides to obtain $x^2 = 12$. Next, take the positive and negative square root of both sides to obtain $x = \pm\sqrt{12} = \pm\sqrt{(3)(4)} = \pm 2\sqrt{3}$.

Reflect: Since this quadratic has no terms containing x, it is possible to solve it without factoring or using the quadratic formula.

*0.13. Set Up and Solve: Notice that $-2 + -3 = -5$ and $(-2)(-3) = 6$. Thus, we can factor the equation as $x^2 - 5x + 6 = (x - 3)(x - 2) = 0$. The two roots are $x = 3$ and $x = 2$.

Reflect: Alternatively, we can use the quadratic formula with $a = 1$, $b = -5$, and $c = 6$ to obtain $x = \dfrac{-b \pm \sqrt{b^2 - 4ac}}{2a} = \dfrac{5 \pm \sqrt{(-5)^2 - 4(1)(6)}}{2(1)} = \dfrac{5 \pm 1}{2} = 3$ or 2.

*0.17. Set Up and Solve: Divide the second equation by 3 to obtain $\dfrac{2x}{3} - \dfrac{y}{3} = \dfrac{1}{3}$. Add this result to the first equation to obtain $\left(\dfrac{x}{2} + \dfrac{y}{3}\right) + \left(\dfrac{2x}{3} - \dfrac{y}{3}\right) = 2 + \dfrac{1}{3}$. The y terms cancel and we can clear the fractions by multiplying both sides by 6. Thus, we obtain $3x + 4x = 12 + 2$. Solving for x we obtain $x = 2$. By substituting this value for x into the second equation we obtain $2(2) - y = 1$, so $y = 3$. Thus, the solution is $x = 2$, $y = 3$.

Reflect: We could also solve this equation by solving the second equation for y and substituting the result into the first equation.

*0.21. Set Up and Solve: Let A be the amplitude of the sound and d the distance from the source. Thus, $A \propto \dfrac{1}{d}$

and we have $Ad = (4.8 \times 10^{-6}$ m$)(1.0$ m$)$ or $A = \dfrac{(4.8 \times 10^{-6}$ m$^2)}{d}$. Thus, when $d = 4.0$ m we have $A = 1.2 \times 10^{-6}$ m.

Reflect: Since the distance is increased by a factor of 4, the amplitude is decreased by a factor of 4 (i.e., multiplied by $\dfrac{1}{4}$).

*0.23. Set Up and Solve: Let F be the force of gravity on an object that is a distance d from the earth's center. Thus, we have $F \propto \dfrac{1}{d^2}$ or $Fd^2 = k$. Also, when $d = 6.38 \times 10^6$ m we know that $F = 700$ N so we can find the value of k. When the astronaut is 6000 km (6×10^6 m) from the earth's surface he is $6.38 \times 10^6 + 6.00 \times 10^6 = 1.238 \times 10^7$ m from the center of the earth. Thus, we have $F = \dfrac{k}{d^2} = \dfrac{(700 \text{ N})(6.38 \times 10^6 \text{ m})^2}{(1.238 \times 10^7 \text{ m})^2} = 186$ N.

Reflect: We could also solve this problem using the ratio equation $\dfrac{F_2}{F_1} = \left(\dfrac{d_1}{d_2}\right)^2$. Since the astronaut nearly doubles his distance from the earth's center as he moves from the surface into orbit, his weight decreases by nearly a factor of 4 (i.e., $\dfrac{1}{2^2}$).

*0.25. Set Up and Solve: **(a)** $4\log x + \log y - 3\log(x+y) = \log(x^4) + \log y - \log[(x+y)^3] = \log\left[\dfrac{x^4 y}{(x+y)^3}\right]$

(b) $\log(xy + x^2) - \log(xz + yz) + 2\log z = \log\left(\dfrac{x(y+x)}{z(x+y)}\right) + \log(z^2) = \log(xz)$

Reflect: There are multiple ways to combine each of these expressions.

*0.29. Set Up and Solve: **(a)** The circumference of the circle is given by $C = 2\pi r = 2\pi(0.12$ m$) = 0.75$ m and its area is given by $A = \pi r^2 = \pi(0.12$ m$)^2 = 0.045$ m^2.

(b) The surface area of the sphere is given by $A = 4\pi r^2 = 4\pi(0.21$ m$)^2 = 0.55$ m^2 and its volume is given by $\dfrac{4}{3}\pi r^3 = \dfrac{4}{3}\pi(0.21$ m$)^3 = 0.039$ m^3.

(c) The total surface area of the rectangular solid (which consists of three identical pairs of rectangular faces) is $2lw + 2lh + 2wh = 2[(0.18$ m$)(0.15$ m$) + (0.18$ m$)(0.8$ m$) + (0.15$ m$)(0.8$ m$)] = 0.6$ m^2 (rounding to the nearest tenth). Its volume is $V = lwh = (0.18$ m$)(0.15$ m$)(0.8$ m$) = 0.02$ m^3 (rounding to the nearest hundredth).

(d) The total surface area of the cylinder consists of its lateral (side) surface area and the area of its two circular end-caps: $A = 2\pi rh + 2\pi r^2 = 2\pi[(0.18$ m$)(0.33$ m$) + (0.18$ m$)^2] = 0.58$ m^2. The volume of the cylinder is given by $V = \pi r^2 h = \pi(0.18$ m$)^2(0.33$ m$) = 0.034$ m^3.

Reflect: When the lateral surface of a cylinder of radius r and height h is unrolled, it forms a rectangle with dimensions $2\pi r$ by h, so it has a surface area of $A = 2\pi rh$. The area of each of its two circular end-caps is πr^2, so the total surface area of a cylinder is $A = 2\pi rh + 2\pi r^2$.

*0.31. Set Up and Solve: Assume that each step forms a right angle so that a line drawn between the corners of adjacent steps forms a right triangle with a base of 30 cm and a height of h. The angle between the hypotenuse and the base is $36°$. Thus, $\tan 36° = \dfrac{h}{30 \text{ cm}}$, or $h = (30 \text{ cm})\tan 36° = 22 \text{ cm}$.

Reflect: A line drawn from the base of the stairs to the top of the stairs has a slope of $\dfrac{22}{30}$, which is also equal to $\tan 36°$.

MODELS, MEASUREMENTS, AND VECTORS

Problems 1, 5, 9, 13, 17, 19, 23, 27, 31, 33, 35, 39, 41, 43, 47, 49, 53, 55, 59, 61, 65, 67, 69

Solutions to Problems

*1.1. Set Up: We know the following equalities: $1 \text{ mg} = 10^{-3}$ g; $1 \mu g = 10^{-6}$ g; 1 kilohms $= 1000$ ohms; and $1 \text{ milliamp} = 10^{-3}$ amp.

Solve: In each case multiply the quantity to be converted by unity, expressed in different units. Construct an expression for unity so that the units to be changed cancel and we are left with the new desired units.

(a) $(2400 \text{ mg/day}) \left(\dfrac{10^{-3} \text{ g}}{1 \text{ mg}} \right) = 2.40 \text{ g/day} \sqrt{b^2 - 4ac}$

(b) $(120 \ \mu g/day) \left(\dfrac{10^{-6} \text{ g}}{1 \ \mu g} \right) = 1.20 \times 10^{-4} \text{ g/day}$

(c) $(500 \text{ mg/day}) \left(\dfrac{10^{-3} \text{ g}}{1 \text{ mg}} \right) = 0.500 \text{ g/day}$

(d) $(1500 \text{ ohms}) \left(\dfrac{1 \text{ kilohm}}{10^3 \text{ ohms}} \right) = 1.50 \text{ kilohms}$

(e) $(0.020 \text{ amp}) = \left(\dfrac{1 \text{ milliamp}}{10^{-3} \text{ amp}} \right) = 20 \text{ milliamps}$

Reflect: In each case, the number representing the quantity is larger when expressed in the smaller unit. For example, it takes more milligrams to express a mass than to express the mass in grams.

*1.5. Set Up: We need to apply the following conversion equalities: $1000 \text{ g} = 1.00 \text{ kg}$, $100 \text{ cm} = 1.00 \text{ m}$, and $1.00 \text{ L} = 1000 \text{ cm}^3$.

Solve: (a) $(1.00 \text{ g/cm}^3) \left(\dfrac{1.00 \text{ kg}}{1000 \text{ g}} \right) \left(\dfrac{100 \text{ cm}}{1.00 \text{ m}} \right)^3 = 1000 \text{ kg/m}^3$

(b) $(1050 \text{ kg/m}^3) \left(\dfrac{1000 \text{ g}}{1.00 \text{ kg}} \right) \left(\dfrac{1.00 \text{ m}}{1000 \text{ cm}} \right)^3 = 1.05 \text{ g/cm}^3$

(c) $(1.00 \text{ L}) \left(\dfrac{1000 \text{ cm}^3}{1.00 \text{ L}} \right) \left(\dfrac{1.00 \text{ g}}{1.00 \text{ cm}^3} \right) \left(\dfrac{1.00 \text{ kg}}{1000 \text{ g}} \right) = 1.00 \text{ kg};$ $(1.00 \text{ kg}) \left(\dfrac{2.205 \text{ lb}}{1.00 \text{ kg}} \right) = 2.20 \text{ lb}$

Reflect: We could express the density of water as 1.00 kg/L.

*1.9. Set Up: We apply the equalities of 1 km = 0.6214 mi and 1 gal = 3.788 L.

Solve: $(37.5 \text{ mi/gal}) \left(\dfrac{1 \text{ km}}{0.6214 \text{ mi}} \right) \left(\dfrac{1 \text{ gal}}{3.788 \text{ L}} \right) = 15.9 \text{ km/L}$

Reflect: Note how the unit conversion strategy, of cancellation of units, automatically tells us whether to multiply or divide by the conversion factor.

*1.13. Set Up: We apply the basic time relations of 1 h = 60 min and 1 min = 60 s in part (a) and use the result of (a) in part (b). Similarly, apply the result of part (b) in solving part (c).

Solve: (a) $(1 \text{ h}) \left(\dfrac{60 \text{ min}}{1 \text{ h}} \right) \left(\dfrac{60 \text{ s}}{1 \text{ min}} \right) = 3600 \text{ s}$

(b) $(24 \text{ h/day}) \left[(3600 \text{ s})/(1 \text{ h}) \right] = 86.4 \times 10^3 \text{ s/day}$

(c) $(365 \text{ day/yr}) [(86.4 \times 10^3 \text{ s})/(1 \text{ day})] = 31{,}536{,}000 \text{ s/yr} = 3.15 \times 10^7 \text{ s/yr}$

*1.17. Set Up: Calculate the trig function with the maximum and minimum possible values of the angle.

Solve: (a) $\cos 3.5° = 0.9981$ and $\cos 4.5° = 0.9969$, so the range is 0.9969 to 0.9981.

(b) $\sin 3.5° = 0.061$ and $\sin 4.5° = 0.078$, so the range is 0.061 to 0.078.

(c) $\tan 3.5° = 0.061$ and $\tan 4.5° = 0.079$, so the range is 0.061 to 0.079.

*1.19. Set Up: We are given the relation density = mass/volume = m/V where $V = \frac{4}{3}\pi r^3$ for a sphere. From Appendix F, the earth has mass of $m = 5.97 \times 10^{24}$ kg and a radius of $r = 6.38 \times 10^6$ m whereas for the sun at the end of its lifetime, $m = 1.99 \times 10^{30}$ kg and $r = 7500 \text{ km} = 7.5 \times 10^6$ m. The star possesses a radius of $r = 10 \text{ km} = 1.0 \times 10^4$ m and a mass of $m = 1.99 \times 10^{30}$ kg.

Solve: (a) The earth has volume $V = \frac{4}{3}\pi r^3 = \frac{4}{3}\pi (6.38 \times 10^6 \text{ m})^3 = 1.088 \times 10^{21} \text{ m}^3$.

$$\text{density} = \frac{m}{V} = \frac{5.97 \times 10^{24} \text{ kg}}{1.088 \times 10^{21} \text{ m}^3} = (5.49 \times 10^3 \text{ kg/m}^3) \left(\frac{10^3 \text{ g}}{1 \text{ kg}} \right) \left(\frac{1 \text{ m}}{10^2 \text{ cm}} \right)^3 = 5.49 \text{ g/cm}^3$$

(b) $V = \frac{4}{3}\pi r^3 = \frac{4}{3}\pi (7.5 \times 10^6 \text{ m})^3 = 1.77 \times 10^{21} \text{ m}^3$

$$\text{density} = \frac{m}{V} = \frac{1.99 \times 10^{30} \text{ kg}}{1.77 \times 10^{21} \text{ m}^3} = (1.1 \times 10^9 \text{ kg/m}^3) \left(\frac{1 \text{ g/cm}^3}{1000 \text{ kg/m}^3} \right) = 1.1 \times 10^6 \text{ g/cm}^3$$

(c) $V = \frac{4}{3}\pi r^3 = \frac{4}{3}\pi (1.0 \times 10^4 \text{ m})^3 = 4.19 \times 10^{12} \text{ m}^3$

$$\text{density} = \frac{m}{V} = \frac{1.99 \times 10^{30} \text{ kg}}{4.19 \times 10^{12} \text{ m}^3} = (4.7 \times 10^{17} \text{ kg/m}^3) \left(\frac{1 \text{ g/cm}^3}{1000 \text{ kg/m}^3} \right) = 4.7 \times 10^{14} \text{ g/cm}^3$$

Reflect: For a fixed mass, the density scales as $1/r^3$. Thus, the answer to (c) can also be obtained from (b) as

$$(1.1 \times 10^6 \text{ g/cm}^3) \left(\frac{7.50 \times 10^6 \text{ m}}{1.0 \times 10^4 \text{ m}} \right)^3 = 4.7 \times 10^{14} \text{ g/cm}^3.$$

*1.23. Set Up: The mass can be calculated as the product of the density and volume. The volume of the washer is the volume V_d of a solid disk of radius r_d minus the volume V_h of the disk-shaped hole of radius r_h. In general, the volume of a disk of radius r and thickness t is $\pi r^2 t$. We also need to apply the unit conversions $1 \text{ m}^3 = 10^6 \text{ cm}^3$ and $1 \text{ g/cm}^3 = 10^3 \text{ kg/m}^3$.

Solve: The volume of the washer is:

$$V = V_d - V_h = \pi(r_d{}^2 - r_h{}^2)t = \pi[(2.25 \text{ cm})^2 - (0.625 \text{ cm})^2](0.150 \text{ cm}) = 2.20 \text{ cm}^3$$

The density of the washer material is $8600 \text{ kg/m}^3[(1 \text{ g/cm}^3)/(10^3 \text{ kg/m}^3)] = 8.60 \text{ g/cm}^3$. Finally, the mass of the washer is: $\text{mass} = (\text{density})(\text{volume}) = (8.60 \text{ g/cm}^3)(2.20 \text{ cm}^3) = 18.9 \text{ g}$.

Reflect: This mass corresponds to a weight of about 0.7 oz, a reasonable value for a washer.

*1.27. Set Up: Estimate that we blink 10 times per minute and a typical lifetime is 80 years. Convert this into blinks per lifetime using $1 \text{ y} = 365 \text{ days}$, $1 \text{ day} = 24 \text{ h}$, $1 \text{ h} = 60 \text{ min}$.

Solve: The number of blinks is $(10 \text{ per min})\left(\dfrac{60 \text{ min}}{1 \text{ h}}\right)\left(\dfrac{24 \text{ h}}{1 \text{ day}}\right)\left(\dfrac{365 \text{ days}}{1 \text{ y}}\right)(80 \text{ y/lifetime}) = 4 \times 10^8$

Reflect: Our estimate of the number of blinks per minute can be off by a factor of two but our calculation is surely accurate to a power of 10.

*1.31. Set Up: An average middle-aged (40 year-old) adult at rest has a heart rate of roughly 75 beats per minute. To calculate the number of beats in a lifetime, use the current average lifespan of 80 years. The volume of blood pumped during this interval is then the volume per beat multiplied by the total beats.

Solve: $N_\text{beats} = (75 \text{ beats/min})\left(\dfrac{60 \text{ min}}{1 \text{ h}}\right)\left(\dfrac{24 \text{ h}}{1 \text{ day}}\right)\left(\dfrac{365 \text{ days}}{\text{yr}}\right)\left(\dfrac{80 \text{ yr}}{\text{lifespan}}\right) = 3 \times 10^9 \text{ beats/lifespan}$

$$V_\text{blood} = (50 \text{ cm}^3/\text{beat})\left(\dfrac{1 \text{ L}}{1000 \text{ cm}^3}\right)\left(\dfrac{1 \text{ gal}}{3.788 \text{ L}}\right)\left(\dfrac{3 \times 10^9 \text{ beats}}{\text{lifespan}}\right) = 4 \times 10^7 \text{ gal/lifespan}$$

*1.33. Set Up: The cost would equal the number of dollar bills required; the surface area of the U.S. divided by the surface area of a single dollar bill. By drawing a rectangle on a map of the U.S., the approximate area is 2600 mi by 1300 mi or $3,380,000 \text{ mi}^2$. This estimate is within 10 percent of the actual area, $3,794,083 \text{ mi}^2$. The population is roughly 3.0×10^8 while the area of a dollar bill, as measured with a ruler, is approximately $6\frac{1}{8}$ in. by $2\frac{5}{8}$ in.

Solve: $A_\text{U.S.} = (3,380,000 \text{ mi}^2)[(5280 \text{ ft})/(1 \text{ mi})]^2[(12 \text{ in.})/(1 \text{ ft})]^2 = 1.4 \times 10^{16} \text{ in.}^2$

$$A_\text{bill} = (6.125 \text{ in.})(2.625 \text{ in.}) = 16.1 \text{ in.}^2$$

$$\text{Total cost} = N_\text{bills} = A_\text{U.S.}/A_\text{bill} = (1.4 \times 10^{16} \text{ in.}^2)/(16.1 \text{ in.}^2/\text{bill}) = 9 \times 10^{14} \text{ bills}$$

$$\text{Cost per person} = (9 \times 10^{14} \text{ dollars})/(3.0 \times 10^8 \text{ persons}) = 3 \times 10^6 \text{ dollars/person}$$

Reflect: The actual cost would be somewhat larger, because the land isn't flat.

*1.35. Set Up: The sum with the largest magnitude is when the two displacements are parallel and the sum with the smallest magnitude is when the two displacements are antiparallel.

Solve: The orientations of the displacements that give the desired sum are shown in the figure below.

Reflect: The orientations of the two displacements can be chosen such that the sum has any value between 0.6 m and 4.2 m.

*1.39. Set Up: Use components to add the two forces. Take the $+x$-direction to be forward and the $+y$-direction to be upward.

Solve: The second force has components $F_{2x} = F_2 \cos 32.4° = 433$ N and $F_{2y} = F_2 \sin 32.4° = 275$ N. The first force has components $F_{1x} = 725$ N and $F_{1y} = 0$.

$$F_x = F_{1x} + F_{2x} = 1158 \text{ N} \quad \text{and} \quad F_y = F_{1y} + F_{2y} = 275 \text{ N}$$

The resultant force is 1190 N in the direction 13.4° above the forward direction.

Reflect: Since the two forces are not in the same direction the magnitude of their vector sum is less than the sum of their magnitudes.

*1.41. Set Up: In each case, create a sketch (see the figure below) showing the components and the resultant to determine the quadrant in which the resultant vector $\vec{A}$ lies. The component vectors add to give the resultant.

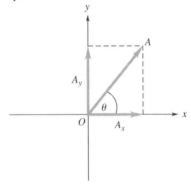

(a)

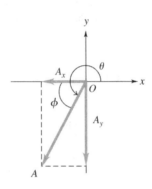

(b)

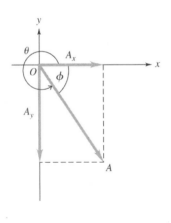

(c)

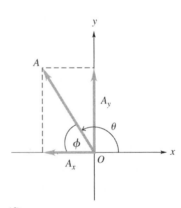

(d)

Solve: (a) $A = \sqrt{A_x{}^2 + A_y{}^2} = \sqrt{(4.0 \text{ m})^2 + (5.0 \text{ m})^2} = 6.4 \text{ m}; \quad \tan\theta = \dfrac{A_y}{A_x} = \dfrac{5.0 \text{ m}}{4.0 \text{ m}}$ and $\theta = 51°$.

(b) $A = \sqrt{(-3.0 \text{ km})^2 + (-6.0 \text{ km})^2} = 6.7 \text{ km}; \quad \tan\theta = \dfrac{-6.0 \text{ m}}{-3.0 \text{ m}}$ and $\theta = 243°$.

(My calculator gives $\phi = 63°$ and $\theta = \phi + 180°$.)

(c) $A = \sqrt{(9.0 \text{ m/s})^2 + (-17 \text{ m/s})^2} = 19 \text{ m/s}; \quad \tan\theta = \dfrac{-17 \text{ m/s}}{9 \text{ m/s}}$ and $\theta = 298°$.

(My calculator gives $\phi = -62°$ and $\theta = \phi + 360°$.)

(d) $A = \sqrt{(-8.0 \text{ N})^2 + (12 \text{ N})^2} = 14$ N. $\tan\theta = \dfrac{12 \text{ N}}{-8.0 \text{ N}}$ and $\theta = 124°$.

(My calculator gives $\phi = -56°$ and $\theta = \phi + 180°$.)

Reflect: The signs of the components determine the quadrant in which the resultant lies.

*1.43. Set Up: In each case, use a sketch (see the figure below) showing the components and the resultant in order to determine the quadrant in which the resultant vector $\overline{A}$ lies. The component vectors add to give the resultant.

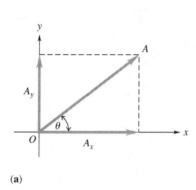

(a)

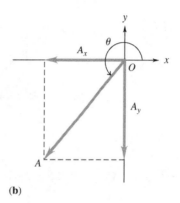

(b)

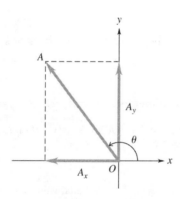

(c)

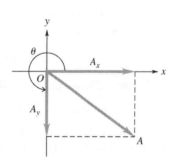

(d)

Solve: (a) $A = \sqrt{A_x^2 + A_y^2} = \sqrt{(8.0 \text{ lb})^2 + (6.0 \text{ lb})^2} = 10.0$ lb; $\tan\theta = \dfrac{A_y}{A_x} = \dfrac{6.0 \text{ lb}}{8.0 \text{ lb}}$ and $\theta = 37°$.

(b) $A = \sqrt{(-24 \text{ m/s})^2 + (-31 \text{ m/s})^2} = 39$ m/s; $\tan\theta = \dfrac{-31 \text{ m/s}}{-24 \text{ m/s}}$ and $\theta = 232°$

(My calculator gives $\phi = 52°$ and $\theta = \phi + 180°$.)

(c) $A = \sqrt{(-1500 \text{ km})^2 + (2000 \text{ km})^2} = 2500$ km; $\tan\theta = \dfrac{2000 \text{ km}}{-1500 \text{ km}}$ and $\theta = 127°$

(My calculator gives $\phi = -53°$ and $\theta = \phi + 180°$.)

(d) $A = \sqrt{(71.3 \text{ N})^2 + (-54.7 \text{ N})^2} = 89.9$ N; $\tan\theta = \dfrac{-54.7 \text{ N}}{71.3 \text{ N}}$ and $\theta = 323°$

(My calculator gives $\phi = -37°$ and $\theta = \phi + 360°$.)

*1.47. Set Up: We know that the two force vectors, $\vec{A}$ and $\vec{B}$, have the same magnitude $(A = B)$ and form a right angle; thus, the two forces and their resultant form an isosceles right triangle. Use coordinates for which the resultant force is parallel to the $+y$-direction and find the magnitude of A and B by setting the y-component of the resultant equal to 620 N.

Solve: (a) The two forces and their resultant are shown below.

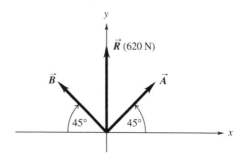

(b) $R_y = A_y + B_y = A\sin 45° + B\sin 135° = 2A\sin 45° = 620$ N. Thus, $A = \dfrac{620 \text{ N}}{2\sin 45°} = 440$ N.

Reflect: Note that $R_x = A_x + B_x = A\cos 45° + B\cos 135° = 0$.

*1.49. Set Up: Use coordinates for which $+x$ is east and $+y$ is north. Each of the professor's displacement vectors make an angle of 0° or 180° with one of these axes. The components of his total displacement can thus be calculated directly from $R_x = A_x + B_x + C_x$ and $R_y = A_y + B_y + C_y$.

Solve: (a) $R_x = A_x + B_x + C_x = 0 + (-4.75 \text{ km}) + 0 = -4.75$ km $= 4.75$ km west; $R_y = A_y + B_y + C_y = 3.25$ km $+ 0$ $+ (-1.50 \text{ km}) = 1.75$ km $= 1.75$ km north; $R = \sqrt{R_x^2 + R_y^2} = 5.06$ km; $\theta = \tan^{-1}(R_y/R_x) = \tan^{-1}[(+1.75)/(-4.75)] =$ $-20.2°$; $\phi = 180° - 20.2° = 69.8°$ west of north

(b) From the scaled sketch in the figure below, the graphical sum agrees with the calculated values.

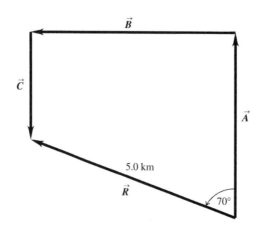

Reflect: The magnitude of his resultant displacement is very different from the distance he traveled, which is 159.50 km.

*1.53. Set Up: Estimate 12 breaths per minute. We know $1\,\text{day} = 24\,\text{h}$, $1\,\text{h} = 60\,\text{min}$, and $1000\,\text{L} = 1\,\text{m}^3$.

Solve: (a) $(12\ \text{breaths/min})\left(\dfrac{60\ \text{min}}{1\ \text{h}}\right)\left(\dfrac{24\ \text{h}}{1\ \text{day}}\right) = 17{,}280$ breaths/day. The volume of air breathed in one day is

$\left(\tfrac{1}{2}\ \text{L/breath}\right)(17{,}280\ \text{breaths/day}) = 8640\ \text{L} = 8.64\ \text{m}^3$.

The mass of air breathed in one day is the density of air times the volume of air breathed: $m = (1.29\ \text{kg/m}^3)(8.64\ \text{m}^3) = 11.1\ \text{kg}$. As 20% of this quantity is oxygen, the mass of oxygen breathed in 1 day is $(0.20)(11.1\ \text{kg}) = 2.2\ \text{kg} = 2200\ \text{g}$.

(b) $8.64\ \text{m}^3$ and $V = l^3$, so $l = V^{1/3} = 2.1\ \text{m}$.

Reflect: A person could not survive one day in a closed tank of this size because the exhaled air is breathed back into the tank and thus reduces the percent of oxygen in the air in the tank. That is, a person cannot extract all of the oxygen from the air in an enclosed space.

*1.55. Set Up: The volume V of blood pumped during each heartbeat is the total volume of blood in the body divided by the number of heartbeats in 1.0 min. We will need to apply $1\ \text{L} = 1000\ \text{cm}^3$.
Solve: The number of heartbeats in 1.0 min is 75. The volume of blood is thus:

$$V = \frac{5.0\ \text{L}}{75\ \text{heartbeats}} = 6.7 \times 10^{-2}\ \text{L} = 67\ \text{cm}^3.$$

*1.59. Set Up: Estimate the volume of each object. The mass, m, of an object is equal to its density times its volume: the volume of a sphere of radius r is $V = \tfrac{4}{3}\pi r^3$. The volume of a cylinder of radius r and length l is $V = \pi r^2 l$. The density of water is $1000\ \text{kg/m}^3$.

Solve: (a) Estimate the volume as that of a sphere of diameter 10 cm: $V = 5.2 \times 10^{-4}\ \text{m}^3$.
$m = (0.98)(1000\ \text{kg/m}^3)(5.2 \times 10^{-4}\ \text{m}^3) = 0.5\ \text{kg}$.

(b) Approximate as a sphere of radius $r = 0.25\ \mu\text{m}$ (probably an overestimate): $V = 6.5 \times 10^{-20}\ \text{m}^3$. So
$m = (0.98)(1000\ \text{kg/m}^3)(6.5 \times 10^{-20}\ \text{m}^3) = 6 \times 10^{-17}\ \text{kg} = 6 \times 10^{-14}\ \text{g}$.

(c) Estimate the volume as that of a cylinder of length 1 cm and radius 3 mm: $V = \pi r^2 l = 2.8 \times 10^{-7}\ \text{m}^3$. So
$m = (0.98)(1000\ \text{kg/m}^3)(2.8 \times 10^{-7}\ \text{m}^3) = 3 \times 10^{-4}\ \text{kg} = 0.3\ \text{g}$.
Reflect: The mass is directly proportional to the volume.

*1.61. Set Up: Use coordinates for which $+x$ is east and $+y$ is north. The spelunker's vector displacements are: $\vec{A} = 180\ \text{m}$, $0°$ of west; $\vec{B} = 210\ \text{m}$, $45°$ east of south; $\vec{C} = 280\ \text{m}$, $30°$ east of north; and the unknown displacement $\vec{D}$. The vector sum of these four displacements is zero.
Solve: $A_x + B_x + C_x + D_x = -180\ \text{m} + (210\ \text{m})(\sin 45°) + (280\ \text{m})(\sin 30°) + D_x = 0$ and $D_x = -108\ \text{m}$.

$A_y + B_y + C_y + D_y = (-210\ \text{m})(\cos 45°) + (280\ \text{m})(\cos 30°) + D_y = 0$ and $D_y = -94\ \text{m}$.

$$D = \sqrt{D_x{}^2 + D_y{}^2} = 143\ \text{m};\ \theta = \tan^{-1}[(-94\ \text{m})/(-108)] = 41°;\ \vec{D} = 143\ \text{m},\ 41°\ \text{south of west}.$$

This result is confirmed by the figure below.

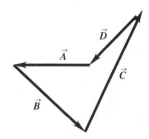

Reflect: We always add vectors by separately adding their x and y components.

*1.65. Set Up: Referring to the vector diagram in the figure below, the resultant weight of the forearm and the weight, $\vec{R}$, is 132.5 N upward while the biceps' force, $\vec{B}$, acts at 43° from the $+y$ axis. The force of the elbow is thus found as $\vec{E} = \vec{R} - \vec{B}$.

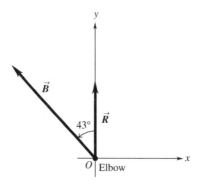

Solve: The components are: $E_x = R_x - B_x = 0 - (-232 \text{ N})(\sin 43°) = 158 \text{ N}$;

$E_y = R_y - B_y = 132.5 \text{ N} - (232 \text{ N})(\cos 43°) = -37 \text{ N}$. The elbow force is thus

$$E = \sqrt{(158 \text{ N})^2 + (-37 \text{ N})^2} = 162 \text{ N}$$

and acts at $\theta = \tan^{-1}[(-37)/158] = -13°$ or 13° below the horizontal.

Reflect: The force exerted by the elbow is larger than the total weight of the arm and the object it is carrying.

Solutions to Passage Problems

*1.67. Set Up: The total volume is equal to the number of alveoli times the average volume of a single alveolus. Note that $1 \, \mu\text{m} = 10^{-6}$ m and $1 \text{ m}^3 = 10^3$ L.

Solve: $(480 \times 10^6 \text{ alveoli}) \left(\dfrac{4.2 \times 10^6 \, \mu\text{m}^3}{\text{alveolus}} \right) \left(\dfrac{10^{-6} \text{ m}}{1 \, \mu\text{m}} \right)^3 \left(\dfrac{10^3 \text{ L}}{1 \text{ m}^3} \right) = 2.0$ L. The correct answer is C.

*1.69. Set Up: The graph shows a nearly horizontal line, with no systematic upward or downward trend. Thus, the average volume of the alveoli is not dependent on the total lung volume.

Solve: The total lung volume must be equal to the average volume of the alveoli times the number of alveoli; thus, as the total volume of the lungs increases, the volume of the individual alveoli remains constant and the number of alveoli increases. The correct answer is C.

2

MOTION ALONG A STRAIGHT LINE

Problems 1, 5, 7, 11, 15, 17, 19, 25, 27, 29, 35, 37, 39, 43, 45, 51, 53, 55, 59, 61, 65, 67, 69, 73, 75, 79, 83, 85

Solutions to Problems

*2.1. Set Up: Let the $+x$ direction be to the right in the figure.

Solve: **(a)** The lengths of the segments determine the distance of each point from O:

$$x_A = -5 \text{ cm}, \quad x_B = +45 \text{ cm}, \quad x_C = +15 \text{ cm}, \quad \text{and } x_D = -5 \text{ cm}.$$

(b) The displacement is Δx; the sign of Δx indicates its direction. The distance is always positive.

(i) A to B: $\Delta x = x_B - x_A = +45 \text{ cm} - (-5 \text{ cm}) = +50 \text{ cm}$. Distance is 50 cm.

(ii) B to C: $\Delta x = x_C - x_B = +15 \text{ cm} - 45 \text{ cm} = -30 \text{ cm}$. Distance is 30 cm.

(iii) C to D: $\Delta x = x_D - x_C = -5 \text{ cm} - 15 \text{ cm} = -20 \text{ cm}$. Distance is 20 cm.

(iv) A to D: $\Delta x = x_D - x_A = 0$. Distance $= 2(AB) = 100$ cm.

Reflect: When the motion is always in the same direction during the interval the magnitude of the displacement and the distance traveled are the same. In (iv) the ant travels to the right and then to the left and the magnitude of the displacement is less than the distance traveled.

*2.5. Set Up: $x_A = 0$, $x_B = 3.0$ m, $x_C = 9.0$ m. $t_A = 0$, $t_B = 1.0$ s, $t_C = 5.0$ s.

Solve: **(a)** $v_{\text{av-}x} = \dfrac{\Delta x}{\Delta t}$

$$A \text{ to } B: \ v_{\text{av-}x} = \frac{\Delta x}{\Delta t} = \frac{x_B - x_A}{t_B - t_A} = \frac{3.0 \text{ m}}{1.0 \text{ s}} = 3.0 \text{ m/s}$$

$$B \text{ to } C: \ v_{\text{av-}x} = \frac{x_C - x_B}{t_C - t_B} = \frac{6.0 \text{ m}}{4.0 \text{ s}} = 1.5 \text{ m/s}$$

$$A \text{ to } C: \ v_{\text{av-}x} = \frac{x_C - x_A}{t_C - t_A} = \frac{9.0 \text{ m}}{5.0 \text{ s}} = 1.8 \text{ m/s}$$

(b) The velocity is always in the same direction ($+x$-direction), so the distance traveled is equal to the displacement in each case, and the average speed is the same as the magnitude of the average velocity.

Reflect: The average speed is different for different time intervals.

*2.7. Set Up: The positions x_t at time t are: $x_0 = 0$, $x_1 = 1.0$ m, $x_2 = 4.0$ m, $x_3 = 9.0$ m, $x_4 = 16.0$ m.

Solve: (a) The distance is $x_3 - x_1 = 8.0$ m.

(b) $v_{\text{av-}x} = \dfrac{\Delta x}{\Delta t}$. **(i)** $v_{\text{av,}x} = \dfrac{x_1 - x_0}{1.0 \text{ s}} = 1.0$ m/s; **(ii)** $v_{\text{av,}x} = \dfrac{x_2 - x_1}{1.0 \text{ s}} = 3.0$ m/s; **(iii)** $v_{\text{av,}x} = \dfrac{x_3 - x_2}{1.0 \text{ s}} = 5.0$ m/s;

(iv) $v_{\text{av,}x} = \dfrac{x_4 - x_3}{1.0 \text{ s}} = 7.0$ m/s; **(v)** $v_{\text{av,}x} = \dfrac{x_4 - x_0}{4.0 \text{ s}} = 4.0$ m/s

Reflect: In successive 1 s time intervals the boulder travels greater distances and the average velocity for the intervals increases from one interval to the next.

*2.11. Set Up: 1.0 century $= 100$ yr. 1 km $= 10^5$ cm.

Solve: (a) $d = vt = (5.0 \text{ cm/yr})(100 \text{ yr}) = 500$ cm $= 5.0$ m

(b) $t = \dfrac{d}{t} = \dfrac{550 \times 10^5 \text{ cm}}{5.0 \text{ cm/yr}} = 1.1 \times 10^7$ yr

*2.15. Set Up: The velocity of the truck relative to the road is equal to the velocity of the truck relative to me plus my velocity relative to the road: $v_{\text{T/R}} = v_{\text{T/M}} + v_{\text{M/R}}$. Assume that all numbers are good to two significant figures.

Solve: (a) As seen by me, the truck moves *backwards* 25 meters during 5.5 seconds. Thus, the truck's velocity relative to me is $v_{\text{T/M}} = \dfrac{-25 \text{ m}}{5.5 \text{ s}} = -4.55$ m/s. My velocity relative to the road is $v_{\text{M/R}} = (110 \text{ km/h})(1000 \text{ m/km})(1 \text{ h/3600 s}) = 30.6$ m/s.

Thus, the velocity of the truck relative to the road is $v_{\text{T/R}} = v_{\text{T/M}} + v_{\text{M/R}} = -4.55$ m/s $+ 30.6$ m/s $= 26$ m/s.

(b) The distance the truck moves (relative to the road) is $\Delta x = (26 \text{ m/s})(5.5 \text{ s}) = 143$ m $= 1.4 \times 10^2$ m.

Reflect: Note that the distance that I move is $(30.6 \text{ m/s})(5.5 \text{ s}) = 168$ m, which is 25 meters further than the distance that the truck travels.

*2.17. Set Up: Use the normal driving time to find the distance. Use this distance to find the time on Friday.

Solve: $\Delta x = v_{\text{av,}x} \Delta t = (105 \text{ km/h})(1.33 \text{ h}) = 140$ km. Then on Friday $\Delta t = \dfrac{\Delta x}{v_{\text{av,}x}} = \dfrac{140 \text{ km}}{70 \text{ km/h}} = 2.00$ h. The increase in time is $2.00 \text{ h} - 1.33 \text{ h} = 0.67 \text{ h} = 40$ min.

Reflect: A smaller average speed corresponds to a longer travel time when the distance is the same.

*2.19. Set Up: The instantaneous velocity is the slope of the tangent to the x versus t graph.

Solve: (a) The velocity is zero where the graph is horizontal; point IV.

(b) The velocity is constant and positive where the graph is a straight line with positive slope; point I.

(c) The velocity is constant and negative where the graph is a straight line with negative slope; point V.

(d) The slope is positive and increasing at point II.

(e) The slope is positive and decreasing at point III.

*2.25. Set Up: The acceleration a_x equals the slope of the v_x versus t curve.

Solve: The qualitative graphs of acceleration as a function of time are given in the figure below.

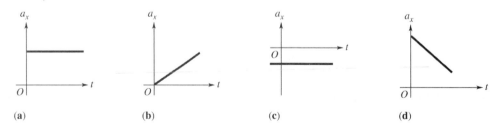

The acceleration can be described as follows: **(a)** positive and constant, **(b)** positive and increasing, **(c)** negative and constant, **(d)** positive and decreasing.

Reflect: When v_x and a_x have the same sign then the speed is increasing. In (c) the velocity and acceleration have opposite signs and the speed is decreasing.

*2.27. Set Up: Assume constant acceleration. $v_{0x} = 88$ ft/s, $v_x = 110$ ft/s, and $t = 3.50$ s. Let $x_0 = 0$.

Solve: (a) $v_x = v_{0x} + a_x t$ and $a_x = \dfrac{v_x - v_{0x}}{t} = \dfrac{110 \text{ ft/s} - 88 \text{ ft/s}}{3.50 \text{ s}} = 6.3 \text{ ft/s}^2$.

(b) $x = x_0 + v_{0x}t + \frac{1}{2}a_x t^2 = (88 \text{ ft/s})(3.50 \text{ s}) + \frac{1}{2}(6.3 \text{ ft/s}^2)(3.50 \text{ s})^2 = 347$ ft

*2.29. Set Up: Take the $+y$ direction to be upward. For part (a) we assume that the cat is in freefall with $a_y = -g$. Since the cat falls a known distance, we can find its final velocity using $v_y^2 = v_{0y}^2 + 2a_y(y - y_0)$. For parts (b) and (c) we assume that the cat has a constant (but unknown) acceleration due to its interaction with the floor. We may use the equations for constant acceleration.

Solve: (a) Solving for v_y we obtain

$$v_y = \pm\sqrt{v_{0y}^2 + 2a_y(y - y_0)}$$

Here we set $\Delta y = (-4.0 \text{ ft})(1 \text{ m}/3.28 \text{ ft}) = -1.22$ m, $a_y = -g$, and $v_y = 0$.

$$v_y = -\sqrt{v_{0y}^2 + 2a_y(y - y_0)}$$
$$= -\sqrt{-2g(-1.22 \text{ m})} = -4.89 \text{ m/s} = -4.9 \text{ m/s}.$$

Where we choose the negative root since the cat is falling. The *speed* of the cat just before impact is the magnitude of its velocity, which is 4.9 m/s.

(b) During its impact with the floor, the cat is brought to rest over a distance of 12 cm. Thus, we have $v_{0y} = -4.89$ m/s, $v_y = 0$, and $\Delta y = -0.12$ m. Solving $\Delta y = \dfrac{1}{2}(v_y + v_{0y})t$ for time we obtain:

$$t = \frac{2\Delta y}{(v_y + v_{0y})} = \frac{2(-0.12 \text{ m})}{(0 + -4.89 \text{ m/s})} = 0.049 \text{ s}.$$

(c) Solving $v_y^2 = v_{0y}^2 + 2a_y(y - y_0)$ for a_y we obtain $a_y = \dfrac{(v_y^2 - v_{0y}^2)}{2\Delta y} = \dfrac{0^2 - (-4.89 \text{ m/s})^2}{2(-0.12 \text{ m})} = 99.6 \text{ m/s}^2$. Since this answer is only accurate to two significant figures, we can write it as 1.0×10^2 m/s^2 or approximately 10 g's.

Reflect: During freefall the cat has a negative velocity and a negative acceleration—so it is speeding up. In contrast, during impact with the ground the cat has a negative velocity and a positive acceleration—so it is slowing down.

*2.35. Set Up: Let $+x$ be in the direction of motion of the bullet. $v_{0x} = 0$, $x_0 = 0$, $v_x = 335$ m/s, and $x = 0.127$ m.

Solve: (a) $v_x^2 = v_{0x}^2 + 2a_x(x - x_0)$ and

$$a_x = \frac{v_x^2 - v_{0x}^2}{2(x - x_0)} = \frac{(335 \text{ m/s})^2 - 0}{2(0.127 \text{ m})} = 4.42 \times 10^5 \text{ m/s}^2 = 4.51 \times 10^4 \, g$$

(b) $v_x = v_{0x} + a_x t$ so $t = \dfrac{v_x - v_{0x}}{a_x} = \dfrac{335 \text{ m/s} - 0}{4.42 \times 10^5 \text{ m/s}^2} = 0.758$ ms

Reflect: The acceleration is very large compared to g. In (b) we could also use $(x - x_0) = \left(\dfrac{v_{0x} + v_x}{2}\right)t$ to calculate

$$t = \frac{2(x - x_0)}{v_x} = \frac{2(0.127 \text{ m})}{335 \text{ m/s}} = 0.758 \text{ ms}$$

*2.37. Set Up: Let $+x$ be the direction the car is moving. We can use the equations for constant acceleration.

Solve: (a) From Eq. (2.13), with $v_{0x} = 0$, $a_x = \dfrac{v_x^2}{2(x-x_0)} = \dfrac{(20 \text{ m/s})^2}{2(120 \text{ m})} = 1.67 \text{ m/s}^2$.

(b) Using Eq. (2.14), $t = 2(x-x_0)/v_x = 2(120 \text{ m})/(20 \text{ m/s}) = 12 \text{ s}$.

(c) $(12 \text{ s})(20 \text{ m/s}) = 240$ m.

Reflect: The average velocity of the car is half the constant speed of the traffic, so the traffic travels twice as far.

*2.39. Set Up: $0.250 \text{ mi} = 1320 \text{ ft}$. $60.0 \text{ mph} = 88.0 \text{ ft/s}$. Let $+x$ be the direction the car is traveling.

Solve: (a) braking: $v_{0x} = 88.0 \text{ ft/s}$, $x-x_0 = 146 \text{ ft}$, $v_x = 0$. $v_x^2 = v_{0x}^2 + 2a_x(x-x_0)$ gives

$$a_x = \frac{v_x^2 - v_{0x}^2}{2(x-x_0)} = \frac{0-(88.0 \text{ ft/s})^2}{2(146 \text{ ft})} = -26.5 \text{ ft/s}^2$$

Speeding up: $v_{0x} = 0$, $x-x_0 = 1320 \text{ ft}$, $t = 19.9 \text{ s}$. $x-x_0 = v_{0x}t + \frac{1}{2}a_x t^2$ gives

$$a_x = \frac{2(x-x_0)}{t^2} = \frac{2(1320 \text{ ft})}{(19.9 \text{ s})^2} = 6.67 \text{ ft/s}^2$$

(b) $v_x = v_{0x} + a_x t = 0 + (6.67 \text{ ft/s}^2)(19.9 \text{ s}) = 133 \text{ ft/s} = 90.5 \text{ mph}$

(c) $t = \dfrac{v_x - v_{0x}}{a_x} = \dfrac{0 - 88.0 \text{ ft/s}}{-26.5 \text{ ft/s}^2} = 3.32 \text{ s}$

Reflect: The magnitude of the acceleration while braking is much larger than when speeding up. That is why it takes much longer to go from 0 to 60 mph than to go from 60 mph to 0.

*2.43. Set Up: The volume of a cylinder of radius R and height H is given by $V = \pi R^2 H$. We know the ratio of the heights of the two tanks and their volumes. From this information we can determine the ratio of their radii.

Solve: We take the ratio of the volumes of the two tanks: $\dfrac{V_{\text{large}}}{V_{\text{small}}} = \dfrac{218}{150} = \dfrac{\pi R_{\text{large}}^2 H_{\text{large}}}{\pi R_{\text{small}}^2 H_{\text{small}}} = 1.20 \left(\dfrac{R_{\text{large}}}{R_{\text{small}}}\right)^2$, where we

have used $\dfrac{H_{\text{large}}}{H_{\text{small}}} = 1.20$. Solving for the ratio of the radii we obtain $\dfrac{R_{\text{large}}}{R_{\text{small}}} = \sqrt{\left(\dfrac{1}{1.20}\right)\left(\dfrac{218}{150}\right)} = 1.10$. Thus, the larger radius is 10% larger than the smaller radius.

Reflect: All of the ratios used are dimensionless, and independent of the units used for measurement.

*2.45. Set Up: $a_A = a_B$, $x_{0A} = x_{0B} = 0$, $v_{0x,A} = v_{0x,B} = 0$, and $t_A = 2t_B$.

Solve: (a) $x = x_0 + v_{0x}t + \frac{1}{2}a_x t^2$ gives $x_A = \frac{1}{2}a_A t_A^2$ and $x_B = \frac{1}{2}a_B t_B^2$. $a_A = a_B$ gives $\dfrac{x_A}{t_A^2} = \dfrac{x_B}{t_B^2}$ and

$$x_B = \left(\frac{t_B}{t_A}\right)^2 x_A = \left(\frac{1}{2}\right)^2 (250 \text{ km}) = 62.5 \text{ km}$$

(b) $v_x = v_{0x} + a_x t$ gives $a_A = \dfrac{v_A}{t_A}$ and $a_B = \dfrac{v_B}{t_B}$. Since $a_A = a_B$, $\dfrac{v_A}{t_A} = \dfrac{v_B}{t_B}$ and

$$v_B = \left(\frac{t_B}{t_A}\right)v_A = \left(\frac{1}{2}\right)(350 \text{ m/s}) = 175 \text{ m/s}.$$

Reflect: v_x is proportional to t and for $v_{0x} = 0$, x is proportional to t^2.

*2.51. Set Up: Take $+y$ upward. $v_y = 0$ at the maximum height. $a_y = -0.379g = -3.71 \text{ m/s}^2$.

Solve: Consider the motion from the maximum height back to the initial level. For this motion $v_{0y} = 0$ and $t = 4.25$ s. $y = y_0 + v_{0y}t + \frac{1}{2}a_yt^2 = \frac{1}{2}(-3.71 \text{ m/s}^2)(4.25 \text{ s})^2 = -33.5$ m

The ball went 33.5 m above its original position.

(b) Consider the motion from just after it was hit to the maximum height. For this motion $v_y = 0$ and $t = 4.25$ s.

$v_y = v_{0y} + a_yt$ gives $v_{0y} = -a_yt = -(-3.71 \text{ m/s}^2)(4.25 \text{ s}) = 15.8$ m/s.

(c) The graphs are sketched in the figure below.

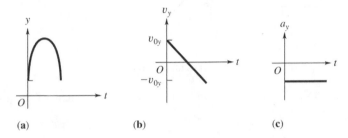

(a)   (b)   (c)

Reflect: The answers can be checked several ways. For example, $v_y = 0$, $v_{0y} = 15.8$ m/s, and $a_y = -3.7$ m/s^2 in $v_y^2 = v_{0y}^2 + 2a_y(y - y_0)$ gives

$$y - y_0 = \frac{v_y^2 - v_{0y}^2}{2a_y} = \frac{0 - (15.8 \text{ m/s})^2}{2(-3.71 \text{ m/s}^2)} = 33.6 \text{ m},$$

which agrees with the height calculated in **(a)**.

*2.53. Set Up: Take $+y$ upward. $a_y = -9.80$ m/s^2. The initial velocity of the sandbag equals the velocity of the balloon, so $v_{0y} = +5.00$ m/s. When the balloon reaches the ground, $y - y_0 = -40.0$ m. At its maximum height the sandbag has $v_y = 0$.

Solve: (a) $t = 0.250$ s:

$$y - y_0 = v_{0y}t + \frac{1}{2}a_yt^2 = (5.00 \text{ m/s})(0.250 \text{ s}) + \frac{1}{2}(-9.80 \text{ m/s}^2)(0.250 \text{ s})^2 = 0.94 \text{ m}.$$

The sandbag is 40.9 m above the ground.

$$v_y = v_{0y} + a_yt = +5.00 \text{ m/s} + (-9.80 \text{ m/s}^2)(0.250 \text{ s}) = 2.55 \text{ m/s}.$$

$t = 1.00$ s:

$$y - y_0 = (5.00 \text{ m/s})(1.00 \text{ s}) + \frac{1}{2}(-9.80 \text{ m/s}^2)(1.00 \text{ s})^2 = 0.10 \text{ m}.$$

The sandbag is 40.1 m above the ground.

$$v_y = v_{0y} + a_yt = +5.00 \text{ m/s} + (-9.80 \text{ m/s}^2)(1.00 \text{ s}) = -4.80 \text{ m/s}.$$

(b) $y - y_0 = -40.0$ m, $v_{0y} = 5.00$ m/s, $a_y = -9.80$ m/s^2. $y - y_0 = v_{0y}t + \frac{1}{2}a_yt^2$ gives $-40.0 \text{ m} = (5.00 \text{ m/s}) \, t -$ $(4.90 \text{ m/s}^2)t^2$. $(4.90 \text{ m/s}^2)t^2 - (5.00 \text{ m/s})t - 40.0 \text{ m} = 0$ and

$$t = \frac{1}{9.80}\left(5.00 \pm \sqrt{(-5.00)^2 - 4(4.90)(-40.0)}\right) \text{ s} = (0.51 \pm 2.90) \text{ s}.$$

t must be positive, so $t = 3.41$ s.

(c) $v_y = v_{0y} + a_yt = +5.00 \text{ m/s} + (-9.80 \text{ m/s}^2)(3.41 \text{ s}) = -28.4 \text{ m/s}$

(d) $v_{0y} = 5.00$ m/s, $a_y = -9.80$ m/s^2, $v_y = 0$. $v_y^2 = v_{0y}^2 + 2a_y(y - y_0)$ gives

$$y - y_0 = \frac{v_y^2 - v_{0y}^2}{2a_y} = \frac{0 - (5.00 \text{ m/s})^2}{2(-9.80 \text{ m/s}^2)} = 1.28 \text{ m}.$$

The maximum height is 41.3 m above the ground.

(e) The graphs of a_y, v_y, and y versus t are given in the figure below. Take $y = 0$ at the ground.

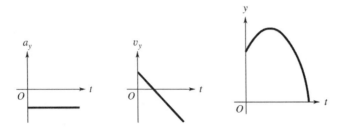

*2.55. Set Up: $a_M = 0.170 a_E$. Take $+y$ to be upward and $y_0 = 0$.

Solve: (a) $v_{0E} = v_{0M}$. $v_y^2 = v_{0y}^2 + 2a_y(y - y_0)$ with $v_y = 0$ at the maximum height gives $2a_y y = -v_{0y}^2$, so $a_M y_M = a_E y_E$.

$$y_M = \left(\frac{a_E}{a_M}\right) y_E = \left(\frac{1}{0.170}\right)(12.0 \text{ m}) = 70.6 \text{ m}$$

(b) Consider the time to the maximum height on the earth. The total travel time is twice this. First solve for v_{0y}, with $v_y = 0$ and $y = 12.0$ m. $v_y^2 = v_{0y}^2 + 2a_y(y - y_0)$ gives

$$v_{0y} = \sqrt{-2(-a_E)y} = \sqrt{-2(-9.8 \text{ m/s}^2)(12.0 \text{ m})} = 15.3 \text{ m/s}.$$

Then $v_y = v_{0y} + a_y t$ gives

$$t = \frac{v_y - v_{0y}}{a_y} = \frac{0 - 15.3 \text{ m/s}}{-9.8 \text{ m/s}^2} = 1.56 \text{ s}.$$

The total time is $2(1.56 \text{ s}) = 3.12$ s. Then, on the moon $v_y = v_{0y} + a_y t$ with $v_{0y} = 15.3$ m/s, $v_y = 0$, and $a = -1.666$ m/s^2 gives

$$t = \frac{v_y - v_{0y}}{a_y} = \frac{0 - 15.3 \text{ m/s}}{-1.666 \text{ m/s}^2} = 9.18 \text{ s}.$$

The total time is 18.4 s. It takes 15.3 s longer on the moon.

Reflect: The maximum height is proportional to $1/a$, so the height on the moon is greater. Since the acceleration is the rate of change of the speed, the wrench loses speed at a slower rate on the moon and it takes more time for its speed to reach $v = 0$ at the maximum height. In fact, $t_M / t_E = a_E / a_M = 1/0.170 = 5.9$, which agrees with our calculated times. But to find the difference in the times we had to solve for the actual times, not just their ratios.

*2.59. Set Up: Use subscripts f and s to refer to the faster and slower stones, respectively. Take $+y$ to be upward and $y_0 = 0$ for both stones. $v_{0f} = 3v_{0s}$. When a stone reaches the ground, $y = 0$.

Solve: (a) $y = y_0 + v_{0y}t + \frac{1}{2}a_y t^2$ gives $a_y = -\frac{2v_{0y}}{t}$. Since both stones have the same a_y, $\frac{v_{0f}}{t_f} = \frac{v_{0s}}{t_s}$ and

$$t_s = t_f\left(\frac{v_{0s}}{v_{0f}}\right) = \left(\tfrac{1}{3}\right)10 \text{ s} = 3.3 \text{ s}$$

(b) Since $v_y = 0$ at the maximum height, then $v_y^2 = v_{0y}^2 + 2a_y(y - y_0)$ gives $a_y = -\dfrac{v_{0y}^2}{2y}$. Since both have the

same a_y, $\dfrac{v_{0f}^2}{y_f} = \dfrac{v_{0s}^2}{y_s}$ and $y_f = y_s\left(\dfrac{v_{0f}}{v_{0s}}\right)^2 = 9H$.

Reflect: The faster stone reaches a greater height so it travels a greater distance than the slower stone and takes more time to return to the ground.

*2.61. Set Up: $\vec{v}_{T/E} = 65$ mph, north. $\vec{v}_{VW/E} = 42$ mph, south. Let $+y$ be north.

Solve: $\vec{v}_{T/E} = \vec{v}_{T/VW} + \vec{v}_{VW/E}$ and $\vec{v}_{T/VW} = \vec{v}_{T/E} - \vec{v}_{VW/E}$

(a) $(v_{T/VW})_y = (v_{T/E})_y - (v_{VW/E})_y = 65$ mph $-(-42$ mph$) = 107$ mph. Relative to the VW, the Toyota is traveling

north at 107 mph. $\vec{v}_{VW/T} = -\vec{v}_{T/VW}$. Relative to the Toyota the VW is traveling south at 107 mph.

(b) The answers are the same as in **(a)**.

*2.65. Set Up: 1 light year $= (3.00 \times 10^8$ m/s$)(3.156 \times 10^7$ s$) = 9.47 \times 10^{15}$ m

Solve: $t = \dfrac{d}{v} = \dfrac{(4.25 \text{ light years})(9.47 \times 10^{15} \text{ m/light year})}{1000 \times 10^3 \text{ m/s}} = 4.02 \times 10^{10}$ s $= 1300$ yr

*2.67. Set Up: The average speed is the total distance traveled divided by the total time. The elapsed time is the distance traveled divided by the average speed. The total distance traveled is 20 mi. With an average speed of 8 mi/h

for 10 mi, the time for that first 10 miles is $\dfrac{10 \text{ mi}}{8 \text{ mi/h}} = 1.25$ h.

Solve: (a) An average speed of 4 mi/h for 20 mi gives a total time of $\dfrac{20 \text{ mi}}{4 \text{ mi/h}} = 5.0$ h. The second 10 mi must be

covered in 5.0 h $- 1.25$ h $= 3.75$ h. This corresponds to an average speed of $\dfrac{10 \text{ mi}}{3.75 \text{ h}} = 2.7$ mi/h.

(b) An average speed of 12 mi/h for 20 mi gives a total time of $\dfrac{20 \text{ mi}}{12 \text{ mi/h}} = 1.67$ h. The second 10 mi must be covered

in 1.67 h $- 1.25$ h $= 0.42$ h. This corresponds to an average speed of $\dfrac{10 \text{ mi}}{0.42 \text{ h}} = 24$ mi/h.

(c) An average speed of 16 mi/h for 20 mi gives a total time of $\dfrac{20 \text{ mi}}{16 \text{ mi/h}} = 1.25$ h. But 1.25 h was already spent during

the first 10 miles and the second 10 miles would have to be covered in zero time. This is not possible and an average speed of 16 mi/h for the 20-mile ride is not possible.

Reflect: The average speed for the total trip is not the average of the average speeds for each 10-mile segment. The rider spends a different amount of time traveling at each of the two average speeds.

*2.69. Set Up: Let $+y$ to be upward and $y_0 = 0$. $a_M = a_E/6$. At the maximum height $v_y = 0$.

Solve: (a) $v_y^2 = v_{0y}^2 + 2a_y(y - y_0)$. Since $v_y = 0$ and v_{0y} is the same for both rocks, $a_M y_M = a_E y_E$ and

$$y_E = \left(\frac{a_M}{a_E}\right)y_M = H/6$$

(b) $v_y = v_{0y} + a_y t$. $a_M t_M = a_E t_E$ and $t_M = \left(\dfrac{a_E}{a_M}\right)t_E = 6(4.0 \text{ s}) = 24.0$ s

Reflect: On the moon, where the acceleration is less, the rock reaches a greater height and takes more time to reach that maximum height.

*2.73. Set Up: Let $+y$ be downward. The egg has $v_{0y} = 0$ and $a_y = 9.80 \text{ m/s}^2$. Find the distance the professor walks during the time t it takes the egg to fall to the height of his head. At this height, the egg has $y - y_0 = 44.2 \text{ m}$.

Solve: $y - y_0 = v_{0y}t + \frac{1}{2}a_yt^2$ gives

$$t = \sqrt{\frac{2(y - y_0)}{a_y}} = \sqrt{\frac{2(44.2 \text{ m})}{9.80 \text{ m/s}^2}} = 3.00 \text{ s}.$$

The professor walks a distance $x - x_0 = v_{0x}t = (1.20 \text{ m/s})(3.00 \text{ s}) = 3.60 \text{ m}$. Release the egg when your professor is 3.60 m from the point directly below you.

Reflect: Just before the egg lands its speed is $(9.80 \text{ m/s}^2)(3.00 \text{ s}) = 29.4 \text{ m/s}$. It is traveling much faster than the professor.

*2.75. Set Up: Since air resistance is ignored, the boulder is in free-fall and has a constant downward acceleration of magnitude 9.80 m/s^2. Apply the constant acceleration equations to the motion of the boulder. Take $+y$ to be upward.

Solve: (a) $v_{0y} = +40.0 \text{ m/s}$, $v_y = +20.0 \text{ m/s}$, $a_y = -9.80 \text{ m/s}^2$. $v_y = v_{0y} + a_yt$ gives

$$t = \frac{v_y - v_{0y}}{a_y} = \frac{20.0 \text{ m/s} - 40.0 \text{ m/s}}{-9.80 \text{ m/s}^2} = +2.04 \text{ s}.$$

(b) $v_y = -20.0 \text{ m/s}$. $t = \frac{v_y - v_{0y}}{a_y} = \frac{-20.0 \text{ m/s} - 40.0 \text{ m/s}}{-9.80 \text{ m/s}^2} = +6.12 \text{ s}.$

(c) $y - y_0 = 0$, $v_{0y} = +40.0 \text{ m/s}$, $a_y = -9.80 \text{ m/s}^2$. $y - y_0 = v_{0y}t + \frac{1}{2}a_yt^2$ gives $t = 0$ and

$$t = -\frac{2v_{0y}}{a_y} = -\frac{2(40.0 \text{ m/s})}{-9.80 \text{ m/s}^2} = +8.16 \text{ s}.$$

(d) $v_y = 0$, $v_{0y} = +40.0 \text{ m/s}$, $a_y = -9.80 \text{ m/s}^2$. $v_y = v_{0y} + a_yt$ gives $t = \frac{v_y - v_{0y}}{a_y} = \frac{0 - 40.0 \text{ m/s}}{-9.80 \text{ m/s}^2} = 4.08 \text{ s}.$

(e) The acceleration is 9.80 m/s^2, downward, at all points in the motion.

(f) The graphs are sketched in the following figure.

Reflect: We have $v_y = 0$ at the maximum height. The time to reach the maximum height is half the total time in the air, so the answer in part (d) is half the answer in part (c). Also note that $2.04 \text{ s} < 4.08 \text{ s} < 6.12 \text{ s}$. The boulder is going upward until it reaches its maximum height, and after the maximum height it is traveling downward.

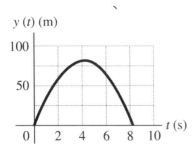
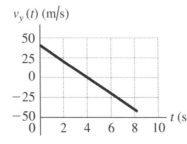
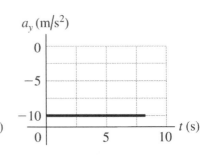

*2.79. Set Up: Take $+y$ to be upward. There are two periods of constant acceleration: $a_y = +2.50 \text{ m/s}^2$ while the engines fire and $a_y = -9.8 \text{ m/s}^2$ after they shut off. Constant acceleration equations can be applied within each period of constant acceleration.

Solve: (a) Find the speed and height at the end of the first 20.0 s. $a_y = +2.50 \text{ m/s}^2$, $v_{0y} = 0$, and $y_0 = 0$.

$v_y = v_{0y} + a_y t = (2.50 \text{ m/s}^2)(20.0 \text{ s}) = 50.0 \text{ m/s}$ and $y = y_0 + v_{0y}t + \frac{1}{2}a_y t^2 = \frac{1}{2}(2.50 \text{ m/s}^2)(20.0 \text{ s})^2 = 500 \text{ m}$.

Next consider the motion from this point to the maximum height. $y_0 = 500 \text{ m}$, $v_y = 0$, $v_{0y} = 50.0 \text{ m/s}$, and $a_y = -9.8 \text{ m/s}^2$. $v_y^2 = v_{0y}^2 + 2a_y(y - y_0)$ gives

$$y - y_0 = \frac{v_y^2 - v_{0y}^2}{2a_y} = \frac{0 - (50.0 \text{ m/s})^2}{2(-9.8 \text{ m/s}^2)} = +128 \text{ m},$$

so $y = 628 \text{ m}$. The duration of this part of the motion is obtained from $v_y = v_{0y} + a_y t$:

$$t = \frac{v_y - v_{0y}}{a_y} = \frac{-50 \text{ m/s}}{-9.8 \text{ m/s}^2} = 5.10 \text{ s}$$

(b) At the highest point, $v_y = 0$ and $a_y = 9.8 \text{ m/s}^2$, downward.

(c) Consider the motion from the maximum height back to the ground. $a_y = -9.8 \text{ m/s}^2$, $v_{0y} = 0$, $y = 0$, and $y_0 = 628 \text{ m}$. $y = y_0 + v_{0y}t + \frac{1}{2}a_y t^2$ gives

$$t = \sqrt{\frac{2(y - y_0)}{a_y}} = 11.3 \text{ s}.$$

The total time the rocket is in the air is $20.0 \text{ s} + 5.10 \text{ s} + 11.3 \text{ s} = 36.4 \text{ s}$. $v_y = v_{0y} + a_y t = (-9.8 \text{ m/s}^2)(11.3 \text{ s}) = -111 \text{ m/s}$. Just before it hits the ground the rocket will have speed 111 m/s.

Reflect: We could calculate the time of free fall directly by considering the motion from the point of engine shutoff to the ground: $v_{0y} = 50.0 \text{ m/s}$, $y - y_0 = 500 \text{ m}$ and $a_y = -9.8 \text{ m/s}^2$. $y - y_0 = v_{0y}t + \frac{1}{2}a_y t^2$ gives $t = 16.4 \text{ s}$, which agrees with a total time of 36.4 s.

*2.83. Set Up: Let t_{fall} be the time for the rock to fall to the ground and let t_s be the time it takes the sound to travel from the impact point back to you. $t_{\text{fall}} + t_s = 10.0 \text{ s}$. Both the rock and sound travel a distance d that is equal to the height of the cliff. Take $+y$ downward for the motion of the rock. The rock has $v_{0y} = 0$ and $a_y = 9.80 \text{ m/s}^2$.

Solve: (a) For the rock, $y - y_0 = v_{0y}t + \frac{1}{2}a_y t^2$ gives $t_{\text{fall}} = \sqrt{\dfrac{2d}{9.80 \text{ m/s}^2}}$.

For the sound, $t_s = \dfrac{d}{330 \text{ m/s}} = 10.0 \text{ s}$. Let $\alpha^2 = d$. $0.00303\alpha^2 + 0.4518\alpha - 10.0 = 0$. $\alpha = 19.6$ and $d = 384 \text{ m}$.

(b) You would have calculated $d = \frac{1}{2}(9.80 \text{ m/s}^2)(10.0 \text{ s})^2 = 490 \text{ m}$. You would have overestimated the height of the cliff. It actually takes the rock less time than 10.0 s to fall to the ground.

Reflect: Once we know d we can calculate that $t_{\text{fall}} = 8.8 \text{ s}$ and $t_s = 1.2 \text{ s}$. The time for the sound of impact to travel back to you is 12% of the total time and cannot be neglected. The rock has speed 86 m/s just before it strikes the ground.

Solutions to Passage Problems

*2.85. Set Up: Assuming that the aorta and arteries are circular in cross-sectional area, we can use $A = \pi r^2 = \pi\left(\dfrac{d}{2}\right)^2$. Let d_a be the diameter of the aorta and d_b be the diameter of each branch.

Solve: Since the combined area of the two arteries is equal to that of the aorta we have $\pi\left(\dfrac{d_b}{2}\right)^2 + \pi\left(\dfrac{d_b}{2}\right)^2 = \pi\left(\dfrac{d_a}{2}\right)^2$, which reduces to $2d_b^{\ 2} = d_a^{\ 2}$. Thus, we have $d_b = d_a/\sqrt{2}$. The correct answer is B.

MOTION IN A PLANE

Problems 3, 5, 9, 11, 13, 19, 21, 25, 29, 31, 33, 37, 39, 41, 45, 47, 51, 53, 57, 59, 61, 65, 67, 69

Solutions to Problems

*3.3. Set Up: Coordinates of point A are $(2.0 \text{ m}, 1.0 \text{ m})$ and for point B they are $(10.0 \text{ m}, 6.0 \text{ m})$.
Solve: (a) At A, $x = 2.0 \text{ m}$, $y = 1.0 \text{ m}$.

(b) $\vec{r}$ and its components are shown in Figure (a) below. $r = \sqrt{x^2 + y^2} = 2.2 \text{ m}$; $\tan\theta = \dfrac{y}{x}$ and $\theta = 26.6°$, counter-clockwise from the $+x$-axis.

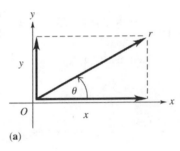

(a)

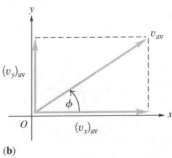

(b)

(c) $(v_x)_{\text{av}} = \dfrac{\Delta x}{\Delta t} = \dfrac{10.0 \text{ m} - 2.0 \text{ m}}{1.50 \text{ s}} = 5.3 \text{ m/s}$; $(v_y)_{\text{av}} = \dfrac{\Delta y}{\Delta t} = \dfrac{6.0 \text{ m} - 1.0 \text{ m}}{1.50 \text{ s}} = 3.3 \text{ m/s}$

(d) $\vec{v}_{av}$ and its components are shown in Figure (b) above. $v_{av} = \sqrt{(v_x)_{av}{}^2 + (v_y)_{av}{}^2} = 6.2$ m/s; $\tan\phi = \dfrac{(v_y)_{av}}{(v_x)_{av}}$ and

$\phi = 32°$, counterclockwise from the +x-axis.

Reflect: The displacement of the dragonfly is in the direction of $\vec{v}_{av}$.

*3.5. Set Up: The coordinates of each point are: A, $(-50\text{ m}, 0)$; B, $(0, +50\text{ m})$; C, $(+50\text{ m}, 0)$; D, $(0, -50\text{ m})$. At each point the velocity is tangent to the circular path, as shown in the figure below. The components (v_x, v_y) of the velocity at each point are: A, $(0, +6.0\text{ m/s})$; B, $(+6.0\text{ m/s}, 0)$; C, $(0, -6.0\text{ m/s})$; D, $(-6.0\text{ m/s}, 0)$.

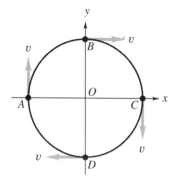

Solve: (a) A to B The time for one full lap is

$$t = \frac{2\pi r}{v} = \frac{2\pi(50\text{ m})}{6.0\text{ m/s}} = 52.4\text{ s}.$$

A to B is one-quarter lap and takes $\frac{1}{4}(52.4\text{ s}) = 13.1\text{ s}$.

$$(v_x)_{av} = \frac{\Delta x}{\Delta t} = \frac{0-(-50\text{ m})}{13.1\text{ s}} = 3.8\text{ m/s}; \ (v_y)_{av} = \frac{\Delta y}{\Delta t} = \frac{+50\text{ m}-0}{13.1\text{ s}} = 3.8\text{ m/s}$$

$$(a_x)_{av} = \frac{\Delta v_x}{\Delta t} = \frac{6.0\text{ m/s}-0}{13.1\text{ s}} = 0.46\text{ m/s}^2; \ (a_y)_{av} = \frac{\Delta v_y}{\Delta t} = \frac{0-6.0\text{ m/s}}{13.1\text{ s}} = -0.46\text{ m/s}^2$$

(b) A to C $t = \frac{1}{2}(52.4\text{ s}) = 26.2\text{ s}$

$$(v_x)_{av} = \frac{\Delta x}{\Delta t} = \frac{+50\text{ m}-(-50\text{ m})}{26.2\text{ s}} = 3.8\text{ m/s}; \ (v_y)_{av} = \frac{\Delta y}{\Delta t} = 0$$

$$(a_x)_{av} = \frac{\Delta v_x}{\Delta t} = 0; \ (a_y)_{av} = \frac{\Delta v_y}{\Delta t} = \frac{-6.0\text{ m/s}-6.0\text{ m/s}}{26.2\text{ s}} = -0.46\text{ m/s}^2$$

(c) C to D $t = \frac{1}{4}(52.4\text{ s}) = 13.1\text{ s}$

$$(v_x)_{av} = \frac{\Delta x}{\Delta t} = \frac{0-50\text{ m}}{13.1\text{ s}} = -3.8\text{ m/s}; \ (v_y)_{av} = \frac{\Delta y}{\Delta t} = \frac{-50\text{ m}-0}{13.1\text{ s}} = -3.8\text{ m/s}$$

$$(a_x)_{av} = \frac{\Delta v_x}{\Delta t} = \frac{-6.0\text{ m/s}-0}{13.1\text{ s}} = -0.46\text{ m/s}^2; \ (a_y)_{av} = \frac{\Delta v_y}{\Delta t} = \frac{0-(-6.0\text{ m/s})}{13.1\text{ s}} = 0.46\text{ m/s}^2$$

(d) A to A $\Delta x = \Delta y = 0$ so $(v_x)_{av} = (v_y)_{av} = 0$, and $\Delta v_x = \Delta v_y = 0$ so $(a_x)_{av} = (a_y)_{av} = 0$

(e) For A to B, $v_{av} = \sqrt{(v_x)_{av}{}^2 + (v_y)_{av}{}^2} = \sqrt{(3.8\text{ m/s})^2 + (3.8\text{ m/s})^2} = 5.4\text{ m/s}$. The speed is constant so the average speed is 6.0 m/s. The average speed is larger than the magnitude of the average velocity because the distance traveled is larger than the displacement.

(f) Velocity is a vector, with both magnitude and direction. The magnitude of the velocity is constant but its direction is changing.

Reflect: For this motion the acceleration describes the rate of change of the direction of the velocity.

*3.9. Set Up: Take $+y$ downward, so $a_x = 0$, $a_y = +9.80$ m/s^2 and $v_{0y} = 0$. When the ball reaches the floor, $y - y_0 = 0.750$ m.

Solve: (a) $y - y_0 = v_{0y}t + \frac{1}{2}a_yt^2$ gives $t = \sqrt{\dfrac{2(y - y_0)}{a_y}} = \sqrt{\dfrac{2(0.750 \text{ m})}{9.80 \text{ m/s}^2}} = 0.391$ s.

(b) $x - x_0 = v_{0x}t + \frac{1}{2}a_xt^2$ gives $v_{0x} = \dfrac{x - x_0}{t} = \dfrac{1.40 \text{ m}}{0.391 \text{ s}} = 3.58$ m/s. Since $v_{0y} = 0$, $v_0 = v_{0x} = 3.58$ m/s.

(c) $v_x = v_{0x} = 3.58$ m/s. $v_y = v_{0y} + a_yt = (9.80 \text{ m/s}^2)(0.391 \text{ s}) = 3.83$ m/s. $v = \sqrt{v_x^2 + v_y^2} = 5.24$ m/s.

$$\tan\theta = \frac{|v_y|}{|v_x|} = \frac{3.83 \text{ m/s}}{3.58 \text{ m/s}}$$

and $\theta = 46.9°$. The final velocity of the ball has magnitude 5.24 m/s and is directed at 46.9° below the horizontal.

Reflect: The time for the ball to reach the floor is the same as if it had been dropped from a height of 0.750 m; the horizontal component of velocity has no effect on the vertical motion.

*3.11. Set Up: The ball moves with projectile motion with an initial velocity that is horizontal and has magnitude v_0. In both cases, the height h of the table and v_0 have the same values; but the acceleration due to gravity changes from $g_E = 9.80$ m/s^2 on earth to g_X on planet X. Let $+x$ be horizontal and in the direction of the initial velocity of the marble and let $+y$ be upward. $v_{0x} = v_0$, $v_{0y} = 0$, $a_x = 0$, $a_y = -g$, where g is either g_E or g_X.

Solve: Use the vertical motion to find the time in the air: $y - y_0 = -h$. $y - y_0 = v_{0y}t + \frac{1}{2}a_yt^2$ gives $t = \sqrt{\dfrac{2h}{g}}$. Then

$x - x_0 = v_{0x}t + \frac{1}{2}a_xt^2$ gives $x - x_0 = v_{0x}t = v_0\sqrt{\dfrac{2h}{g}}$. $x - x_0 = D$ on earth and $2.76D$ on Planet X. $(x - x_0)\sqrt{g} = v_0\sqrt{2h}$,

which is constant, so $D\sqrt{g_E} = 2.76D\sqrt{g_X}$. $g_X = \dfrac{g_E}{(2.76)^2} = 0.131g_E = 1.28$ m/s^2.

Reflect: On Planet X the acceleration due to gravity is less, it takes the ball longer to reach the floor, and it travels farther horizontally.

*3.13. Set Up: Take $+y$ to be downward. $a_x = 0$, $a_y = +9.80$ m/s^2. $v_{0x} = v_0$, $v_{0y} = 0$. The car travels 21.3 m $-$ 1.80 m $= 19.5$ m downward during the time it travels 61.0 m horizontally.

Solve: Use the vertical motion to find the time in the air:

$$y - y_0 = v_{0y}t + \frac{1}{2}a_yt^2 \text{ gives } t = \sqrt{\frac{2(y - y_0)}{a_y}} = \sqrt{\frac{2(19.5 \text{ m})}{9.80 \text{ m/s}^2}} = 1.995 \text{ s}$$

Then $x - x_0 = v_{0x}t + \frac{1}{2}a_xt^2$ gives $v_0 = v_{0x} = \dfrac{x - x_0}{t} = \dfrac{61.0 \text{ m}}{1.995 \text{ s}} = 30.6$ m/s.

(b) $v_x = 30.6$ m/s since $a_x = 0$. $v_y = v_{0y} + a_yt = -19.6$ m/s. $v = \sqrt{v_x^2 + v_y^2} = 36.3$ m/s.

Reflect: We calculate the final velocity by calculating its x and y components.

*3.19. Set Up: The horizontal displacement when the ball returns to its original height is

$$R = \frac{v_0^2 \sin(2\theta_0)}{g}.$$

At its maximum height $v_y = 0$. $g = 32$ ft/s^2. $a_x = 0$, $a_y = -g$.

Solve: (a) $v_0 = \sqrt{\dfrac{Rg}{\sin(2\theta_0)}} = \sqrt{\dfrac{(375 \text{ ft})(32 \text{ ft/s}^2)}{\sin(60.0°)}} = 118 \text{ ft/s}$

(b) $v_{0y} = v_0 \sin\theta_0 = (118 \text{ ft/s})\sin 30.0° = 59 \text{ ft/s}.$ $v_y^2 = v_{0y}^2 + 2a_y(y - y_0).$

$$y - y_0 = \frac{v_y^2 - v_{0y}^2}{2a_y} = \frac{0 - (59 \text{ ft/s})^2}{2(-32 \text{ ft/s}^2)} = 54.4 \text{ ft}$$

Reflect: At the maximum height $v_y = 0$. But $v \neq 0$ there because the ball still has its constant horizontal component of velocity. The horizontal range equation,

$$R = \frac{v_0^2 \sin(2\theta_0)}{g},$$

can be used only when the initial and final points of the motion are at the same elevation.

*3.21. Set Up: Use coordinates with the origin at the ground and $+y$ upward. $a_x = 0$, $a_y = -9.80 \text{ m/s}^2$. At the maximum height $v_y = 0$.

Solve: (a) $v_y^2 = v_{0y}^2 + 2a_y(y - y_0)$ gives

$$v_y = \sqrt{-2a_y(y - y_0)} = \sqrt{-2(-9.80 \text{ m/s}^2)(0.587 \text{ m})} = 3.39 \text{ m/s}$$

$$v_{0y} = v_0 \sin\theta_0 \text{ so } v_0 = \frac{v_{0y}}{\sin\theta_0} = \frac{3.39 \text{ m/s}}{\sin 58.0°} = 4.00 \text{ m/s}$$

(b) Use the vertical motion to find the time in the air. When the froghopper has returned to the ground, $y - y_0 = 0$.

$$y - y_0 = v_{0y}t + \tfrac{1}{2}a_y t^2 \text{ gives } t = -\frac{2v_{0y}}{a_y} = -\frac{2(3.39 \text{ m/s})}{-9.80 \text{ m/s}^2} = 0.692 \text{ s}$$

Then $x - x_0 = v_{0x}t + \tfrac{1}{2}a_x t^2 = (v_0 \cos\theta_0)t = (4.00 \text{ m/s})(\cos 58.0°)(0.692 \text{ s}) = 1.47 \text{ m}$

Reflect: $v_y = 0$ when $t = -\dfrac{v_{0y}}{a_y} = -\dfrac{3.39 \text{ m/s}}{-9.80 \text{ m/s}^2} = 0.346 \text{ s}$. The total time in the air is twice this.

*3.25. Set Up: Example 3.5 gives $R = \dfrac{v_0^2 \sin 2\theta_0}{g}$ and $t = \dfrac{2v_0 \sin\theta_0}{g}$.

Solve: (a) The maximum range occurs when $\sin 2\theta_0 = 1$, $2\theta_0 = 90°$ and $\theta_0 = 45°$.

(b) $R_{max} = \dfrac{v_0^2}{g} = \dfrac{(25.0 \text{ m/s})^2}{9.80 \text{ m/s}^2} = 63.8 \text{ m}$

(c) $t = \dfrac{2(25.0 \text{ m/s})\sin 45°}{g} = 3.61 \text{ s}$

Reflect: The time that the balloon is in the air can also be calculated from $t = \dfrac{\Delta x}{v_{0x}} = \dfrac{63.8 \text{ m}}{(25.0 \text{ m/s})\cos 45°} = 3.61 \text{ s}$.

*3.29. Set Up: We can determine the initial velocity of the bottle rocket from its maximum vertical height by using the equation $h = \dfrac{v_0^2 \sin^2\theta_0}{2g}$, which is derived in Example 3.5. The maximum range of the projectile can be determined from $R = \dfrac{v_0^2 \sin 2\theta_0}{g}$, which is also derived in Example 3.5.

Solve: When fired vertically we have $\theta_0 = 90°$ and so $h = \dfrac{v_0^2 \sin^2 90°}{2g} = \dfrac{v_0^2}{2g}$. Solving for v_0 we obtain $v_0 = \sqrt{2gh}$.

The maximum range occurs when $\sin 2\theta_0 = 1$ so $\theta_0 = 45°$.

Substituting this value into $R = \dfrac{v_0^2 \sin 2\theta_0}{g}$ we obtain $R = \dfrac{(2gh)\sin 90°}{g} = 2h = 50.0$ m.

Reflect: When a projectile is fired at a fixed initial speed without air resistance, the maximum range that it can achieve is twice the maximum height that it achieves when fired vertically.

*3.31. Set Up: The acceleration is a_{rad}. $R = 0.75$ m.

Solve: $a_{rad} = \dfrac{v^2}{R}$ so $v = \sqrt{Ra_{rad}} = \sqrt{(0.75 \text{ m})(9.8 \text{ m/s}^2)} = 2.7$ m/s

*3.33. Set Up: The radius of the earth's orbit is $r = 1.50 \times 10^{11}$ m and its orbital period is $T = 365.3$ days $= 3.16 \times 10^7$ s. For mercury, $r = 5.79 \times 10^{10}$ m and $T = 88.0$ days $= 7.60 \times 10^6$ s.

Solve: (a) $v = \dfrac{2\pi r}{T} = 2.98 \times 10^4$ m/s

(b) $a_{rad} = \dfrac{v^2}{r} = 5.91 \times 10^{-3} \text{ m/s}^2$.

(c) $v = 4.79 \times 10^4$ m/s, and $a_{rad} = 3.96 \times 10^{-2} \text{ m/s}^2$.

*3.37. Set Up: $R = 0.070$ m. For 3.0 rev/s, the period T (time for one revolution) is $T = (1.0 \text{ s})/(3.0 \text{ rev}) = 0.333$ s.

Solve: $a_{rad} = \dfrac{4\pi^2 R}{T^2} = \dfrac{4\pi^2 (0.070 \text{ m})}{(0.333 \text{ s})^2} = 25 \text{ m/s}^2 = 2.5g$

Reflect: The acceleration is large and the force on the fluid must be 2.5 times its weight.

*3.39. Set Up: Apply the relative velocity relation. The relative velocities are $\vec{v}_{C/E}$, the canoe relative to the earth, $\vec{v}_{R/E}$, the velocity of the river relative to the earth and $\vec{v}_{C/R}$, the velocity of the canoe relative to the river.

Solve: $\vec{v}_{C/E} = \vec{v}_{C/R} + \vec{v}_{R/E}$ and therefore $\vec{v}_{C/R} = \vec{v}_{C/E} - \vec{v}_{R/E}$. The velocity components of $\vec{v}_{C/R}$ are -0.50 m/s $+$ $(0.40 \text{ m/s})/\sqrt{2}$, east and $(0.40 \text{ m/s})/\sqrt{2}$, south, for a velocity relative to the river of 0.36 m/s, at $52.5°$ south of west.

Reflect: The velocity of the canoe relative to the river has a smaller magnitude than the velocity of the canoe relative to the earth.

*3.41. Set Up: The relative velocities are the water relative to the earth, $\vec{v}_{W/E}$, the boat relative to the water, $\vec{v}_{B/W}$, and the boat relative to the earth, $\vec{v}_{B/E} \cdot \vec{v}_{B/E}$ is due east, $\vec{v}_{W/E}$ is due south and has magnitude 2.0 m/s. $\vec{v}_{B/W} = 4.2$ m/s. $\vec{v}_{B/E} = \vec{v}_{B/W} + \vec{v}_{W/E}$. The velocity addition diagram is given in the figure below.

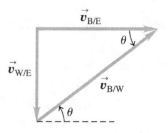

Solve: (a) Find the direction of $\vec{v}_{B/W}$. $\sin\theta = \dfrac{v_{W/E}}{v_{B/W}} = \dfrac{2.0 \text{ m/s}}{4.2 \text{ m/s}}$. $\theta = 28.4°$, north of east.

(b) $v_{B/E} = \sqrt{v_{B/W}^2 - v_{W/E}^2} = \sqrt{(4.2 \text{ m/s})^2 - (2.0 \text{ m/s})^2} = 3.7 \text{ m/s}$

(c) $t = \dfrac{800 \text{ m}}{v_{B/E}} = \dfrac{800 \text{ m}}{3.7 \text{ m/s}} = 216 \text{ s}.$

Reflect: It takes longer to cross the river in this problem than it did in Problem 3.40. In the direction straight across the river (east) the component of his velocity relative to the earth is less than 4.2 m/s.

*3.45. Set Up: For the motion of the football take $+y$ upward. $v_{0x} = v_0 \cos 40.0° = 9.19$ m/s, $v_{0y} = v_0 \sin 40.0° = 7.71$ m/s, $a_x = 0$, and $a_y = -9.80$ m/s^2. Use the y component motion to find the time the football is in the air. This same time applies to the horizontal motion.

Solve: When the ball returns to the ground, $y - y_0 = 0$. $y - y_0 = v_{0y}t + \frac{1}{2}a_y t^2$ gives $0 = v_{0y}t + \frac{1}{2}a_y t^2$ and

$$t = -\frac{2v_{0y}}{a_y} = -\frac{2(7.71 \text{ m/s})}{-9.80 \text{ m/s}^2} = 1.57 \text{ s}.$$

Then $x - x_0 = v_{0x}t + \frac{1}{2}a_x t^2 = (9.19 \text{ m/s})(1.57 \text{ s}) = 14.4$ m. In the 1.57 s that the ball is in the air the second player must travel $30.0 \text{ m} - 14.4 \text{ m} = 15.6$ m. Therefore, his (constant) velocity must be

$$v_x = \frac{x - x_0}{t} = \frac{15.6 \text{ m}}{1.57 \text{ s}} = 9.94 \text{ m/s}.$$

*3.47. Set Up: Take $+y$ downward. $v_{0x} = 64.0$ m/s, $v_{0y} = 0$.

Solve: Use the vertical motion to find the time in the air: $y - y_0 = v_{0y}t + \frac{1}{2}a_y t^2$ with $y - y_0 = 90.0$ m gives

$$t = \sqrt{\frac{2(y - y_0)}{a_y}} = \sqrt{\frac{2(90.0 \text{ m})}{9.80 \text{ m/s}^2}} = 4.29 \text{ s}.$$

Then $x - x_0 = (64.0 \text{ m/s})(4.29 \text{ s}) = 275$ m.

*3.51. Set Up: Example 3.5 gives $R = \dfrac{v_0^2 \sin 2\theta_0}{g}$.

Solve: We have the maximum range when $\theta_0 = 45°$ so $R_{max} = \dfrac{v_0^2}{g_{mars}} = \dfrac{(70.0 \text{ m/s})^2}{3.71 \text{ m/s}^2} = 1.32 \times 10^3$ m.

Reflect: The maximum range is inversely proportional to g. Thus, the range on the earth would be $(1320 \text{ m})\left(\dfrac{g_{mars}}{g_{earth}}\right) = 500$ m, ignoring air resistance.

*3.53. Set Up: Use coordinates with the origin at the boy and with $+y$ downward. The ball has $v_{0y} = 0$, $v_{0y} = 0$, $v_{0x} = 8.50$ m/s, $a_x = 0$ and $a_y = 9.80$ m/s^2.

Solve: (a) The dog must travel horizontally the same distance the ball travels horizontally, so the dog must have speed 8.50 m/s.

(b) Use the vertical motion of the ball to find its time in the air. $y - y_0 = v_{0y}t + \frac{1}{2}a_y t^2$ gives

$$t = \sqrt{\frac{2(y - y_0)}{a_y}} = \sqrt{\frac{2(12.0 \text{ m})}{9.80 \text{ m/s}^2}} = 1.56 \text{ s}$$

Then $x - x_0 = v_{0x}t + \frac{1}{2}a_x t^2 = (8.50 \text{ m/s})(1.56 \text{ s}) = 13.3$ m

*3.57. Set Up: Let $+y$ be downward. $a_x = 0$, $a_y = +9.80$ m/s^2. $v_{0x} = v_0 \cos\theta_0 = 5.36$ m/s, $v_{0y} = v_0 \sin\theta_0 = 4.50$ m/s.

Solve: Use the vertical motion to find the time in the air: $y - y_0 = v_{0y}t + \frac{1}{2}a_yt^2$ with $y - y_0 = 14.0$ m gives 14.0 m $= (4.50$ m/s$)t + (4.9$ m/s$^2)t^2$. The quadratic formula gives

$$t = \frac{1}{2(4.9)}\left(-4.50 \pm \sqrt{(4.50)^2 - 4(4.9)(-14.0)}\right) \text{ s.}$$

The positive root is $t = 1.29$ s. Then $x - x_0 = v_{0x}t + \frac{1}{2}a_xt^2 = (5.36$ m/s$)(1.29$ s$) = 6.91$ m.

*3.59. Set Up: The water moves in projectile motion. Let $x_0 = y_0 = 0$ and take $+y$ to be positive. $a_x = 0$, $a_y = -g$.

Solve: The equations of motions are $y = (v_0\sin\theta_0)t - \frac{1}{2}gt^2$ and $x = (v_0\cos\theta_0)t$. When the water goes in the tank for the *minimum* velocity, $y = 2D$ and $x = 6D$. When the water goes in the tank for the *maximum* velocity, $y = 2D$ and $x = 7D$. In both cases we have $\theta_0 = 45°$, so $\sin\theta_0 = \cos\theta_0 = \sqrt{2}/2$.

To reach the *minimum* distance: $6D = \frac{\sqrt{2}}{2}v_0 t$, and $2D = \frac{\sqrt{2}}{2}v_0 t - \frac{1}{2}gt^2$. Solving the first equation for t gives

$t = \frac{6D\sqrt{2}}{v_0}$. Substituting this into the second equation gives $2D = 6D - \frac{1}{2}g\left(\frac{6D\sqrt{2}}{v_0}\right)^2$. Solving this for v_0 gives

$v_0 = 3\sqrt{gD}$.

To reach the *maximum* distance: $7D = \frac{\sqrt{2}}{2}v_0 t$, and $2D = \frac{\sqrt{2}}{2}v_0 t - \frac{1}{2}gt^2$. Solving the first equation for t gives

$t = \frac{7D\sqrt{2}}{v_0}$. Substituting this into the second equation gives $2D = 7D - \frac{1}{2}g\left(\frac{7D\sqrt{2}}{v_0}\right)^2$. Solving this for v_0 gives

$v_0 = \sqrt{49gD/5} = 3.13\sqrt{gD}$, which, as expected, is larger than the previous result.

Reflect: A launch speed of $v_0 = \sqrt{6}\sqrt{gD} = 2.45\sqrt{gD}$ is required for a horizontal range of $6D$. The minimum speed required is greater than this, because the water must be at a height of at least $2D$ when it reaches the front of the tank.

*3.61. Set Up: For circular motion the acceleration has magnitude

$$a_{\text{rad}} = \frac{v^2}{R}$$

and is directed toward the center of the circle. The period T (time for one revolution) is

$$T = \frac{2\pi R}{v}.$$

Solve: (a) $a_{\text{rad}} = \frac{v^2}{R} = \frac{(7.00 \text{ m/s})^2}{14.0 \text{ m}} = 3.50$ m/s^2, upward.

(b) $a_{\text{rad}} = 3.50$ m/s^2, downward.

(c) $T = \frac{2\pi R}{v} = \frac{2\pi(14.0 \text{ m})}{7.00 \text{ m/s}} = 12.6$ s.

*3.65. Set Up: Use coordinates with $+y$ upward. Both the shell and the projectile have $a_x = 0$ and $a_y = -9.80$ m/s^2 as they move through the air.

Solve: (a) Find the maximum height of the shell: $v_{0y} = v_0\sin\theta_0 = 120$ m/s and $v_y = 0$

$$v_y{}^2 = v_{0y}{}^2 + 2a_y(y - y_0) \text{ gives } y - y_0 = \frac{v_y{}^2 - v_{0y}{}^2}{2a_y} = \frac{0 - (120 \text{ m/s})^2}{2(-9.80 \text{ m/s}^2)} = 735 \text{ m}$$

At its maximum height the shell has $v_x = v_{0x} = v_0 \cos 53° = 90.3$ m/s and $v_y = 0$. Relative to the ground the projectile has velocity components

$$v_{0x} = 90.3 \text{ m/s} + (100.0 \text{ m/s})\cos 30.0° = 176.9 \text{ m/s} \text{ and } v_{0y} = (100.0 \text{ m/s})\sin 30.0° = 50.0 \text{ m/s}$$

Find the maximum height of the projectile: $y_0 = 735$ m. $v_y = 0$. $v_y^2 = v_{0y}^2 + 2a_y(y - y_0)$ gives

$$y - y_0 = \frac{v_y^2 - v_{0y}^2}{2a_y} = \frac{0 - (50.0 \text{ m/s})^2}{2(-9.80 \text{ m/s}^2)} = 128 \text{ m}$$

and the maximum height is $y = 735 \text{ m} + 128 \text{ m} = 863$ m.

(b) Find the time for the shell to reach its maximum height: $v_y = v_{0y} + a_y t$ gives

$$t = \frac{v_y - v_{0y}}{a_y} = \frac{0 - 120 \text{ m/s}}{-9.80 \text{ m/s}^2} = 12.2 \text{ s}$$

During this time the shell has traveled horizontally a distance

$$x - x_0 = v_{0x}t = (v_0 \cos\theta_0)t = (150 \text{ m/s})(\cos 53°)(12.2 \text{ s}) = 1100 \text{ m}$$

Find the time for the projectile to reach the ground from the point where it is launched. It starts at $y_0 = 735$ m and ends up at $y = 0$, so $y - y_0 = -735$ m. $y - y_0 = v_{0y}t + \frac{1}{2}a_y t^2$ gives $-735 \text{ m} = (50.0 \text{ m/s})t - (4.90 \text{ m/s}^2)t^2$ and $t = 18.4$ s. During this time the projectile has a horizontal displacement $x - x_0 = v_{0x}t = (176.9 \text{ m/s})(18.4 \text{ s}) = 3255$ m. It is at $x = x_0 + 3255 \text{ m} = 1100 \text{ m} + 3255 \text{ m} = 4360$ m from the original launch point when it reaches the ground.

Reflect: If the shell didn't fire the projectile its horizontal range would be

$$R = \frac{v_0^2 \sin(2\theta_0)}{g} = \frac{(150 \text{ m/s})^2 (\sin 106°)}{9.80 \text{ m/s}^2} = 2210 \text{ m}.$$

The projectile gets an extra boost when it is fired and travels farther than this.

Solutions to Passage Problems

*3.67. Set Up: We may use the equation for the range of a projectile derived in Example 3.5: $R = \frac{v_0^2 \sin 2\theta_0}{g}$. Assume that v_0 is the same on both the earth and the moon. The actual value of θ_0 is irrelevant as long as it is the same on both the earth and the moon.

Solve: We compare the range on the moon to that on the earth: $\frac{R_{moon}}{R_{earth}} = \frac{v_0^2 \sin 2\theta_0/g_{moon}}{v_0^2 \sin 2\theta_0/g_{earth}} = \frac{g_{earth}}{g_{moon}} = 6$. Thus, the range on the moon is greater than that on the earth by a factor of 6. The correct answer is B.

*3.69. Set Up: We may use the equation for the maximum height of a projectile derived in Example 3.5: $h = \frac{v_0^2 \sin^2 \theta_0}{2g}$. Assume that v_0 is the same on both the earth and the moon. The actual value of θ_0 is irrelevant as long as it is the same on both the earth and the moon.

Solve: We compare the maximum height of the projectile on the moon to that on the earth:

$\frac{h_{moon}}{h_{earth}} = \frac{v_0^2 \sin^2 \theta_0/2g_{moon}}{v_0^2 \sin^2 \theta_0/2g_{earth}} = \frac{g_{earth}}{g_{moon}} = 6$. Thus, the maximum height of the projectile on the moon is greater than that on the earth by a factor of 6. The correct answer is B.

NEWTON'S LAWS OF MOTION

4

Problems 3, 5, 9, 11, 15, 19, 21, 23, 27, 29, 31, 35, 39, 41, 43, 47, 49, 51, 55, 59, 61

Solutions to Problems

*4.3. Set Up: The force and its horizontal and vertical components are shown in the figure below. The force is directed at an angle of $20.0° + 30.0° = 50.0°$ above the horizontal.

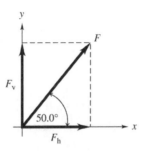

Solve: $F_h = F \cos 50.0° = 241$ N; $F_v = F \sin 50.0° = 287$ N.

Reflect: We could also find the components of $\vec{F}$ in the directions parallel and perpendicular to the incline but the problem asks for horizontal and vertical components.

*4.5. Set Up: Let $F_1 = 985$ N, $F_2 = 788$ N, and $F_3 = 411$ N. The angles θ that each force makes with the $+x$-axis are $\theta_1 = 31°$, $\theta_2 = 122°$, and $\theta_3 = 233°$.

Solve: **(a)** $F_{1x} = F_1 \cos \theta_1 = 844$ N; $F_{1y} = F_1 \sin \theta_1 = 507$ N

$$F_{2x} = F_2 \cos \theta_2 = -418 \text{ N}; \quad F_{2y} = F_2 \sin \theta_2 = 668 \text{ N}$$
$$F_{3x} = F_3 \cos \theta_3 = -247 \text{ N}; \quad F_{3y} = F_3 \sin \theta_3 = -328 \text{ N}$$

(b) $R_x = F_{1x} + F_{2x} + F_{3x} = 179$ N; $R_y = F_{1y} + F_{2y} + F_{3y} = 847$ N

$$R = \sqrt{R_x^2 + R_y^2} = 886 \text{ N}; \quad \tan \theta = \frac{R_y}{R_x} \text{ so } \theta = 78.1°$$

$\vec{R}$ and its components are shown in the figure below.

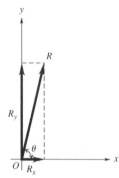

Reflect: Adding the forces as vectors gives a very different result from adding their magnitudes.

*4.9. Set Up: Take $+x$ to be in the direction in which the cheetah moves. $v_{0x} = 0$.

Solve: (a) $v_x = v_{0x} + a_x t$ so $a_x = \dfrac{v_x - v_{0x}}{t} = \dfrac{20.1 \text{ m/s} - 0}{2.0 \text{ s}} = 10.05 \text{ m/s}^2$

$$F_x = ma_x = (68 \text{ kg})(10.05 \text{ m/s}^2) = 680 \text{ N}$$

(b) The force is exerted on the cheetah by the ground.

Reflect: The net force on the cheetah is in the same direction as the acceleration of the cheetah.

*4.11. Set Up: Let $+x$ be the direction of the force. $\sum F_x = 80.0 \text{ N}$. Use the information about the motion to find the acceleration and then use $\sum F_x = ma_x$ to calculate m.

Solve: $x - x_0 = 11.0 \text{ m}$, $t = 5.00 \text{ s}$, $v_{0x} = 0$. $x - x_0 = v_{0x}t + \frac{1}{2}a_x t^2$ gives

$$a_x = \frac{2(x - x_0)}{t^2} = \frac{2(11.0 \text{ m})}{(5.00 \text{ s})^2} = 0.880 \text{ m/s}^2.$$

$$m = \frac{\sum F_x}{a_x} = \frac{80.0 \text{ N}}{0.880 \text{ m/s}^2} = 90.9 \text{ kg}.$$

Reflect: The mass determines the amount of acceleration produced by a given force.

*4.15. Set Up: $1 \mu g = 10^{-6} \text{ g} = 10^{-9} \text{ kg}$. $1 \text{ mg} = 10^{-3} \text{ g} = 10^{-6} \text{ kg}$. $1 \text{ N} = 0.2248 \text{ lb}$. $g = 9.80 \text{ m/s}^2 = 32.2 \text{ ft/s}^2$.

Solve: (a) $m = 210 \mu g = 2.10 \times 10^{-7} \text{ kg}$. $w = mg = (2.10 \times 10^{-7} \text{ kg})(9.80 \text{ m/s}^2) = 2.06 \times 10^{-6} \text{ N}$

(b) $m = 12.3 \text{ mg} = 1.23 \times 10^{-5} \text{ kg}$. $w = mg = (1.23 \times 10^{-5} \text{ kg})(9.80 \text{ m/s}^2) = 1.21 \times 10^{-4} \text{ N}$

(c) $(45 \text{ N})\left(\dfrac{0.2248 \text{ lb}}{1 \text{ N}}\right) = 10.1 \text{ lb}$. $m = \dfrac{w}{g} = \dfrac{45 \text{ N}}{9.80 \text{ m/s}^2} = 4.6 \text{ kg}$

*4.19. Set Up: The weight of an object depends on its mass and the value of g at its location and is independent of the motion of the object.

Solve: (a) 138 N **(b)** 138 N **(c)** $w = mg = (138 \text{ kg})(9.80 \text{ m/s}^2) = 1350 \text{ N}$, for both **(a)** and **(b)**. **(d)** They would be the same, 138 N.

Reflect: The weight of an object is the gravitational force exerted on it by the earth.

*4.21. Set Up: On Earth, $g_E = 9.80 \text{ m/s}^2$.

Solve: (a) $w_E = mg_E$ so $m = \dfrac{w_E}{g_E} = \dfrac{85.2 \text{ N}}{9.80 \text{ m/s}^2} = 8.69 \text{ kg}$. The mass m is independent of the location of the object.

$$w_M = mg_M \text{ so } g_M = \frac{w_M}{g} = \frac{32.2 \text{ N}}{8.69 \text{ kg}} = 3.71 \text{ m/s}^2.$$

(b) $m = 8.69 \text{ kg}$, the same at either location.

Reflect: The object weighs more on earth because the acceleration due to gravity is greater there.

*4.23. Set Up: Use coordinates for which the $+x$-direction is the direction the car is traveling initially. $m = 1750$ kg. When the car has stopped, $v_x = 0$.

Solve: (a) $f = 0.25mg$ and is in the $-x$-direction.

$$\Sigma F_x = ma_x \text{ gives } -f = ma_x \text{ and } a_x = \frac{-f}{m} = \frac{-0.25mg}{m} = -0.25g = -2.45 \text{ m/s}^2$$

The acceleration is 2.45 m/s^2, directed opposite to the motion.

(b) $v_{0x} = 110$ km/h $= 30.6$ m/s, $v_x = 0$, $a_x = -2.45$ m/s^2. $v_x^2 = v_{0x}^2 + 2a_x(x - x_0)$ so

$$x - x_0 = \frac{v_x^2 - v_{0x}^2}{2a_x} = \frac{0 - (30.6 \text{ m/s})^2}{2(-2.45 \text{ m/s}^2)} = 191 \text{ m}$$

Reflect: The stopping distance is proportional to the square of the initial speed.

*4.27. Set Up: The reaction forces in Newton's third law are always between a pair of objects. In Newton's second law all the forces act on a single object. Let $+y$ be downward. $m = w/g$.

Solve: The reaction to the upward normal force on the passenger is the downward normal force, also of magnitude 620 N, that the passenger exerts on the floor. The reaction to the passenger's weight is the gravitational force that the passenger exerts on the earth, upward and also of magnitude 650 N. $\dfrac{\Sigma F_y}{m} = a_y$ gives $a_y = \dfrac{650 \text{ N} - 620 \text{ N}}{(650 \text{ N})/(9.80 \text{ m/s}^2)} = 0.452 \text{ m/s}^2$. The

passenger's acceleration is 0.452 m/s^2, downward.

Reflect: There is a net downward force on the passenger and the passenger has a downward acceleration.

*4.29. Set Up: The brakes cause a friction force from the road that is directed opposite to the motion.

Solve: (a) The free-body diagram is shown in Figure (a) below. w is the gravity force, n is the normal force, and f is the friction force. The friction force is directed opposite to the direction the car is traveling.

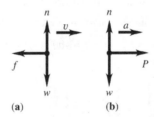

(a)   (b)

(b) The free-body diagram is shown in Figure (b) above. The force P is applied to the passenger by the car seat and is in the direction the car is moving.

Reflect: In each case there is a net horizontal force on the object and this force produces an acceleration in the direction of the net force.

*4.31. Set Up: Identify the forces on each object. In each case the forces are the noncontact force of gravity (the weight) and the forces applied by objects that are in contact with each crate. Each crate touches the floor and the other crate, and some object applies $\vec{F}$ to crate A.

Solve: (a) The free-body diagrams for each crate are given in the figure below.

F_{AB} (the force on m_A due to m_B) and F_{BA} (the force on m_B due to m_A) form an action-reaction pair.

(b) Since there is no horizontal force opposing F, any value of F, no matter how small, will cause the crates to accelerate to the right. The weight of the two crates acts at a right angle to the horizontal and is in any case balanced by the upward force of the surface on them.

Reflect: Crate B is accelerated by F_{BA} and crate A is accelerated by the net force $F - F_{AB}$. The greater the total weight of the two crates, the greater their total mass and the smaller will be their acceleration.

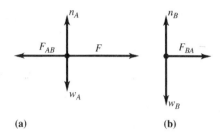

(a)   (b)

*4.35. Set Up: Call the blocks A and B, with $w_A = 850$ N and $w_B = 750$ N. Since the worker pulls to the right, the friction force on each block is to the left. Let $\vec{F}$ be the force the worker applies and let T be the tension in the rope. **Solve:** The free-body diagrams are shown in the figure below. The rope pulls to the right on block A and to the left on block B.

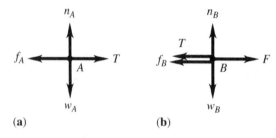

(a)   (b)

Reflect: The force $\vec{F}$ acts only on block B.

*4.39. Set Up: Take $+y$ to be upward. The mass of the bucket is $m_B = w_B/g = 28.1$ kg and the mass of the chain is $m_C = w_C/g = 12.8$ kg. The chain also has an upward acceleration of 2.50 m/s^2.
Solve: (a) Consider the chain and bucket as a combined object of mass $m_{tot} = 28.1$ kg $+ 12.8$ kg $= 40.9$ kg and weight $w_{tot} = 400$ N. The free-body diagram is shown in Figure (a) below. T_t is the tension in the top link. $\Sigma F_y = ma_y$ gives $T_t - w_{tot} = m_{tot}a$ and $T_t = w_{tot} + m_{tot}a = 400$ N $+ (40.9$ kg$)(2.50$ m/s$^2) = 502$ N

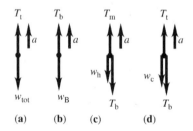

(a)   (b)   (c)   (d)

(b) Apply $\Sigma F_y = ma_y$ to the bucket. The free-body diagram (Figure (b) above) gives $T_b - w_B = m_B a$ so $T_b = w_B + m_B a =$ 275 N $+ (28.1$ kg$)(2.50$ m/s$^2) = 345$ N
(c) Apply $\Sigma F_y = ma_y$ to the bottom half of the chain. The free-body diagram is shown in Figure (c) above. The bottom half of the chain has mass $m_h = 6.4$ kg and weight $w_h = 62.5$ N. T_m is the tension in the middle link of the chain. $\Sigma F_y = ma_y$ gives $T_m - w_h - T_b = m_h a$.

$$T_m = w_h + T_b + m_h a = 62.5 \text{ N} + 345 \text{ N} + (6.4 \text{ kg})(2.5 \text{ m/s}^2) = 424 \text{ N}$$

Reflect: Part (b) can also be done by applying $\Sigma F_y = ma_y$ to the chain. The free-body diagram is given in Figure (d) above.
$\Sigma F_y = ma_y$ gives $T_t - w_c - T_b = m_c a$ and $T_b = T_t - w_c - m_c a = 502$ N $- 125$ N $- (12.8$ kg$)(2.50$ m/s$^2) = 345$ N, which agrees with the value obtained previously. The tension increases from the bottom to the top of the chain.

*4.41. Set Up: Identify the forces on the chair. The floor exerts a normal force and a friction force. Let $+y$ be upward and let $+x$ be in the direction of the motion of the chair.

Solve: (a) The free-body diagram for the chair is given in the figure below.
(b) For the chair, $a_y = 0$ so $\Sigma F_y = ma_y$ gives $n - mg - F\sin 37° = 0$ and $n = 142$ N.

Reflect: n is larger than the weight because $\vec{F}$ has a downward component.

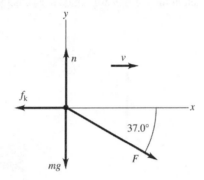

*4.43. Set Up: Your mass is $m = w/g = 63.8$ kg. Both you and the package have the same acceleration as the elevator. Take $+y$ to be upward, in the direction of the acceleration of the elevator.

Solve: (a) Your free-body diagram is shown in Figure (a) below. n is the scale reading. $\Sigma F_y = ma_y$ gives $n - w = ma$ and $n = w + ma = 625$ N $+ (63.8$ kg$)(2.50$ m/s$^2) = 784$ N

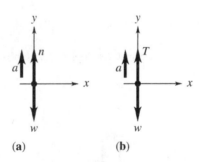

(a)   (b)

(b) The free-body diagram for the package is given in Figure (b) above. $\Sigma F_y = ma_y$ gives $T - w = ma$ so $T = w + ma =$
$(3.85$ kg$)(9.80$ m/s$^2 + 2.50$ m/s$^2) = 47.4$ N
Reflect: The objects accelerate upward so for each object the upward force is greater than the downward force.

*4.47. Set Up: Use constant acceleration equations to calculate the acceleration a_x that would be required. Then use $\Sigma F_x = ma_x$ to find the necessary force. Let $+x$ be the direction of the initial motion of the auto.

Solve: $v_x^2 = v_{0x}^2 + 2a_x(x - x_0)$ with $v_x = 0$ gives $a_x = -\dfrac{v_{0x}^2}{2(x - x_0)}$. The force F is directed opposite to the motion

and $a_x = -\dfrac{F}{m}$. Equating these two expressions for a_x gives

$$F = m\frac{v_{0x}^2}{2(x - x_0)} = (850 \text{ kg})\frac{(12.5 \text{ m/s})^2}{2(1.8 \times 10^{-2} \text{ m})} = 3.7 \times 10^6 \text{ N.}$$

Reflect: A very large force is required to stop such a massive object in such a short distance.

*4.49. Set Up: Apply $\Sigma \vec{F} = m\vec{a}$ to the parachutist. Let $+y$ be upward. $\vec{F}_{air}$ is the force of air resistance.

Solve: (a) $w = mg = (55.0 \text{ kg})(9.80 \text{ m/s}^2) = 539 \text{ N}$

(b) The free-body diagram is given in the figure below. $\Sigma F_y = F_{air} - w = 620 \text{ N} - 539 \text{ N} = 81 \text{ N}$. The net force is upward.

(c) $a_y = \dfrac{\Sigma F_y}{m} = \dfrac{81 \text{ N}}{55.0 \text{ kg}} = 1.5 \text{ m/s}^2$. The acceleration is upward.

Reflect: Both the net force and the acceleration are upward. Since her velocity is downward and her acceleration is upward, her speed decreases.

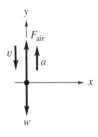

*4.51. Set Up: Let $+y$ be upward. At his maximum height, $v_y = 0$. While he is in the air, $a_y = -9.80 \text{ m/s}^2$. $m = w/g = 90.8 \text{ kg}$

Solve: (a) $v_{0y} = 0$, $a_y = -9.80 \text{ m/s}^2$ and $y - y_0 = 1.2 \text{ m}$. $v_y^2 = v_{0y}^2 + 2a_y(y - y_0)$ gives

$$v_{0y} = \sqrt{-2a_y(y - y_0)} = \sqrt{-2(-9.80 \text{ m/s}^2)(1.2 \text{ m})} = 4.85 \text{ m/s}$$

(b) For the motion while he is pushing against the floor, $v_{0y} = 0$, $v_y = 4.85 \text{ m/s}$, and $t = 0.300 \text{ s}$. $v_y = v_{0y} + a_y$ gives

$a_y = \dfrac{v_y - v_{0y}}{t} = \dfrac{4.85 \text{ m/s} - 0}{0.300 \text{ s}} = 16.2 \text{ m/s}^2$. The acceleration is upward.

(c) The free-body diagram while he is pushing against the floor is given in the figure below. $\vec{F}$ is the vertical force the floor applies to him.

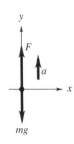

(d) $\Sigma F_y = ma_y$ gives $F - mg = ma$ and $F = m(g + a) = (90.8 \text{ kg})(9.80 \text{ m/s}^2 + 16.2 \text{ m/s}^2) = 2.36 \times 10^3 \text{ N}$. The force he applied to the ground has this same magnitude and is downward.

Reflect: The ground must push upward on him with a force greater than his weight in order to give him an upward acceleration.

*4.55. Set Up: Apply $\Sigma \vec{F} = m\vec{a}$ to the barbell and to the athlete. Use the motion of the barbell to calculate its acceleration. Let $+y$ be upward.

Solve: (a) The free-body diagrams for the barbell and for the athlete are sketched in the figure below.

(b) The athlete's weight is $mg = (90.0 \text{ kg})(9.80 \text{ m/s}^2) = 882 \text{ N}$. The upward acceleration of the barbell is found from

$y - y_0 = v_{0y}t + \frac{1}{2}a_y t^2$. $a_y = \dfrac{2(y - y_0)}{t^2} = \dfrac{2(0.600 \text{ m})}{(1.6 \text{ s})^2} = 0.469 \text{ m/s}^2$. The force needed to lift the barbell is given by

$F_{\text{lift}} - w_{\text{barbell}} = ma_y$. The barbell's mass is $(490 \text{ N})/(9.80 \text{ m/s}^2) = 50.0 \text{ kg}$, so $F_{\text{lift}} = w_{\text{barbell}} + ma = 490 \text{ N} + (50.0 \text{ kg})$

$(0.469 \text{ m/s}^2) = 490 \text{ N} + 23 \text{ N} = 513 \text{ N}$.

The athlete is not accelerating, so $F_{\text{floor}} - F_{\text{lift}} - w_{\text{athlete}} = 0$. Thus, $F_{\text{floor}} = F_{\text{lift}} + w_{\text{athlete}} = 513 \text{ N} + 882 \text{ N} = 1395 \text{ N}$, which is the total force that the ground exerts on the athlete's feet as he lifts the barbell.

Reflect: Since the athlete pushes upward on the barbell with a force greater than its weight, the barbell pushes down on him with a force that is greater than its weight, and so the normal force on the athlete is greater than the total weight, 1372 N, of the athlete plus barbell.

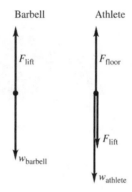

Solutions to Passage Problems

*4.59. Set Up: We can find the magnitude of the viscous force from Newton's second law: $F = ma$. The mass of the bacterium is $10^{-12} \text{ g} = 10^{-15} \text{ kg}$.

Solve: The viscous force needed to stop the bacterium in the required distance is $F = ma = (1 \times 10^{-15} \text{ kg})$ $(5 \times 10^5 \text{ m/s}^2) = 5 \times 10^{-10} \text{ N}$. The correct answer is D.

*4.61. Set Up: According to Newton's Second Law $\sum F_x = ma_x$.

Solve: To change from a constant speed (where the net force is zero) to an acceleration of 0.001 m/s^2, the flagellum must increase the net force to $F = ma = (10^{-15} \text{ kg})(10^{-3} \text{ m/s}^2) = 10^{-18} \text{ N}$. The correct answer is A.

5

APPLICATIONS OF NEWTON'S LAWS

Problems 1, 5, 7, 11, 15, 17, 21, 23, 25, 29, 31, 33, 37, 39, 41, 45, 47, 51, 53, 55, 61, 63, 69, 71, 73, 77, 81, 83, 87, 89, 91

Solutions to Problems

*5.1. Set Up: Constant speed means $a = 0$. Use coordinates where $+y$ is upward. The breaking strength for the rope in each part is (a) 50 N, (b) 4000 N, and (c) 6×10^4 N. The free-body diagram for the bucket plus its contents is given in the figure below.

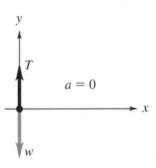

Solve: $\Sigma F_y = ma_y$ gives $T - w = 0$ so $T = w$.

(a) $w = 50$ N, so 35 N of cement. The mass of cement is 3.6 kg.

(b) $w = 4000$ N, so 3985 N of cement. The mass of cement is 410 kg.

(c) $w = 6 \times 10^4$ N, so 6×10^4 N of cement. The mass of cement is 6×10^3 kg.

Reflect: Since $a = 0$, the upward pull of the rope equals the total weight of the object.

*5.5. Set Up: Take $+y$ upward. Apply $\Sigma \vec{F} = m\vec{a}$ to the person. $a = 0$.
Solve: (a) The free-body diagram for the person is given in the figure below. The two sides of the rope each exert a force with vertical component $T \sin \theta$. $\Sigma F_y = ma_y$ gives $2T \sin \theta - w = 0$.

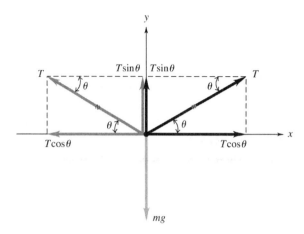

$$T = \frac{mg}{2\sin\theta} = \frac{(90.0 \text{ kg})(9.80 \text{ m/s}^2)}{2\sin 10.0°} = 2.54 \times 10^3 \text{ N}$$

(b) Set $T = 2.50 \times 10^4 \text{ N}$ and solve for θ:

$$\sin\theta = \frac{mg}{2T} = \frac{(90.0 \text{ kg})(9.80 \text{ m/s}^2)}{2(2.50 \times 10^4 \text{ N})} \text{ and } \theta = 1.01°.$$

Reflect: Only the vertical components of the tension on each side of the person act to hold him up. The tension in the rope is much greater than his weight.

*5.7. Set Up: The SI units for force and area are N and m^2. $1 \text{ N} = 1 \text{ kg} \cdot \text{m/s}^2$.

Solve: (a) $\sigma = \dfrac{F_{\text{max}}}{A}$ so σ has units of N/m^2 or $\text{kg/m} \cdot \text{s}^2$

(b) $A = \dfrac{F_{\text{max}}}{\sigma} = \dfrac{755 \text{ N}}{3.0 \times 10^5 \text{ N/m}^2} = 2.5 \times 10^{-3} \text{ m}^2 = 25 \text{ cm}^2$

*5.11. Set Up: Use coordinates with $+y$ upward and $+x$ to the right. Call the objects A and B, with $w_A = 175 \text{ N}$ and $m_B = 32 \text{ kg}$. Label the tensions in the three wires T_1, T_2, and T_3.

Solve: (a) The free-body diagrams for each object are given in the figure below. T_3 has been replaced by its x and y components.

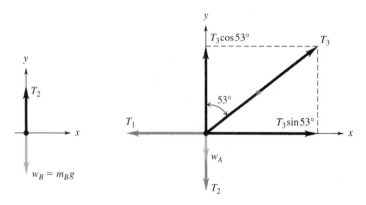

(b) $\Sigma F_y = 0$ for B gives $T_2 - w_B = 0$ and $T_2 = m_B g = 314 \text{ N}$. $\Sigma F_y = 0$ for A gives $T_3 \cos 53° - T_2 - w_A = 0$ so $T_3 = \dfrac{T_2 + w_A}{\cos 53°} = 813 \text{ N}$. $\Sigma F_x = 0$ for A gives $T_3 \sin 53° - T_1 = 0$ and $T_1 = 649 \text{ N}$.

(c) The tensions are unaffected by the length of the wires, so long as the third wire still makes a $53°$ angle with one wall.

Reflect: $T_3 \cos 53°$ equals the combined weight of both objects, 489 N.

*5.15. Set Up: Apply $\Sigma \vec{F} = m\vec{a}$ to each block. $a = 0$. Take $+y$ perpendicular to the incline and $+x$ parallel to the incline.

Solve: The free-body diagrams for each block, A and B, are given in the figure below.

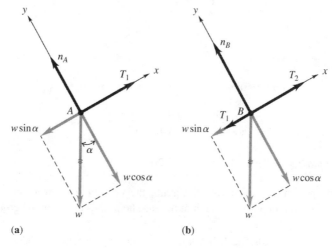

(a)   (b)

(a) For B, $\Sigma F_x = ma_x$ gives $T_1 - w \sin \alpha = 0$ and $T_1 = w \sin \alpha$.

(b) For block A, $\Sigma F_x = ma_x$ gives $T_1 - T_2 - w \sin \alpha = 0$ and $T_2 = 2w \sin \alpha$.

(c) $\Sigma F_y = ma_y$ for each block gives $n_A = n_B = w \cos \alpha$.

(d) For $\alpha \to 0$, $T_1 = T_2 \to 0$ and $n_A = n_B \to w$. For $\alpha \to 90°$, $T_1 = w$, $T_2 = 2w$ and $n_A = n_B = 0$.

*5.17. Set Up: $60g = 588 \text{ m/s}^2$. Use coordinates where the $+x$ direction is in the direction of the acceleration, opposite to the initial velocity of the car.

Solve: (a) The free-body diagram for the person is given in the figure below. F is the force the air bag exerts and n is the normal force from the car seat. $\Sigma F_x = ma_x$ gives $F = ma = (75 \text{ kg})(588 \text{ m/s}^2) = 4.41 \times 10^4 \text{ N}$

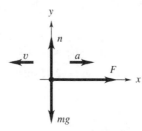

(b) $\dfrac{F}{w} = \dfrac{F}{mg} = \dfrac{4.41 \times 10^4 \text{ N}}{(75 \text{ kg})(9.80 \text{ m/s}^2)} = 60$, so $F = 60w$.

*5.21. Set Up: Take $+y$ to be upward. After he leaves the ground the person travels upward 60 cm and his acceleration is $g = 9.80 \text{ m/s}^2$, downward. His weight is W so his mass is W/g.

Solve: (a) $v_y = 0$ (at the maximum height), $y - y_0 = 0.60 \text{ m}$, $a_y = -9.80 \text{ m/s}^2$. $v_y^2 = v_{0y}^2 + 2a_y(y - y_0)$ gives

$$v_{0y} = \sqrt{-2a_y(y - y_0)} = \sqrt{-2(-9.80 \text{ m/s}^2)(0.60 \text{ m})} = 3.4 \text{ m/s}.$$

(b) The free-body diagram for the person while he is pushing up against the ground is given in the figure below.

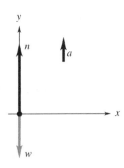

(c) For the jump, $v_{0y} = 0$, $v_y = 3.4$ m/s (from part (a)), $y - y_0 = 0.50$ m. $v_y^2 = v_{0y}^2 + 2a_y(y - y_0)$ gives

$$a_y = \frac{v_y^2 - v_{0y}^2}{2(y - y_0)} = \frac{(3.4 \text{ m/s})^2 - 0}{2(0.50 \text{ m})} = 11.6 \text{ m/s}^2$$

$\Sigma F_y = ma_y$ gives $n - W = ma$. $n = W + ma = W\left(1 + \dfrac{a}{g}\right) = 2.2W$

Reflect: To accelerate the person upward during the jump, the upward force from the ground must exceed the downward pull of gravity. The ground pushes up on him because he pushes down on the ground.

*5.23. Set Up: Take $+y$ to be upward. The fish has the same upward acceleration as the elevator. Let $\vec{F}$ be the upward force exerted on the fish by the spring balance; F is what the balance reads.

Solve: (a) The free-body diagram for the fish is sketched in the figure below.

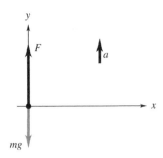

$\Sigma F_y = ma_y$ gives $F - mg = ma$ and

$$m = \frac{F}{g + a} = \frac{60.0 \text{ N}}{9.80 \text{ m/s}^2 + 2.45 \text{ m/s}^2} = 4.90 \text{ kg}.$$

The weight of the fish is $w = mg = 48.0$ N.

(b) $F - mg = ma_y$ and $m = 4.90$ kg from part (a).

$$a = \frac{F - mg}{m} = \frac{35.0 \text{ N} - (4.90 \text{ kg})(9.80 \text{ m/s}^2)}{4.90 \text{ kg}} = -2.65 \text{ m/s}^2$$

The elevator is accelerating downward with $a = 2.65$ m/s^2.

(d) If the cable breaks, $a_y = -9.80$ m/s^2. $F - mg = m(-9.80 \text{ m/s}^2)$ and $F = 0$. The elevator is in free-fall and the balance reads zero.

*5.25. Set Up: The camera is at rest relative to the train car and has the same horizontal acceleration as the car. Use coordinates with $+y$ upward and $+x$ in the direction of motion of the train car.

Solve: (a) The free-body-diagram is given in the figure below. The tension T in the strap has been replaced by its x and y components. $\phi = 12.0°$

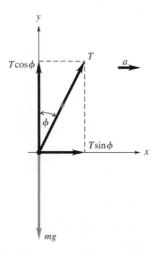

(b) $\Sigma F_x = ma_x$ gives $T\sin\phi = ma$. $\Sigma F_y = ma_y$ gives $T\cos\phi - mg = 0$. Combining these two equations to eliminate T gives $a = g\tan\phi = 2.08 \text{ m/s}^2$. And the y component equation gives

$$T = \frac{mg}{\cos\phi} = \frac{3.00 \text{ N}}{\cos 12.0°} = 3.07 \text{ N}$$

Reflect: As a increases there must be a larger horizontal component of T and the angle ϕ increases.

*5.29. Set Up: Let $m_1 = 50.0$ kg and $m_2 = 30.0$ kg. The two boxes have the same magnitude of acceleration. For the 30.0 kg box let $+y$ be downward and for the 50.0 kg box use coordinates parallel and perpendicular to the ramp, with $+x$ up the ramp.

Solve: (a) The free-body diagrams are given in the figure below. $\phi = 30.0°$

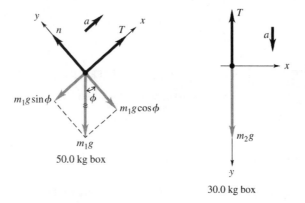

(b) The downward force on the 30.0 kg box is $m_2 g = 294$ N. The force pulling the system in the opposite direction is the component of $m_1 g$ directed down the ramp, $m_1 g\sin\phi = 245$ N. $m_2 g > m_1 g\sin\phi$ and the 30.0 kg box moves downward. Therefore, the 50.0 kg box moves up the ramp.

(c) $\Sigma F_x = ma_x$ applied to m_1 gives $T - m_1 g\sin\phi = m_1 a$. $\Sigma F_y = ma_y$ applied to m_2 gives $m_2 g - T = m_2 a$. Combining these two equations to eliminate T gives

$$a = \frac{m_2 g - m_1 g \sin\phi}{m_1 + m_2} = \frac{294\ \text{N} - 245\ \text{N}}{50.0\ \text{kg} + 30.0\ \text{kg}} = 0.612\ \text{m/s}^2$$

The 50.0 kg box accelerates up the ramp at $0.612\ \text{m/s}^2$ and the 30.0 kg box accelerates downward at $0.612\ \text{m/s}^2$.
Reflect: Only the component of the weight of the 50 kg box that is parallel to the ramp acts to oppose the weight of the 30 kg box. Therefore, the 30 kg box pulls the 50 kg box up the ramp even though its weight is less than the weight of the 50 kg box.

*5.31. Set Up: Since the only vertical forces are n and w, the normal force on the box equals its weight. Static friction is as large as it needs to be to prevent relative motion between the box and the surface, up to its maximum possible value of $f_s^{\text{max}} = \mu_s n$. If the box is sliding then the friction force is $f_k = \mu_k n$.
Solve: **(a)** If there is no applied force, no friction force is needed to keep the box at rest.
(b) $f_s^{\text{max}} = \mu_s n = (0.40)(40.0\ \text{N}) = 16.0\ \text{N}$. If a horizontal force of 6.0 N is applied to the box, then $f_s = 6.0\ \text{N}$ in the opposite direction.
(c) The monkey must apply a force equal to f_s^{max}, 16.0 N.
(d) Once the box has started moving, a force equal to $f_k = \mu_k n = 8.0\ \text{N}$ is required to keep it moving at constant velocity.
Reflect: $\mu_k < \mu_s$ and less force must be applied to the box to maintain its motion than to start it moving.

*5.33. Set Up: The free-body diagram for the two crates treated as a single object, weight w_C, is shown in Figure (a) below. The system doesn't move so the friction force exerted by the roof is static friction. For the heaviest pallet of bricks this force has its maximum possible, $f_s = \mu_s n$. The free-body diagram for the pallet of bricks is given in Figure (b) below.

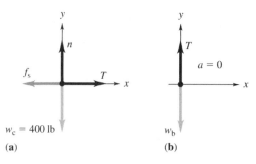

(a)   (b)

Solve: **(a)** For the crates, $\Sigma F_y = ma_y$ gives $n - w_C = 0$ and $n = 400\ \text{lb}$. Then

$$f_s = \mu_s n = (0.666)(400\ \text{lb}) = 266\ \text{lb}.$$

$\Sigma F_x = ma_x$ gives $T - f_s = 0$ and $T = f_s = 266\ \text{lb}$
For the bricks, $\Sigma F_y = ma_y$ gives $T - w_b = 0$ and $w_b = T = 266\ \text{lb}$
(b) For the upper crate the only horizontal force on the crate would be friction. This crate has $a_x = 0$ so $\Sigma F_x = 0$ and the friction force is zero.
Reflect: If some bricks are removed, so the weight of the pallet is reduced, the system remains at rest. The friction force on the crate is equal to the new weight of the pallet and is less than $\mu_s n$.

*5.37. Set Up: Apply $\Sigma \vec{F} = m\vec{a}$ to the box and calculate the normal and friction forces. The coefficient of kinetic friction is the ratio $\dfrac{f_k}{n}$. Let $+x$ be in the direction of motion. $a_x = -0.90\ \text{m/s}^2$. The box has mass 8.67 kg.
Solve: The normal force has magnitude $85\ \text{N} + 25\ \text{N} = 110\ \text{N}$. The friction force, from $F_H - f_k = ma$ is

$$f_k = F_H - ma = 20\ \text{N} - (8.67\ \text{kg})(-0.90\ \text{m/s}^2) = 28\ \text{N}.\quad \mu_k = \frac{28\ \text{N}}{110\ \text{N}} = 0.25.$$

Reflect: The normal force is greater than the weight of the box, because of the downward component of the push force.

*5.39. Set Up: $n = mg$. For constant speed, $a_x = 0$. Apply $\Sigma F_x = ma_x$ with $+x$ in the direction of motion of the box.

Solve: (a) The free-body diagram for the box is given in the figure below. $\Sigma F_x = ma_x$ gives $P - f_k = 0$ and

$P = f_k = \mu_k mg = (0.200)(11.2 \text{ kg})(9.80 \text{ m/s}^2) = 22.0 \text{ N}.$

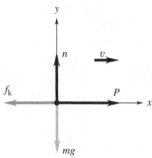

(b) $P = 0$ and $-f_k = ma_x$. $a_x = -\dfrac{f_k}{m} = -\mu_k g = -1.96 \text{ m/s}^2.$

Reflect: In (b) the friction force and the acceleration are directed opposite to the motion. Since $\vec{a}$ and $\vec{v}$ are in opposite directions, the box slows down.

*5.41. Set Up: The free-body diagram for the box is given in the figure below. When the box is either at rest or sliding at constant speed, the acceleration of the box is zero. Use coordinates as shown, with the x axis parallel to the incline. The weight mg has been resolved into its x and y components.

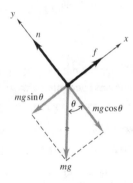

Solve: $\Sigma F_y = ma_y$ so $n - mg\cos\theta = 0$ and $n = mg\cos\theta$. $\Sigma F_x = ma_x$ so $f - mg\sin\theta = 0$

$f = \mu n = \mu mg\cos\theta$ so $\mu mg\cos\theta = mg\sin\theta$ and $\mu = \tan\theta$. At $\theta = \theta_1$, $f = f_s$ and $\mu_s = \tan\theta_1$. At $\theta = \theta_2$, $f = f_k$ and $\mu_k = \tan\theta_2$.

Reflect: Usually $\mu_k < \mu_s$, so $\theta_2 < \theta_1$. If the angle is kept at θ_1 the box will accelerate down the incline once it has been set into motion.

*5.45. Set Up: The free-body diagram for the crate is given in the figure below. Use coordinates with axes parallel and perpendicular to the ramp. Constant speed means $a = 0$. The mass of the crate is $m = w/g = 38.3 \text{ kg}$. The weight mg has been resolved in to its x and y components.

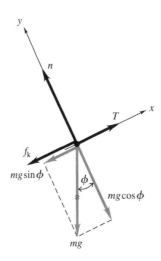

Solve: **(a)** $\Sigma F_y = ma_y$ gives $n = mg\cos\phi$. $f_k = \mu_k n = \mu_k mg\cos\phi$. $\Sigma F_x = ma_x$ with $a_x = 0$ gives $T - mg\sin\phi - f_k = 0$, so

$$T = mg\sin\phi + f_k = (375\text{ N})\sin 33° + (0.250)(375\text{ N})\cos 33° = 283\text{ N}$$

(b) $T = 0$ so $-f_k - mg\sin\phi = ma_x$

$$-\mu_k mg\cos\phi - mg\sin\phi = ma_x \text{ and } a_x = -g(\mu_k\cos\phi + \sin\phi) = -0.754g$$

The acceleration would be $0.754g = 7.39\text{ m/s}^2$, directed down the ramp.

Reflect: In part (a) the force up the incline is 283 N and the force down the incline is 283 N. When the rope breaks $T \to 0$ and the force down the incline remains 283 N. The acceleration down the incline is $(283\text{ N})/(38.3\text{ kg}) = 7.39\text{ m/s}^2$.

*5.47. Set Up: Apply $\Sigma\vec{F} = m\vec{a}$ to the box. When the box is ready to slip, the static friction force has its maximum possible value, $f_s = \mu_s n$. Use coordinates parallel and perpendicular to the ramp.

Solve: **(a)** The normal force will be $w\cos\alpha$ and the component of the gravitational force along the ramp is $w\sin\alpha$. The box begins to slip when $w\sin\alpha > \mu_s w\cos\alpha$, or $\tan\alpha > \mu_s = 0.35$, so slipping occurs at $\alpha = \arctan(0.35) = 19.3°$.

(b) When moving, the friction force along the ramp is $\mu_k w\cos\alpha$, the component of the gravitational force along the ramp is $w\sin\alpha$, so the acceleration is $(w\sin\alpha - w\mu_k\cos\alpha)/m = g(\sin\alpha - \mu_k\cos\alpha) = 0.92\text{ m/s}^2$.

(c) Since $v_{0x} = 0$, $2ax = v^2$, so $v = (2ax)^{1/2}$, or $v = [(2)(0.92\text{m/s}^2)(5\text{ m})]^{1/2} = 3\text{ m/s}$.

Reflect: When the box starts to move, friction changes from static to kinetic and the friction force becomes smaller.

*5.51. Set Up: From Example 5.13 the terminal velocity of the raindrop is $v_T = \sqrt{\dfrac{mg}{D}}$. We know D, so we must find the mass of the raindrop. The density of water is 10^3 kg/m^3 and the volume of a sphere is $V = \dfrac{4}{3}\pi r^3$.

Solve: The volume of the raindrop is $V = \dfrac{4}{3}\pi r^3 = \dfrac{4}{3}\pi(4.15\times10^{-3}\text{ m})^3 = \underline{2.994\times10^{-7}}\text{ m}^3$. The mass of the raindrop is $\dfrac{10^3\text{ kg}}{1\text{ m}^3}\cdot(2.994\times10^{-7}\text{ m}^3) = \underline{2.994\times10^{-4}}\text{ kg}$. Thus, the terminal velocity of the raindrop is

$$v_T = \sqrt{\frac{(2.994\times10^{-4}\text{ kg})(9.80\text{ m/s}^2)}{2.43\times10^{-5}\text{ kg/m}}} = 11.0\text{ m/s}.$$

Reflect: At terminal velocity the upward drag force on the raindrop equals the downward gravitational force it experiences, and the acceleration of the raindrop is zero. A radius of 4 mm represents a very large raindrop, but a speed of 11 m/s (about 25 mph) is reasonable for a raindrop or hailstone of this size.

*5.53. Set Up: Take $+y$ to be downward.

Solve: (a) $F_D = Dv^2$. The F_D versus v graph is sketched in Figure (a) below.

(b) $F_{net} = -F_D + mg = -Dv^2 + mg$. See Figure (b) below.

(c) $v = 0$ at $t = 0$ and $v \to v_t$ as t increases. When $v = v_t$, $F_{net} = 0$. See Figure (c) below.

(d) See Figure (d) below.

(e) $a = F_{net}/m$; graph has same shape as F_{net} versus t. See Figure (e) below.

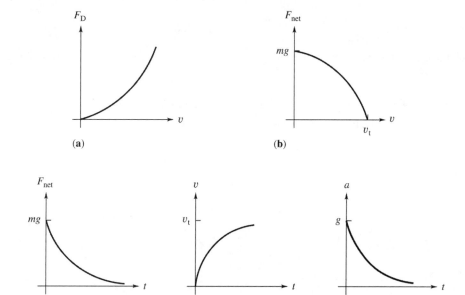

*5.55. Set Up: $|F_{spr}| = mg$ so $mg = kx$. x is the change in length of the spring.

Solve: (a) $k = \dfrac{mg}{x} = \dfrac{(0.875\ \text{kg})(9.80\ \text{m/s}^2)}{0.0240\ \text{m}} = 357\ \text{N/m}$

(b) $m = \dfrac{kx}{g} = \dfrac{(357\ \text{N/m})(0.0572\ \text{m})}{9.80\ \text{m/s}^2} = 2.08\ \text{kg}$

Reflect: The elongation of the vertical spring is proportional to the mass hung from it. So, the mass in (b) is

$$\left(\frac{5.72\ \text{cm}}{2.40\ \text{cm}} \right)(0.875\ \text{kg}) = 2.08\ \text{kg.}$$

*5.61. Set Up: The magnitude of the spring force is kx. The force diagrams are given in the figure below.

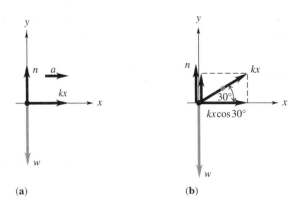

(a)   **(b)**

Solve: (a) $kx = ma$ and $x = \dfrac{ma}{k} = \dfrac{(9.50 \text{ kg})(2.00 \text{ m/s}^2)}{125 \text{ N/m}} = 0.152 \text{ m} = 15.2 \text{ cm}$

(b) $kx\cos 30.0° = ma$ and $x = \dfrac{ma}{k\cos 30.0°} = \dfrac{0.152 \text{ m}}{\cos 30.0°} = 0.176 \text{ m} = 17.6 \text{ cm}$

Reflect: Only the horizontal component of the spring force accelerates the sled, so a greater spring force is needed in part (b).

*5.63. Set Up: Take $+y$ downward. (a) Assume the hip is in free-fall. (b) The free-body diagram for the person is given in the figure below. It is assumed that the whole mass of the person has the same acceleration as her hip.

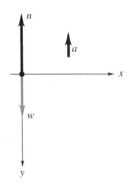

Solve: (a) $v_{0y} = 0$, $y - y_0 = 1.0 \text{ m}$, $a_y = +9.80 \text{ m/s}^2$. $v_y^2 = v_{0y}^2 + 2a_y(y - y_0)$ gives

$$v_y = \sqrt{2a_y(y - y_0)} = \sqrt{2(9.80 \text{ m/s}^2)(1.0 \text{ m})} = 4.4 \text{ m/s}$$

(b) $v_{0y} = 4.4 \text{ m/s}$, $y - y_0 = 0.020 \text{ m}$, $v_y = 1.3 \text{ m/s}$. $v_y^2 = v_{0y}^2 + 2a_y(y - y_0)$ gives

$$a_y = \frac{v_y^2 - v_{0y}^2}{2(y - y_0)} = \frac{(1.3 \text{ m/s})^2 - (4.4 \text{ m/s})^2}{2(0.020 \text{ m})} = -440 \text{ m/s}^2$$

The acceleration is 440 m/s^2, upward. $\Sigma F_y = ma_y$ gives $w - n = -ma$ and

$n = w + ma = m(a + g) = (55 \text{ kg})(440 \text{ m/s}^2 + 9.80 \text{ m/s}^2) = 25{,}000 \text{ N}$

(c) $v_y = v_{0y} + a_y t$ gives $t = \dfrac{v_y - v_{0y}}{a_y} = \dfrac{1.3 \text{ m/s} - 4.4 \text{ m/s}}{-440 \text{ m/s}^2} = 7.0 \text{ ms}$

Reflect: When the velocity change occurs over a small distance the acceleration is large.

*5.69. Set Up: The free-body diagrams for the rocket (weight w_r) and astronaut (weight W) are given in Figures (a) and (b) below. F_T is the thrust and n is the normal force the rocket exerts on the astronaut. The speed of sound is 331 m/s.

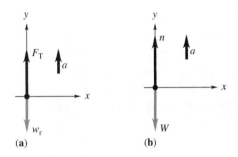

(a)   (b)

(a) Apply $\Sigma F_y = ma_y$ to the rocket: $F_T - w_r = ma$. $a = 4g$ and $w_r = mg$, so
$$F = m(5g) = (2.25 \times 10^6 \text{ kg})(5)(9.80 \text{ m/s}^2) = 1.10 \times 10^8 \text{ N}$$

(b) Apply $\Sigma F_y = ma_y$ to the astronaut: $n - W = ma$. $a = 4g$ and $m = \dfrac{W}{g}$, so
$$n = W + \left(\frac{W}{g}\right)(4g) = 5W$$

(c) $v_0 = 0$, $v = 331$ m/s and $a = 4g = 39.2$ m/s^2. $v = v_0 + at$ gives
$$t = \frac{v - v_0}{a} = \frac{331 \text{ m/s}}{39.2 \text{ m/s}^2} = 8.4 \text{ s}$$

*5.71. Set Up: $a = 0$ for both objects. The tension in the vertical wire is w. Apply $\Sigma \vec{F} = m\vec{a}$ to the knot where the wires are joined and then to block A. The static friction force on block A will be whatever is required to keep block A at rest.

Solve: The free-body diagram for the knot is given in Figure (a) below. $\Sigma F_y = ma_y$ gives $T_1 \sin 45° = w$. $\Sigma F_x = ma_x$ gives $T_2 \cos 45.0° = T_3$. $\sin 45.0° = \cos 45.0°$ and $T_3 = w$. The free-body diagram for block A is given in Figure (b) below. $f_s = T_3 = w = 12.0$ N

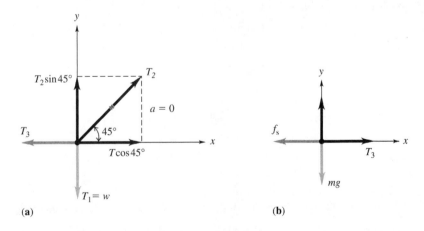

(a)   (b)

(b) $f_s^{\text{max}} = \mu_s n = \mu_s mg = (0.25)(60.0 \text{ N}) = 15.0 \text{ N}$. Then $w = f_s^{\text{max}} = 15.0$ N.

Reflect: The friction force found in part (a) is less than the maximum possible friction force.

*5.73. Set Up: The unstretched length L_0 of the tendon is 20.0 cm. The weight W suspended from the end of the tendon equals the force applied by the tendon.

Solve: (a) $x = L - L_0$ so Hooke's law says $W = kx = kL - kL_0$. The graph of W versus L is of this form.

(b) k is the slope of W versus x, so $k = \dfrac{15.0 \text{ N}}{0.500 \text{ m} - 0.200 \text{ m}} = 50.0$ N/m

(c) $W = kx = (50.0 \text{ N/m})(0.080 \text{ m}) = 4.0$ N

*5.77. Set Up: When the spring is stretched an amount x the spring force has magnitude kx and is directed to the right in the figure in the problem. The free-body diagram for the weight is given in Figure (a) below. When the spring is compressed a distance $|x|$ the spring force has magnitude $k|x|$ and is directed to the left. The free-body diagram for this case is given in Figure (b) below.

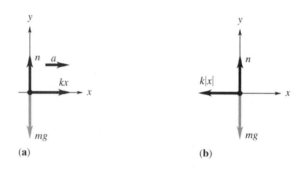

(a)   (b)

Solve: (a) $\Sigma F_x = ma_x$ gives $kx = ma$ and $k = \dfrac{ma}{x} = \dfrac{(1.10 \text{ kg})(2.50 \text{ m/s}^2)}{0.0110 \text{ m}} = 250$ N/m

(b) $\Sigma F_x = ma_x$ gives $-k|x| = ma_x$ and $a_x = -\dfrac{k|x|}{m} = -\dfrac{(250 \text{ N/m})(0.0230 \text{ m})}{1.10 \text{ kg}} = -5.23$ m/s^2. The acceleration is

5.23 m/s^2, in the backward direction.

*5.81. Set Up: Apply $\Sigma \vec{F} = m\vec{a}$ to the brush. Constant speed means $a = 0$. Target variables are two of the forces on the brush. Note that the normal force exerted by the wall is horizontal, since it is perpendicular to the wall. The kinetic friction force exerted by the wall is parallel to the wall and opposes the motion, so it is vertically downward. The free-body diagram is given in the figure below.

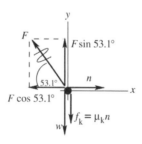

Solve: Using $\Sigma F_x = ma_x$, we obtain $n - F\cos 53.1° = 0$, so $n = F\cos 53.1°$; thus, $f_k = \mu_k n = \mu_k F\cos 53.1°$.

Using $\Sigma F_y = ma_y$ we obtain $F\sin 53.1° - w - f_k = 0$, so $F\sin 53.1° - w - \mu_k F\cos 53.1° = 0$, which factors to give

$F(\sin 53.1° - \mu_k \cos 53.1°) = w$. Solving for F we obtain $F = \dfrac{w}{\sin 53.1° - \mu_k \cos 53.1°}$.

(a) $F = \dfrac{w}{\sin 53.1° - \mu_k \cos 53.1°} = \dfrac{12.0 \text{ N}}{\sin 53.1° - (0.150)\cos 53.1°} = 16.9 \text{ N}$

(b) $n = F \cos 53.1° = (16.9 \text{ N})\cos 53.1° = 10.1 \text{ N}$

Reflect: In the absence of friction, we have $w = F \sin 53.1°$, which agrees with our expression.

*5.83. Set Up: The normal force is horizontal, perpendicular to the wall and the friction force is upward, parallel to the wall.
Solve: (a) The free-body diagram is shown in the figure below.

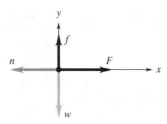

(b) $f = W = 20 \text{ N}$. Note: $n = F = 50 \text{ N}$. The maximum possible static friction force is $f_s = \mu_s n = (0.50)(50 \text{ N}) = 25 \text{ N}$ and the actual friction force is less than this.

(c) Constant speed means $a = 0$. Moving means $f = f_k = \mu_k n$. $f_k = W$ so $\mu_k n = W$ and $n = \dfrac{W}{\mu_k} = \dfrac{20 \text{ N}}{0.20} = 100 \text{ N}$.

*5.87. Set Up: The block has the same horizontal acceleration a as the cart. Let $+x$ be to the right and $+y$ be upward. To find the minimum acceleration required, set the static friction force equal to its maximum value, $\mu_s n$.
Solve: The free-body diagram for the block is given in the figure below.

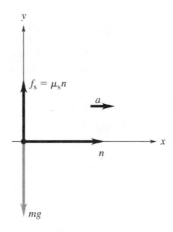

$\Sigma F_x = ma_x$ gives $n = ma$. $\Sigma F_y = ma_y$ gives $f_s - mg = 0$. $\mu_s n = mg$ and $\mu_s ma = mg$. $a = \dfrac{g}{\mu_s}$.

Reflect: The smaller μ_s is the greater a must be to prevent slipping. Increasing a increases the normal force n and that increases the maximum f_s for a given μ_s.

Solutions to Passage Problems

*5.89. Set Up: Consider the analysis of Example 5.6 and use an x-axis that is parallel to the hill: this example shows that the net force on the toboggan (without the sail) is equal to $w \sin \alpha$ and is directed downhill.

Solve: Once the toboggan is moving at a constant speed, the net force on it must be zero. We know that the sum of the gravitational force and normal force is $w \sin \alpha$ (directed downhill), so the sail must provide an equal force directed uphill. The correct answer is B.

*5.91. Set Up: Consider the analysis of Example 5.11 and use an x-axis that is parallel to the hill.

Solve: Since the wind blows parallel to the hill, the force of the wind has no y-component and we have $\sum F_y = n + (-w \cos \alpha) = 0$ so $N = w \cos \alpha$ as in Example 5.11. In the x-direction we have two changes: (i) since the sled is moving uphill, friction is directed downhill (in the $+x$ direction), and (ii) the magnitude of the wind force (which acts in the negative x-direction) must be subtracted. Thus, we have (for constant speed motion) $\sum F_x = w \sin \alpha + \mu_k w \cos \alpha - f_{\text{wind}} = ma_x = 0$. We can solve this for the force of the wind to obtain $f_{\text{wind}} = w \sin \alpha + \mu_k w \cos \alpha$. The correct answer is D.

CIRCULAR MOTION AND GRAVITATION

Problems 1, 3, 5, 9, 11, 15, 17, 19, 23, 27, 29, 33, 37, 39, 43, 47, 49, 53, 55, 57, 59

Solutions to Problems

*6.1. Set Up: The path shows the relative size of the radius of curvature R of each part of the track. $R_B > R_C$. A straight line is an arc of a circle of infinite radius, so $R_A \to \infty$.

Solve: $a_{rad} = \dfrac{v^2}{R}$. At A, $a_{rad} = 0$. $a_{rad, B} < a_{rad, C}$. In each case, $\vec{a}_{rad}$ is directed toward the center of the circular path. The net force is proportional to $\vec{a}_{rad}$ and is shown in the figure below.

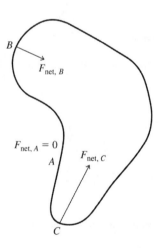

Reflect: The road must exert a greater friction force on the car when it makes tighter turns that have smaller radii of curvature.

*6.3. Set Up: Each hand travels in a circle of radius 0.750 m and has mass $(0.0125)(52 \text{ kg}) = 0.65$ kg and weight 6.4 N. The period for each hand is $T = (1.0 \text{ s})/(2.0) = 0.50$ s. Let $+x$ be toward the center of the circular path.

(a) The free-body diagram for one hand is given in the figure below. $\vec{F}$ is the force exerted on the hand by the wrist. This force has both horizontal and vertical components.

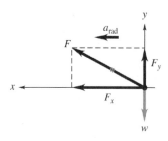

(b) $a_{rad} = \dfrac{4\pi^2 R}{T^2} = \dfrac{4\pi^2(0.750 \text{ m})}{(0.50 \text{ s})^2} = 118 \text{ m/s}^2$

$\Sigma F_x = ma_x$ gives $F_x = ma_{rad} = (0.65 \text{ kg})(118 \text{ m/s}^2) = 77 \text{ N}$

(c) $\dfrac{F}{w} = \dfrac{77 \text{ N}}{6.4 \text{ N}} = 12$. The horizontal force from the wrist is 12 times the weight of the hand.

Reflect: The wrist must also exert a vertical force on the hand equal to its weight.

*6.5. Set Up: The person moves in a circle of radius $R = 3.00 \text{ m} + (5.00 \text{ m})\sin 30.0° = 5.50 \text{ m}$. The acceleration of the person is $a_{rad} = v^2/R$, directed horizontally to the left in the figure in the problem. The time for one revolution is the period $T = \dfrac{2\pi R}{v}$.

Solve: (a) The free-body diagram is given in the figure below. $\vec{F}$ is the force applied to the seat by the rod.

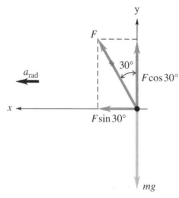

(b) $\Sigma F_y = ma_y$ gives $F\cos 30.0° = mg$ and $F = \dfrac{mg}{\cos 30.0°}$. $\Sigma F_x = ma_x$ gives $F\sin 30.0° = m\dfrac{v^2}{R}$. Combining these two equations gives

$$v = \sqrt{Rg\tan\theta} = \sqrt{(5.50 \text{ m})(9.80 \text{ m/s}^2)\tan 30.0°} = 5.58 \text{ m/s}.$$

Then the period is $T = \dfrac{2\pi R}{v} = \dfrac{2\pi(5.50 \text{ m})}{5.58 \text{ m/s}} = 6.19 \text{ s}.$

(c) The net force is proportional to m so in $\Sigma\vec{F} = m\vec{a}$ the mass divides out and the angle for a given rate of rotation is independent of the mass of the passengers.

Reflect: The person moves in a horizontal circle so the acceleration is horizontal. The net inward force required for circular motion is produced by a component of the force exerted on the seat by the rod.

*6.9. Set Up: At the safe speed no friction force is required. Consider a car of mass m. The free-body diagram for the car is given in the figure below. Use coordinates where $+x$ is toward the center of the horizontal turn. Each turn is one-quarter of a circle so the radius R is given by $0.25 \text{ mi} = \frac{1}{4}(2\pi R)$. $9°12'$ is $9.2°$.

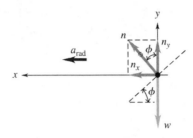

Solve: $R = 4(0.25 \text{ mi})/2\pi = 0.1592 \text{ mi} = 256 \text{ m}$

$\sum F_y = ma_y$ gives $n_y = w$ so $n = \dfrac{mg}{\cos\phi}$

$\sum F_x = ma_x$ gives $n_x = ma_{\text{rad}}$ so $n\sin\phi = m\dfrac{v^2}{R}$

$\left(\dfrac{mg}{\cos\phi}\right)\sin\phi = m\dfrac{v^2}{R}$ and $v = \sqrt{Rg\tan\phi} = \sqrt{(256 \text{ m})(9.80 \text{ m/s}^2)\tan 9.2°} = 20.0 \text{ m/s} = 44.7 \text{ mph}$

Reflect: Race cars travel *much* faster than this. The track must exert a large friction force on the cars to maintain their circular motion.

*6.11. Set Up: The period is $T = 60.0 \text{ s}$ and $T = \dfrac{2\pi R}{v}$. The apparent weight of a person is the normal force exerted on him by the seat he is sitting on. His acceleration is $a_{\text{rad}} = v^2/R$, directed toward the center of the circle. The passenger has mass $m = w/g = 90.0 \text{ kg}$.

Solve: (a) $v = \dfrac{2\pi R}{T} = \dfrac{2\pi(50 \text{ m})}{60.0 \text{ s}} = 5.24 \text{ m/s}$. Note that $a_{\text{rad}} = \dfrac{v^2}{R} = \dfrac{(5.24 \text{ m/s})^2}{50.0 \text{ m}} = 0.549 \text{ m/s}^2$.

(b) The free-body diagram for the person at the top of his path is given in Figure (a) below. The acceleration is downward, so take $+y$ downward. $\sum F_y = ma_y$ gives $mg - n = ma_{\text{rad}}$.

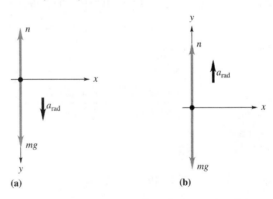

(a)   (b)

$$n = m(g - a_{\text{rad}}) = (90.0 \text{ kg})(9.80 \text{ m/s}^2 - 0.549 \text{ m/s}^2) = 833 \text{ N}.$$

The free-body diagram for the person at the bottom of his path is given in Figure (b) above. The acceleration is upward, so take $+y$ upward. $\sum F_y = ma_y$ gives $n - mg = ma_{\text{rad}}$ and $n = m(g + a_{\text{rad}}) = 931 \text{ N}$.

(c) Apparent weight $= 0$ means $n = 0$ and $mg = ma_{\text{rad}}$.

$$g = \dfrac{v^2}{R} \text{ and } v = \sqrt{gR} = 22.1 \text{ m/s}.$$

The time for one revolution would be

$$T = \frac{2\pi R}{v} = \frac{2\pi(50.0 \text{ m})}{22.1 \text{ m/s}} = 14.2 \text{ s}.$$

Note that $a_{rad} = g$.

(d) $n = m(g + a_{rad}) = 2mg = 2(882 \text{ N}) = 1760 \text{ N}$, twice his true weight.

Reflect: At the top of his path his apparent weight is less than his true weight and at the bottom of his path his apparent weight is greater than his true weight.

*6.15. Set Up: According to the hint, "red out" can occur when the non-gravitational part of the centripetal acceleration exceeds 2.5 g's. Thus, the maximum allowed downward centripetal acceleration is $g + 2.5g = 3.5g$. We know that $a_{rad} = \dfrac{v^2}{r}$. Also, from the appendix we know that 1 km/h = 0.2778 m/s.

Solve: First convert the speed of the plane into m/s: $(320 \text{ km/h})\left(0.2778\dfrac{\text{m/s}}{\text{km/h}}\right) = 88.9 \text{ m/s}$. The minimum allowed radius occurs when the centripetal acceleration has its maximum allowed value of $3.5g = \dfrac{v^2}{r}$. Solving for r we obtain $r = \dfrac{v^2}{3.5g} = \dfrac{(88.9 \text{ m/s})^2}{3.5(9.80 \text{ m/s}^2)} = 230 \text{ m}$.

Reflect: The force of gravity tends to cause all objects to accelerate together and does not contribute to the redistribution of blood necessary for "red out" to occur—it is the non-gravitational part of the acceleration that must be limited below 2.5g.

*6.17. Set Up: A proton has mass $m_p = 1.673 \times 10^{-27}$ kg. 1.0 fm $= 1.0 \times 10^{-15}$ m

Solve: $F_g = G\dfrac{m^2}{r^2} = (6.673 \times 10^{-11} \text{ N} \cdot \text{m}^2/\text{kg}^2)\dfrac{(1.673 \times 10^{-27} \text{ kg})^2}{(1.0 \times 10^{-15} \text{ m})^2} = 1.87 \times 10^{-34} \text{ N}$

*6.19. Set Up: The gravitational force between two objects is $F_g = G\dfrac{m_1 m_2}{r^2}$. The mass of the earth is $m_E = 5.97 \times 10^{24}$ kg and the mass of the sun is $m_S = 1.99 \times 10^{30}$ kg. The distance from the earth to the sun is $r_{SE} = 1.50 \times 10^{11}$ m and the distance from the earth to the moon is $r_{EM} = 3.84 \times 10^8$ m.

Solve: $F_{SM} = G\dfrac{m_S m_M}{r_{SM}^2} = G\dfrac{m_S m_M}{r_{SE}^2}$. $F_{EM} = G\dfrac{m_E m_M}{r_{EM}^2}$.

$$\frac{F_{SM}}{F_{EM}} = \left(\frac{m_S}{m_E}\right)\left(\frac{r_{EM}}{r_{SE}}\right)^2 = \left(\frac{1.99 \times 10^{30} \text{ kg}}{5.97 \times 10^{24} \text{ kg}}\right)\left(\frac{3.84 \times 10^8 \text{ m}}{1.50 \times 10^{11} \text{ m}}\right)^2 = 2.18$$

It is more accurate to say the moon orbits the sun.

Reflect: The sun is farther away but has a much greater mass than the earth.

*6.23. Set Up: The force exerted on the particle by the earth is $w = mg$, where m is the mass of the particle. The force exerted by the 100 kg ball is $F_g = \dfrac{Gm_1 m_2}{r^2}$, where r is the distance of the particle from the center of the ball. $G = 6.67 \times 10^{-11}$ N $\cdot$ m^2/kg^2, $g = 9.80$ m/s^2.

Solve: $F_g = w$ gives $\dfrac{Gmm_{ball}}{r^2} = mg$ and

$$r = \sqrt{\frac{Gm_{ball}}{g}} = \sqrt{\frac{(6.67 \times 10^{-11} \text{ N} \cdot \text{m}^2/\text{kg}^2)(100 \text{ kg})}{9.80 \text{ m/s}^2}} = 2.61 \times 10^{-5} \text{ m} = 0.0261 \text{ mm}.$$ It is not feasible to do this; a 100 kg

ball would have a radius much larger than 0.0261 mm.

Reflect: The gravitational force between ordinary objects is very small. The gravitational force exerted by the earth on objects near its surface is large enough to be important because the mass of the earth is very large.

*6.27. Set Up: The circumference c is related to the radius R_p of the planet by $c = 2\pi R_p$. Take $+y$ downward and use the measured motion to find $a_y = g$, the acceleration due to gravity at the surface of the planet.

$m_E = 5.97 \times 10^{24}$ kg

Solve: (a) $v_{0y} = 0$, $t = 0.811$ s, $y - y_0 = 5.00$ m

$$y = y_0 + v_{0y}t + \frac{1}{2}a_y t^2 \text{ gives } a_y = \frac{2(y - y_0)}{t^2} = \frac{2(5.00 \text{ m})}{(0.811 \text{ s})^2} = 15.2 \text{ m/s}^2$$

$$R_p = \frac{c}{2\pi} = \frac{6.24 \times 10^7 \text{ m}}{2\pi} = 9.93 \times 10^6 \text{ m}$$

$$g = \frac{Gm_p}{R_p{}^2} \text{ so } m_p = \frac{gR_p{}^2}{G} = \frac{(15.2 \text{ m/s}^2)(9.93 \times 10^6 \text{ m})^2}{6.673 \times 10^{-11} \text{ N} \cdot \text{m}^2/\text{kg}^2} = 2.25 \times 10^{25} \text{ kg}$$

(b) $m_p = 3.76 m_E$

Reflect: $R_p = 1.56 R_E$ and $m_p = 3.76 m_E$ so $g_p = \frac{3.76}{(1.56)^2} g_E = 15.2 \text{ m/s}^2$, which checks.

*6.29. Set Up: The mass of the probe on earth is $m = \dfrac{w_E}{g_E} = \dfrac{3120 \text{ N}}{9.80 \text{ m/s}^2} = 318.4$ kg. The radius of Titan is

$R_T = \frac{1}{2}(5150 \text{ km}) = 2.575 \times 10^6$ m.

Solve: $w = G\dfrac{mm_T}{R_T{}^2} = \dfrac{(6.673 \times 10^{-11} \text{ N} \cdot \text{m}^2/\text{kg}^2)(318.4 \text{ kg})(1.35 \times 10^{23} \text{ kg})}{(2.575 \times 10^6 \text{ m})^2} = 433$ N

Reflect: The mass of the probe is independent of its location.

*6.33. Set Up: $T = \dfrac{2\pi r}{v}$. r and v are also related by applying $\Sigma \vec{F} = m\vec{a}$ to the motion of the satellite. The satellite has $a_{rad} = v^2/R$, The only force on the satellite is the gravitational force, $F_g = G\dfrac{m_E m}{r^2}$. $m_E = 5.97 \times 10^{24}$ kg

Solve: (a) and **(c)** The free-body diagram is given in the figure below. $F_g = ma_{rad}$ gives $G\dfrac{m_E m}{r^2} = m\dfrac{v^2}{r}$.

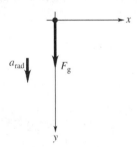

$$G\frac{m_E}{r} = v^2 \text{ and } r = \frac{Gm_E}{v^2} = \frac{(6.673 \times 10^{-11} \text{ N} \cdot \text{m}^2/\text{kg}^2)(5.97 \times 10^{24} \text{ kg})}{(6200 \text{ m/s})^2} = 1.04 \times 10^7 \text{ m}.$$

$$T = \frac{2\pi r}{v} = \frac{2\pi(1.04 \times 10^7 \text{ m})}{6200 \text{ m/s}} = 1.05 \times 10^4 \text{ s} = 176 \text{ min}$$

(b) $a_{\text{rad}} = \dfrac{v^2}{r} = \dfrac{(6200 \text{ m/s})^2}{1.04 \times 10^7 \text{ m}} = 3.70 \text{ m/s}^2$

*6.37. Set Up: The rotational period of the earth is $1 \text{ day} = 8.64 \times 10^4 \text{ s}$. The radius of the satellite's path is $r = h + R_{\text{E}}$, where h is its height above the surface of the earth and $R_{\text{E}} = 6.38 \times 10^6 \text{ m}$. v and r are related by applying $\sum \vec{F} = m\vec{a}$ to the motion of the satellite. $m_{\text{E}} = 5.97 \times 10^{24} \text{ kg}$

Solve: (a) $24 \text{ h} = 8.64 \times 10^4 \text{ s}$

(b)-(c) The free-body diagram for the satellite is given in Figure (a) below. $F_g = ma_{\text{rad}}$ gives $G\dfrac{m_{\text{E}}m}{r^2} = m\dfrac{v^2}{r}$ and

$v^2 = \dfrac{Gm_{\text{E}}}{r}$. $T = \dfrac{2\pi r}{v}$ so $v = \dfrac{2\pi r}{T}$ and $\dfrac{4\pi^2 r^2}{T^2} = \dfrac{Gm_{\text{E}}}{r}$.

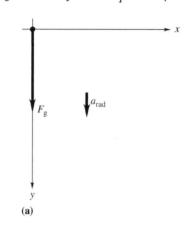

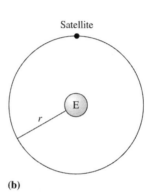

(a)   (b)

$$r = \left(\frac{Gm_{\text{E}}T^2}{4\pi^2}\right)^{1/3} = \left[\frac{(6.673 \times 10^{-11} \text{ N} \cdot \text{m}^2/\text{kg}^2)(5.97 \times 10^{24} \text{ kg})(8.64 \times 10^4 \text{ s})^2}{4\pi^2}\right]^{1/3} = 4.22 \times 10^7 \text{ m}$$

$$h = r - R_{\text{E}} = 3.58 \times 10^7 \text{ m} = 3.58 \times 10^4 \text{ km}$$

(d) $\dfrac{r}{R_{\text{E}}} = 5.6$. The sketch of the earth and orbit of the satellite is given in Figure (b) above.

Reflect: The orbital period is proportional to $r^{3/2}$. To achieve a period as large as 24 h, the satellite is at a large altitude above the earth's surface.

*6.39. Set Up: The period of each satellite is given by equation (6.10). Set up a ratio involving T and r:

$T = \dfrac{2\pi r^{3/2}}{\sqrt{Gm_{\text{p}}}}$ gives $\dfrac{T}{r^{3/2}} = \dfrac{2\pi}{\sqrt{Gm_{\text{p}}}} = \text{constant}$, so $\dfrac{T_1}{r_1^{3/2}} = \dfrac{T_2}{r_2^{3/2}}$.

Solve: $T_2 = T_1 \left(\dfrac{r_2}{r_1}\right)^{3/2} = (6.39 \text{ days})\left(\dfrac{48,000 \text{ km}}{19,600 \text{ km}}\right)^{3/2} = 24.5 \text{ days}$. For the other satellite we have

$T_2 = (6.39 \text{ days})\left(\dfrac{64,000 \text{ km}}{19,600 \text{ km}}\right)^{3/2} = 37.7 \text{ days}$.

Reflect: T increases when r increases.

*6.43. Set Up: $r = 137.5 \text{ m}$. The distance a point on the rim travels in 1 revolution is $2\pi r$.

Solve: $a_{rad} = \dfrac{v^2}{R}$ so $v = \sqrt{ra_{rad}} = \sqrt{(137.5 \text{ m})(9.80 \text{ m/s}^2)} = 36.7$ m/s

In one minute a point on the rim travels a distance $(36.7 \text{ m/s})(60 \text{ s}) = 2202$ m. The number of revolutions in one minute is $\dfrac{2202 \text{ m}}{2\pi r} = 2.55$. The station must turn at 2.55 rpm.

*6.47. Set Up: The person moves in a horizontal circle of radius $r = 2.5$ m. Set the static friction force equal to its maximum value, $f_s = \mu_s n$. The person has an acceleration

$$a_{rad} = \frac{v^2}{r},$$

directed toward the center of the circle. The period is

$$T = \frac{1}{0.60 \text{ rev/s}} = 1.67 \text{ s}$$

and the person has speed

$$v = \frac{2\pi r}{T} = 9.41 \text{ m/s}.$$

Solve: **(a)** The free-body diagram is given in the figure below. The diagram is for when the wall is to the right of her, so the center of the cylinder is to the left.

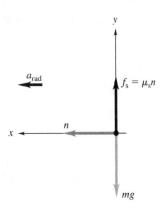

(b) $\Sigma F_x = ma_x$ gives $n = m\dfrac{v^2}{r}$. $\Sigma F_y = ma_y$ gives $f_s = mg$, so $\mu_s n = mg$. Combining these two equations gives

$\mu_s m \dfrac{v^2}{r} = mg$ and

$$\mu_s = \frac{gr}{v^2} = \frac{(9.80 \text{ m/s}^2)(2.5 \text{ m})}{(9.41 \text{ m/s})^2} = 0.28.$$

(c) The mass m of the person divides out of the equation for μ_s; the answer to (b) does not depend on the mass of the person.

Reflect: The greater the rotation rate the larger the normal force exerted by the wall and the larger the friction force. Therefore, for smaller μ_s the rotation rate must be larger.

*6.49. Set Up: density = mass/volume. The volume of a sphere is $\frac{4}{3}\pi R^3$. Use the assumption that the density of Toro is the same as that of earth to calculate the mass of Toro. Then $g_T = G\dfrac{m_T}{R_T^2}$. Apply $\Sigma \vec{F} = m\vec{a}$ to the object to find its speed when it is in a circular orbit around Toro.

Solve: **(a)** $\dfrac{m_E}{\frac{4}{3}\pi R_E^{\,3}} = \dfrac{m_T}{\frac{4}{3}\pi R_T^{\,3}}$ gives

$$m_T = m_E \left(\frac{R_T}{R_E}\right)^3 = (5.97 \times 10^{24}\,\text{kg})\left(\frac{5.0 \times 10^3\,\text{m}}{6.38 \times 10^6\,\text{m}}\right)^3 = 2.9 \times 10^{15}\,\text{kg}.$$

$$g_T = G\frac{m_T}{R_T^{\,2}} = \frac{(6.673 \times 10^{-11}\,\text{N} \cdot \text{m}^2/\text{kg}^2)(2.9 \times 10^{15}\,\text{kg})}{(5.0 \times 10^3\,\text{m})^2} = 7.7 \times 10^{-3}\,\text{m/s}^2$$

(b) The gravity force on the object is mg_T. In a circular orbit just above the surface of Toro, its acceleration is $\dfrac{v^2}{R_T}$.

Then $\Sigma \vec{F} = m\vec{a}$ gives $mg_T = m\dfrac{v^2}{R_T}$ and

$$v = \sqrt{g_T R_T} = \sqrt{(7.7 \times 10^{-3}\,\text{m/s}^2)(5.0 \times 10^3\,\text{m})} = 6.2\,\text{m/s}.$$

A speed of 6.2 m/s corresponds to running 100 m in 16.1 s, which is barely possible for the average person.

*6.53. Set Up: The comet has radius $R = 2.5 \times 10^3\,\text{m}$, volume $V = \frac{4}{3}\pi R^3$ and density $\rho = 2.1 \times 10^3\,\text{kg/m}^3$.
Solve: **(a)** The comet has mass

$$m_C = \rho_C V_C = \rho_C \tfrac{4}{3}\pi R^3 = (2.1 \times 10^3\,\text{kg/m}^3)(\tfrac{4}{3}\pi)(2.5 \times 10^3\,\text{m})^3 = 1.374 \times 10^{14}\,\text{kg}.$$

The gravitational force is

$$F_g = G\frac{mm_C}{r^2} = \frac{(6.673 \times 10^{-11}\,\text{N} \cdot \text{m}^2/\text{kg}^2)(385\,\text{kg})(1.374 \times 10^{14}\,\text{kg})}{(237 \times 10^3\,\text{m})^2} = 6.28 \times 10^{-5}\,\text{N}$$

(b) At the surface of the earth, the earth's gravity force is $w = mg = (385\,\text{kg})(9.80\,\text{m/s}^2) = 3.77 \times 10^3\,\text{N}$. The force exerted by the comet is much less than this.

*6.55. Set Up: The package moves in the arc of a circle so it has acceleration $a_{\text{rad}} = \dfrac{v^2}{r}$, directed toward the center of the curve.
Solve: The free-body diagram for the package is given in the figure below. $\Sigma F_y = ma_y$ gives $T\cos 30.0° = mg$.

$\Sigma F_x = ma_x$ gives $T\sin 30.0° = m\dfrac{v^2}{r}$. Therefore, $T = \dfrac{mg}{\cos 30.0°}$ and $mg\tan 30.0° = m\dfrac{v^2}{r}$.

$$v = \sqrt{gr\tan 30.0°} = \sqrt{(9.80\,\text{m/s}^2)(50.0\,\text{m})\tan 30.0°} = 16.8\,\text{m/s}$$

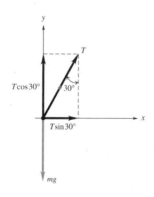

Solutions to Passage Problems

*6.57. Set Up: Assume that the diameter of the space station is 1000 meters so that its radius is 500 meters. For circular motion at a constant speed, the acceleration of the astronaut is $a_r = \dfrac{v^2}{r}$.

Solve: $v = \sqrt{r a_r} = \sqrt{(500 \text{ m})(5 \text{ m/s}^2)} = 50$ m/s. The correct answer is D.

*6.59. Set Up: The radial force required to keep the helmet moving on its circular path is $F = m\dfrac{v^2}{r} = ma_r$.

Solve: $F = ma_r = (2 \text{ kg})(5 \text{ m/s}^2) = 10$ N. The correct answer is B.

WORK AND ENERGY

Problems 3, 5, 7, 9, 13, 15, 19, 21, 23, 25, 29, 31, 33, 35, 39, 43, 45, 49, 51, 55, 57, 61, 63, 65, 69, 71, 75, 79, 83, 87, 89, 93, 95, 97, 101, 105

Solutions to Problems

*7.3. Set Up: Use $W = F_{\parallel}s = (F\cos\phi)s$ with $\phi = 15.0°$.

Solve: $W = (F\cos\phi)s = (180\text{ N})(\cos 15.0°)(300.0\text{ m}) = 5.22 \times 10^4\text{ J}$

Reflect: Since $\cos 15.0° \approx 0.97$, the relatively small angle of $15.0°$ allows the boat to apply approximately 97% of the 180 N force to pulling the skier.

*7.5. Set Up: For parts (a) through (d), identify the appropriate value of ϕ and use the relation $W = F_{\parallel}s = (F\cos\phi)s$. In part (e), apply the relation $W_{\text{net}} = W_{\text{student}} + W_{\text{grav}} + W_n + W_f$.

Solve: (a) Since you are applying a horizontal force, $\phi = 0°$. Thus,

$$W_{\text{student}} = (2.40\text{ N})(\cos 0°)(1.50\text{ m}) = 3.60\text{ J}$$

(b) The friction force acts in the horizontal direction, opposite to the motion, so $\phi = 180°$.

$$W_f = (F_f\cos\phi)s = (0.600\text{ N})(\cos 180°)(1.50\text{ m}) = -0.900\text{ J}$$

(c) Since the normal force acts upward and perpendicular to the tabletop, $\phi = 90°$.

$$W_n = (n\cos\phi)s = (ns)(\cos 90°) = 0.0\text{ J}$$

(d) Since gravity acts downward and perpendicular to the tabletop, $\phi = 270°$.

$$W_{\text{grav}} = (mg\cos\phi)s = (mgs)(\cos 270°) = 0.0\text{ J}$$

(e) $W_{\text{net}} = W_{\text{student}} + W_{\text{grav}} + W_n + W_f = 3.60\text{ J} + 0.0\text{ J} + 0.0\text{ J} - 0.900\text{ J} = 2.70\text{ J}$

Reflect: Whenever a force acts perpendicular to the direction of motion, its contribution to the net work is zero.

*7.7. Set Up: In order to move the crate at constant velocity, the worker must apply a force that equals the force of friction, $F_{\text{worker}} = f_k = \mu_k n$. Each force can be used in the relation $W = F_{\parallel}s = (F\cos\phi)s$ for parts (b) through (d). For part (e), apply the net work relation as $W_{\text{net}} = W_{\text{worker}} + W_{\text{grav}} + W_n + W_f$.

Solve: (a) The magnitude of the force the worker must apply is:

$$F_{\text{worker}} = f_k = \mu_k n = \mu_k mg = (0.25)(30.0\text{ kg})(9.80\text{ m/s}^2) = 74\text{ N}\left(\frac{\pi}{2} - \theta\right)$$

(b) Since the force applied by the worker is horizontal and in the direction of the displacement, $\phi = 0°$ and the work is:

$$W_{worker} = (F_{worker} \cos\phi)s = [(74 \text{ N})(\cos 0°)](4.5 \text{ m}) = +333 \text{ J}$$

(c) Friction acts in the direction opposite of motion, thus $\phi = 180°$ and the work of friction is:

$$W_f = (f_k \cos\phi)s = [(74 \text{ N})(\cos 180°)](4.5 \text{ m}) = -333 \text{ J}$$

(d) Both gravity and the normal force act perpendicular to the direction of displacement. Thus, neither force does any work on the crate and $W_{grav} = W_n = 0.0 \text{ J}$.

(e) Substituting into the net work relation, the net work done on the crate is:

$$W_{net} = W_{worker} + W_{grav} + W_n + W_f = +333 \text{ J} + 0.0 \text{ J} + 0.0 \text{ J} - 333 \text{ J} = 0.0 \text{ J}$$

Reflect: The net work done on the crate is zero because the two contributing forces, F_{worker} and F_f, are equal in magnitude and opposite in direction.

*7.9. Set Up: $W_F = (F \cos\phi)s$, since the forces are constant. We can calculate the total work by summing the work done by each force. The forces are sketched in the figure below.

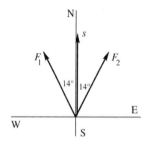

Solve: Using $W_1 = F_1 s \cos\phi_1$ we obtain $W_1 = (1.80\times10^6 \text{ N})(0.75\times10^3 \text{ m})\cos 14°$ so $W_1 = 1.31\times10^9 \text{ J}$. Also, note that $W_2 = F_2 s \cos\phi_2 = W_1$. Thus, $W_{tot} = W_1 + W_2 = 2(1.31\times10^9 \text{ J}) = 2.62\times10^9 \text{ J}$

Reflect: Only the component $F \cos\phi$ of force in the direction of the displacement does work. These components are in the direction of $\vec{s}$ so the forces do positive work.

*7.13. Set Up: Use $K = \frac{1}{2}mv^2$ and solve for m.

Solve: $m = 2K/v^2 = 2(1960 \text{ J})/(965 \text{ m/s})^2 = 4.21\times10^{-3} \text{ kg} = 4.21 \text{ g}$

Reflect: The kinetic energy of an object is proportional to the mass of the object.

*7.15. Set Up: Use $K = \frac{1}{2}mv^2$ to relate v and K. Let v_1 and K_1 be speed and kinetic energy when $v_1 = 10.0$ m/s.

Solve: (a) $K_2 = K_1/2$ and $K_1/K_2 = v_1^2/v_2^2$ so $v_2 = v_1\sqrt{K_2/K_1} = (10.0 \text{ m/s})\sqrt{1/2} = 7.07$ m/s.

(b) Since $v_2 = v_1/2$, we obtain $K_2 = K_1(v_2^2/v_1^2) = K_1(1/4)$. This means that the decrease in K is $3K_1/4$.

*7.19. Set Up: Use the work-kinetic energy theorem: $W_{net} = K_f - K_i = F_{net}s$. Since the net force is due to friction, $F_{net}s = -f_k s = -\mu_k mgs$. Also, since the car stops, $K_f = 0$.

Solve: (a) $W_{net} = K_f - K_i = F_{net}s$ gives $-\frac{1}{2}mv_i^2 = -\mu_k mgs$. Solving for the distance,

$$s = \frac{v_i^2}{2\mu_k g} = \frac{(23.0 \text{ m/s})^2}{2(0.700)(9.80 \text{ m/s}^2)} = 38.6 \text{ m}$$

(b) Since s is proportional to v_i^2, doubling v_i increases s by a factor of 4; s therefore becomes 154 m.

(c) The original kinetic energy was converted into thermal energy by the negative work of friction.

Reflect: To stop the car friction must do negative work equal in magnitude to the initial kinetic energy of the car.

*7.21. Set Up: From the work-energy relation, $W = W_{grav} = \Delta K_{rock}$ or $F_\| s = K_f - K_i$. As the rock rises, the gravitational force, $F = mg$, does work on the rock. Since this force acts in the direction opposite to the motion and displacement, s, the work is negative.

Solve: **(a)** Applying $F_\| s = K_f - K_i$ we obtain:

$$-mgh = \tfrac{1}{2}mv_f^2 - \tfrac{1}{2}mv_i^2$$

Dividing by m and solving for v_i, $v_i = \sqrt{v_f^2 + 2gh}$. Substituting $h = 15.0$ m and $v_f = 25.0$ m/s,

$$v_i = \sqrt{(25.0 \text{ m/s})^2 + 2(9.80 \text{ m/s}^2)(15.0 \text{ m})} = 30.3 \text{ m/s}$$

(b) Solve the same work-energy relation for h. At the maximum height $v_f = 0$.

$$-mgh = \tfrac{1}{2}mv_f^2 - \tfrac{1}{2}mv_i^2$$

$$h = \frac{v_i^2 - v_f^2}{2g} = \frac{(30.3 \text{ m/s})^2 - (0.0 \text{ m/s})^2}{2(9.80 \text{ m/s}^2)} = 46.8 \text{ m}$$

Reflect: Note that the weight of 20 N was never used in the calculations because both gravitational potential and kinetic energy are proportional to mass, m. Thus any object, that attains 25.0 m/s at a height of 15.0 m, must have an initial velocity of 30.3 m/s. As the rock moves upward gravity does negative work and this reduces the kinetic energy of the rock.

*7.23. Set Up: Use $W_{net} = W_f = -f_k s = -\mu_k mgs$ and $W_{net} = K_f - K_i$. The skier stops, so $K_f = 0$.

Solve: **(a)** Setting the two expression for net work equal, $W_{net} = -\mu_k mgs = -\tfrac{1}{2}mv_i^2$. Solving for the coefficient of kinetic friction,

$$\mu_k = \frac{v_i^2}{2gs} = \frac{(12.0 \text{ m/s})^2}{2(9.80 \text{ m/s}^2)(184 \text{ m})} = 3.99 \times 10^{-2}$$

(b) The mass m of the skier divides out and μ_k is independent of m. If v_i is doubled while s is constant then μ_k increases by a factor of 4; $\mu_k = 0.160$.

Reflect: To stop the skier friction does negative work that is equal in magnitude to the initial kinetic energy of the skier.

*7.25. Set Up: Use $W_{on \text{ spring}} = +\tfrac{1}{2}kx^2$ in part (a) and $F_{on \text{ spring}} = kx$ in part (b).

Solve: **(a)** $k = \dfrac{2(W_{on \text{ spring}})}{x^2} = \dfrac{2(8.0 \text{ J})}{(0.025 \text{ m})^2} = 2.6 \times 10^4 \text{ N/m}$.

(b) $F_{on \text{ spring}} = (2.6 \times 10^4 \text{ N/m})(0.025 \text{ m}) = 650 \text{ N}$.

*7.29. Set Up: The work done is the area under the graph of $F_{on \text{ spring}}$ versus x between the initial and final positions.

Solve: **(a)** The area between $x = 0$ and $x = 5.0$ cm is a right triangle. The work is thus calculated as:

$$W_{on \text{ spring}} = \frac{1}{2}(5.0 \times 10^{-2} \text{ m})(250 \text{ N}) = 6.2 \text{ J}$$

(b) Since the area between $x = 2.0$ cm and $x = 7.0$ cm is a trapezoid, the work is calculated as:

$$W_{on \text{ spring}} = \frac{1}{2}(100 \text{ N} + 350 \text{ N})(0.070 \text{ m} - 0.020 \text{ m}) = 11.2 \text{ J}$$

Reflect: The force is larger for $x = 7.0$ cm than for 5.0 cm and the work done in (b) is greater than that done in (a), even though the displacement of the end of the spring is 5.0 cm in each case.

*7.31. Set Up: Use $\Delta U_{grav} = mg(y_f - y_i)$.

Solve: $\Delta U_{grav} = (72 \text{ kg})(9.80 \text{ m/s}^2)(0.60 \text{ m}) = 420$ J.

Reflect: This gravitational potential energy comes from elastic potential stored in his tensed muscles.

*7.33. Set Up: Only the spring does work and Eq. (7.15) applies. Also, we have $a = \dfrac{F}{m} = \dfrac{-kx}{m}$, where F is the force the spring exerts on the mass. Let point 1 be the initial position of the mass against the compressed spring, so $K_1 = 0$ and $U_1 = 11.5$ J. Let point 2 be the point where the mass leaves the spring, so $U_{el,2} = 0$.

Solve: **(a)** $K_1 + U_{el,1} = K_2 + U_{el,2}$ gives $U_{el,1} = K_2$. Thus, $\frac{1}{2}mv_2^2 = U_{el,1}$ and $v_2 = \sqrt{\dfrac{2U_{el,1}}{m}} = \sqrt{\dfrac{2(11.5 \text{ J})}{2.50 \text{ kg}}} = 3.03$ m/s.

K is largest when U_{el} is least and this is when the mass leaves the spring. The mass achieves its maximum speed of 3.03 m/s as it leaves the spring and then slides along the surface with constant speed.

(b) The acceleration is greatest when the force on the mass is the greatest, and this is when the spring has its maximum compression. Since $U_{el} = \frac{1}{2}kx^2$ we have $x = -\sqrt{\dfrac{2U_{el}}{k}} = -\sqrt{\dfrac{2(11.5 \text{ J})}{2500 \text{ N/m}}} = -0.0959$ m. The minus sign indicates compression. Finally, we have $F = -kx = ma_x$ and $a_x = -\dfrac{kx}{m} = -\dfrac{(2500 \text{ N/m})(-0.0959 \text{ m})}{2.50 \text{ kg}} = 95.9$ m/s^2.

Reflect: If the end of the spring is displaced to the left when the spring is compressed, then a_x in part (b) is to the right, and vice versa.

*7.35. Set Up: Use $F_{on\ tendon} = kx$. In part (a), $F_{on\ tendon}$ equals mg, the weight of the object suspended from it. In part (b), also apply $U_{el} = \frac{1}{2}kx^2$ to calculate the stored energy.

Solve: **(a)** $k = \dfrac{F_{on\ tendon}}{x} = \dfrac{(0.250 \text{ kg})(9.80 \text{ m/s}^2)}{0.0123 \text{ m}} = 199$ N/m

(b) $x = \dfrac{F_{on\ tendon}}{k} = \dfrac{138 \text{ N}}{199 \text{ N/m}} = 0.693 \text{ m} = 69.3$ cm; $U_{el} = \frac{1}{2}(199 \text{ N/m})(0.693 \text{ m})^2 = 47.8$ J

Reflect: The 250 g object has a weight of 2.45 N. The 138 N force is much larger than this and stretches the tendon a much greater distance.

*7.39. Set Up: The change in gravitational potential energy is $\Delta U_{grav} = mg(y_f - y_i)$, while the increase in kinetic energy, for a zero initial velocity, is $\Delta K = \frac{1}{2}mv_f^2$. Set the food energy, expressed in joules, equal to the mechanical energies developed.

Solve: **(a)** The food energy equals $mg(y_f - y_i)$, so

$$y_f - y_i = \dfrac{(140 \text{ food calories})(4186 \text{ J/1 food calorie})}{(65 \text{ kg})(9.80 \text{ m/s}^2)} = 920 \text{ m}$$

(b) The mechanical energy would be 20% of the results of part (a): $\Delta y = (0.20)(920 \text{ m}) = 180$ m.

*7.43. Set Up: For part (a), $U_f - U_i = mg(y_f - y_i)$. Take $y_i = 2425$ ft $= 739.1$ m and $y_f = 0$. For part (b), use $K_f + U_f = K_i + U_i$ with $K_i = U_f = 0$ and $K_f = \frac{1}{2}mv_f^2$; this gives $\frac{1}{2}mv_f^2 = U_i$.

Solve: **(a)** $U_i = -mg(y_f - y_i) = -(1.00 \text{ kg})(9.80 \text{ m/s}^2)(0.00 \text{ m} - 739.1 \text{ m}) = 7.24 \times 10^3$ J

(b) The final kinetic energy of each kilogram is equal to the initial potential energy, $K_f = U_i = 7.24 \times 10^3$ J while the corresponding speed is

$$v_f = \sqrt{\frac{2U_i}{m}} = \sqrt{\frac{2(7.24 \times 10^3 \text{ J})}{1.00 \text{ kg}}} = 120 \text{ m/s}.$$

By comparison, for a 70 kg person,

$$v_f = \sqrt{\frac{2(7.24 \times 10^3 \text{ J})}{70 \text{ kg}}} = 14 \text{ m/s} = 32 \text{ mph}.$$

(c) To double K_f, double U_i; since mg is constant, this means h must be doubled. You therefore need $h = 4850$ ft. To double v_f, you must increase h by a factor of 4 $(2v_f = 2\sqrt{2U_i/m} = \sqrt{8gh})$, resulting in $h = 9700$ ft.

*7.45. Set Up: Let θ be the angle of the initial speed above the horizontal. In parts (a)-(c), θ has values of $90°$, $-90°$, and $0°$, respectively. Use $K_f + U_f = K_i + U_i$ with $K_f = \frac{1}{2}(w/g)v_f^2$ and $K_i = \frac{1}{2}(w/g)v_0^2$. Let $y_i = h$ and $y_f = 0$; this gives $\frac{1}{2}(w/g)v_f^2 = \frac{1}{2}(w/g)v_0^2 + wh$.

Solve: **(a)-(d)** $v_f = \sqrt{v_0^2 + 2gh}$. **(e)** Since K and U are both proportional to the weight $w = mg$, w divides out of the expression and v_f is unaffected by a change in weight.

Reflect: The initial kinetic energy depends only on the initial speed and is independent of the direction of the initial velocity.

*7.49. Set Up: Only gravity does work, so apply Eq. (7.11). We have $v_i = 0$, so $\frac{1}{2}mv_f^2 = mg(y_i - y_f)$.

Solve: Tarzan is lower than his original height by a distance $y_i - y_f = l(\cos 30° - \cos 45°)$ so his speed is $v = \sqrt{2gl(\cos 30° - \cos 45°)} = 7.9$ m/s, a bit quick for conversation.

Reflect: The result is independent of Tarzan's mass.

*7.51. Set Up: Apply $K_f + U_f = K_i + U_i$ with $K_i = U_f = 0$, $K_f = K$, and $U_i = \frac{1}{2}kx^2$; the result is $K = \frac{1}{2}kx^2$. Since $\dfrac{K}{x^2}$ is constant, $\dfrac{K_1}{x_1^2} = \dfrac{K_2}{x_2^2}$.

Solve: $K_2 = K_1\left(\dfrac{x_2^2}{x_1^2}\right) = K\left(\dfrac{10.0 \text{ cm}}{5.0 \text{ cm}}\right)^2 = 4K$

*7.55. Set Up: Apply $E_{mech} = K + U$ with $U_i = U_f$ and $K_f = 0$; therefore, the lost mechanical energy is $K_i = \frac{1}{2}mv_i^2$.

Solve: $K_i = \frac{1}{2}(1.50 \text{ kg})(13.0 \text{ m/s})^2 = 127$ J

Reflect: Part of the initial mechanical energy is converted to thermal energy by the negative work done by friction. We did not use the stopping time in our solution.

*7.57. Set Up: Use $E_{mech} = K + U$ with $y_f = 0$, $y_i = h$, and $K_i = U_f = 0$. Conservation of energy thus becomes: $E_{mech,i} - E_{mech,f} = mgy_i - \frac{1}{2}mv_f^2$.

Solve: $E_{mech,i} - E_{mech,f} = (12.0 \times 10^{-3} \text{ kg})[(9.80 \text{ m/s}^2)(2.50 \text{ m}) - \frac{1}{2}(3.20 \text{ m/s})^2] = 0.233$ J

Reflect: In the absence of air resistance an object released from rest has a speed of 7.00 m/s after it has fallen 2.50 m. The speed of the ball is much less than this because of the negative work done by friction.

*7.61. Set Up: Use $E_{\text{mech,f}} = E_{\text{mech,i}} - f_k s$ to solve for f_k. There is no change in height so $U_i = U_f$ and the energy relation becomes $K_f = K_i - f_k s$.

Solve: (a) First solve for the kinetic energy values:

$$K_i = \tfrac{1}{2}mv_i^2 = \tfrac{1}{2}(375 \text{ kg})(4.5 \text{ m/s})^2 = 3800 \text{ J}$$

$$K_f = \tfrac{1}{2}mv_f^2 = \tfrac{1}{2}(375 \text{ kg})(3.0 \text{ m/s})^2 = 1690 \text{ J}$$

Substituting,

$$f_k = \frac{K_i - K_f}{s} = \frac{3800 \text{ J} - 1690 \text{ J}}{7.0 \text{ m}} = 300 \text{ N}$$

(b) (i) $(K_i - K_f)/K_i = [(3800 \text{ J} - 1690 \text{ J})/3800 \text{ J}] \times 100\% = 56\%$

(b) (ii) $(v_f - v_i)/v_i = [(1.5 \text{ m/s})/(4.5 \text{ m/s})] \times 100\% = 33\%$

Reflect: Friction does negative work and reduces the kinetic energy of the object.

*7.63. Set Up: For part (a) use $P = \dfrac{W}{\Delta t}$ to solve for W, the energy the bulb uses. Then set this value equal to $\tfrac{1}{2}mv^2$ for part (b), and solve for the speed. In part (c), equate the W from part (a) to $U_{\text{grav}} = mgh$ and solve for the height.

Solve: (a) $W = P\Delta t = (100 \text{ W})(3600 \text{ s}) = 3.6 \times 10^5 \text{ J}$

(b) $K = 3.6 \times 10^5 \text{ J}$ so $v = \sqrt{\dfrac{2K}{m}} = \sqrt{\dfrac{2(3.6 \times 10^5 \text{ J})}{70 \text{ kg}}} = 100 \text{ m/s}$

(c) $U_{\text{grav}} = 3.6 \times 10^5 \text{ J}$ so $h = \dfrac{U_{\text{grav}}}{mg} = \dfrac{3.6 \times 10^5 \text{ J}}{(70 \text{ kg})(9.80 \text{ m/s}^2)} = 520 \text{ m}$

Reflect: (b) Olympic runners achieve speeds up to approximately 36 m/s, or roughly one = third the result calculated. **(c)** The tallest tree on record, a redwood, stands 364 ft or 110 m, or 4.7 times smaller than the result.

*7.65. Set Up: Since the electricity cost rate is given in cents per kWh, each power must be multiplied by the corresponding time usage in determining the cost: $\text{cost} = Pt \times (\text{cost}/Pt) = (\text{kWh})[\text{cents}/(\text{kWh})]$.

Solve: (a) $\text{cost} = Pt \times \left(\dfrac{\text{cost}}{Pt}\right) = \left[(10.0 \text{ hp})(8.00 \text{ hr}) \times \left(\dfrac{746 \text{ W}}{\text{hp}}\right)\left(\dfrac{\text{kW}}{1000 \text{ W}}\right)\right]\left(\dfrac{7.35 \cancel{c}}{\text{kWh}}\right) = 439\cancel{c} = \$4.39$

(b) $\text{cost} = Pt \times \left(\dfrac{\text{cost}}{Pt}\right) = \left[(75 \text{ W})\left(\dfrac{24.0 \text{ hr}}{\text{day}}\right)\left(\dfrac{\text{kW}}{1000 \text{ W}}\right)\right]\left(\dfrac{7.35 \cancel{c}}{\text{kWh}}\right) = 13\dfrac{\cancel{c}}{\text{day}} = \$0.13 \text{ per day}$

Reflect: Note that the cost of running the motor for an entire day is $3(\$4.39) = \$13.2$; thus, the cost of running the motor is $(\$13.2)/(\$0.13) = 102 \approx 100$ times the cost of running the light.

*7.69. Set Up: From Problem 7.68, the U.S. consumes $3.2 \times 10^{11} \text{ W}$ of power per year. The area required in part (a) may then be calculated as $A = (P_{\text{U.S.}})/(P_{\text{radiation}}/\text{m}^2)$. However, for an array efficiency of 25%, the radiative power rate must be multiplied by 0.25 to account for power losses. For part (b), since the array is square, $A = L^2$ and the length of a side is simply $L = \sqrt{A}$.

Solve: (a) $A = (P_{\text{U.S.}})/(P_{\text{radiation}}/\text{m}^2) = \dfrac{3.2 \times 10^{11} \text{ W}}{0.25(1.0 \times 10^3 \text{ W/m}^2)}\left(\dfrac{\text{km}^2}{10^6 \text{ m}^2}\right) = 1.3 \times 10^3 \text{ km}^2$

(b) $L_{\text{metric}} = \sqrt{A} = \sqrt{1.3 \times 10^3 \text{ km}^2} = 36 \text{ km}$

$$L_{\text{English}} = (36 \text{ km})\left(\frac{1 \text{ mi}}{1.609 \text{ km}}\right) = 22 \text{ mi}$$

Reflect: An array of this size is approximately equivalent to the state of Rhode Island and thus feasible.

*7.71. Set Up: Use $1.00 \text{ hp} = 746 \text{ W}$ to convert the given units. In part (c), also apply the energy relation $W = P \, \Delta t$.

Solve: (a) $(100 \text{ W})\left(\dfrac{1.00 \text{ hp}}{746 \text{ W}}\right) = 0.134 \text{ hp}$

(b) $(75 \text{ hp})\left(\dfrac{746 \text{ W}}{1.00 \text{ hp}}\right) = 5.6 \times 10^4 \text{ W}$

(c) $(2.00 \text{ W})\left(\dfrac{1.00 \text{ hp}}{746 \text{ W}}\right) = 2.68 \times 10^{-3} \text{ hp}$

(d) $(3.92 \times 10^{26} \text{ W})\left(\dfrac{1.00 \text{ hp}}{746 \text{ W}}\right) = 5.25 \times 10^{23} \text{ hp}$

(e) $(25 \text{ hp})\left(\dfrac{746 \text{ W}}{1.00 \text{ hp}}\right) = 1.865 \times 10^4 \text{ W}; \quad W = (1.865 \times 10^4 \text{ W})(1.5 \text{ h}) = 28 \text{ kWh}$

*7.75. Set Up: For part (a), the work performed each day is $W = W_{\text{grav}} = mgh$ where $h = 1.63 \text{ m}$. The mass of the blood is calculated using the density, $\rho = m/V = 1050 \text{ kg/m}^3$, and the known volume of $V = 7500 \text{ L}$. The power output is then found from $P = W_{\text{grav}}/t$.

Solve: (a) The work performed in a single day is:

$$W = W_{\text{grav}} = mgh = \rho V g h = \left(1050\frac{\text{kg}}{\text{m}^3}\right)\left(7500 \text{ L} \times \frac{1 \text{ m}^3}{1000 \text{ L}}\right)\left(9.80\frac{\text{m}}{\text{s}^2}\right)(1.63 \text{ m}) = 1.26 \times 10^5 \text{ J}$$

(b) $P = \dfrac{W_{\text{grav}}}{t} = \dfrac{(1.26 \times 10^5 \text{ J/day})}{(24 \text{ hr/day})(3600 \text{ s/hr})} = 1.46 \text{ J/s} = 1.46 \text{ W}$

(c) Reflect: The heart actually puts out more power than the value calculated because two forms of energy were not considered; the heart also provides kinetic energy to the blood and generates thermal energy by working against the friction of the vein and vessel walls.

*7.79. Set Up: Conservation of energy says the decrease in potential energy equals the gain in kinetic energy. Since the two animals are equidistant from the axis, they each have the same speed v.

Solve: One mass rises while the other falls, so the net loss of potential energy is $(0.500 \text{ kg} - 0.200 \text{ kg})(9.80 \text{ m/s}^2)(0.400 \text{ m}) = 1.176 \text{ J}$. This is the sum of the kinetic energies of the animals and is equal to $\frac{1}{2}m_{\text{tot}}v^2$, and $v = \sqrt{\dfrac{2(1.176 \text{ J})}{(0.700 \text{ kg})}} = 1.83 \text{ m/s}$.

Reflect: The mouse gains both gravitational potential energy and kinetic energy. The rat's gain in kinetic energy is less than its decrease of potential energy, and the energy difference is transferred to the mouse.

*7.83. Set Up: We know that $P_{\text{av}} = F_{\|}v_{\text{av}}$. We can find the average velocity, $v_{\text{av}} = \dfrac{0 + 6.00 \text{ m/s}}{2} = 3.00 \text{ m/s}$, and use $F = ma$ to calculate the force.

Solve: Your friend's average acceleration is $a = \dfrac{v - v_0}{t} = \dfrac{6.00 \text{ m/s}}{3.00 \text{ s}} = 2.00 \text{ m/s}^2$. Since there are no other horizontal forces acting, the force you exert on her is given by $F_{net} = ma = (65.0 \text{ kg})(2.00 \text{ m/s}^2) = 130 \text{ N}$. $P_{av} = (130 \text{ N}) \times (3.00 \text{ m/s}) = 390 \text{ W}$.

Reflect: We could also use the work-energy theorem: $W = K_2 - K_1 = \frac{1}{2}(65.0 \text{ kg})(6.00 \text{ m/s})^2 = 1170 \text{ J}$. $P_{av} = \dfrac{W}{t} = \dfrac{1170 \text{ J}}{3.00 \text{ s}} = 390 \text{ W}$, the same as obtained by our other approach.

*7.87. Set Up: For part (a), apply conservation of energy to the motion from point A to point B: $K_B + U_{grav,B} = K_A + U_{grav,A}$ with $K_A = 0$. Defining $y_B = 0$ and $y_A = 13.0$ m, conservation of energy becomes $\frac{1}{2}mv_B^2 = mgy_A$ or $v_B = \sqrt{2gy_A}$. In part (b), the free-body diagram for the roller coaster car at point B is shown in the figure below. $\Sigma F_y = ma_y$ gives $mg + n = ma_{rad}$, where $a_{rad} = v^2/r$. Solving for the normal force results in:

$$n = m\left(\dfrac{v^2}{r} - g\right)$$

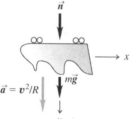

Solve: (a) $v_B = \sqrt{2(9.80 \text{ m/s}^2)(13.0 \text{ m})} = 16.0 \text{ m/s}$

(b) $n = (350 \text{ kg})\left[\dfrac{(16.0 \text{ m/s})^2}{6.0 \text{ m}} - 9.80 \text{ m/s}^2\right] = 1.15 \times 10^4 \text{ N}$

Reflect: The normal force n is the force that the tracks exert on the roller coaster car. The car exerts a force of equal magnitude and opposite direction on the tracks.

*7.89. Set Up: We can calculate the average speed of the horse from the number of trips that the horse makes around a circle of known circumference each hour. The force exerted by the horse can be calculated from its assumed power output by using $P_{av} = F_\parallel v_{av}$. According to the appendix, 1 hp = 746 W = 550 ft $\cdot$ lb/s.

Solve: (a) The average speed of the horse is $\left(\dfrac{144 \text{ trips}}{\text{hour}}\right)\left(\dfrac{24\pi \text{ feet}}{1 \text{ trip}}\right)\left(\dfrac{1 \text{ hour}}{3600 \text{ s}}\right) = 3.0 \text{ ft/s}$. To produce 1 hp, we require a force that satisfies $F_\parallel = \dfrac{P_{av}}{v_{av}} = \dfrac{550 \text{ ft} \cdot \text{lb/s}}{3.0 \text{ ft/s}} = 180 \text{ lb}$.

(b) The work output of the person must be at least her gain in gravitational potential energy, which is $\Delta U = mg\Delta y$. If this work occurs over a time t, the average power output is $P_{av} = \dfrac{mgh}{t} = \dfrac{(70 \text{ kg})(9.80 \text{ m/s}^2)(3.0 \text{ m})}{5.0 \text{ s}} = 410 \text{ W}$. This can also be written as $(410 \text{ W})\left(\dfrac{1 \text{ hp}}{746 \text{ W}}\right) = 0.55 \text{ hp}$.

Reflect: The power output of a person can exceed 1 hp for a brief time.

*7.93. Set Up: Apply $K_1 + U_1 + W_{\text{other}} = K_2 + U_2$ to the motion of the person. Point 1 is where he steps off the platform and point 2 is where he is stopped by the cord. Let $y = 0$ at point 2. $y_1 = 41.0$ m. $W_{\text{other}} = -\frac{1}{2}kx^2$, where $x = 11.0$ m is the amount the cord is stretched at point 2. The cord does negative work.

Solve: $K_1 = K_2 = U_2 = 0$, so $mgy_1 - \frac{1}{2}kx^2 = 0$ and $k = 631$ N/m.

Now apply $F = kx$ to the test pulls:
$F = kx$ so $x = F/k = 0.602$ m.

Reflect: All his initial gravitational potential energy is taken away by the negative work done by the force exerted by the cord, and this amount of energy is stored as elastic potential energy in the stretched cord.

*7.95. Set Up: First apply conservation of energy to find the kinetic energy of the wood as it enters the rough section: $U_1 = K_2$, or $K_2 = mgy_1$. Then determine the distance traveled using the work-energy relation, $W_{\text{nc}} = K_3 + U_3 - K_2 - U_2$ where $K_3 = U_3 = U_2 = 0$ and $W_{\text{nc}} = W_f = -\mu_k mgs$. For part (b), use the same relation and the calculated distance to find the frictional work.

Solve: (a) The work-energy relation reduces to $W_{\text{nc}} = W_f = -K_2$, which gives $-\mu_k mgs = -K_2$. Substituting $K_2 = mgy_1$ and solving for the distance, s:

$$s = \frac{mgy_1}{\mu_k mg} = \frac{y_1}{\mu_k} = \frac{4.0 \text{ m}}{0.20} = 20 \text{ m}$$

(b) $W_{\text{nc}} = W_f = -\mu_k mgs = -(0.20)(2.0 \text{ kg})(9.80 \text{ m/s}^2)(20 \text{ m}) = -78$ J

*7.97. Set Up: For parts (a) and (c), use the energy relation $W = P\,\Delta t$ to find the energy in joules and then convert the respective results to food calorie units.

Solve: (a) Energy $= W = (80 \text{ J/s})(3600 \text{ s/h})(24 \text{ h/day}) = 6.9 \times 10^6$ J/day

$$W = (6.9 \times 10^6 \text{ J/day})\left(\frac{1 \text{ food cal}}{4186 \text{ J}}\right) = 1.6 \times 10^3 \text{ food calories/day}$$

(b) The body's metabolic processes convert food energy to thermal energy.

(c) Energy $= W = [(80 \text{ J/s})(3600 \text{ s/h})(16 \text{ hr}) + (200 \text{ J/s})(3600 \text{ s/h})(8 \text{ h})]$

$$= 1.04 \times 10^7 \text{ J} = 2.48 \times 10^3 \text{ food calories}$$

*7.101. Set Up: $P = F_\| v$ and $1 \text{ m/s} = 3.6$ km/h.

Solve: (a) $F = \dfrac{P}{v} = \dfrac{28.0 \times 10^3 \text{ W}}{(60.0 \text{ km/h})((1 \text{ m/s})/(3.6 \text{ km/h}))} = 1.68 \times 10^3$ N.

(b) The speed is lowered by a factor of one-half, and the resisting force is lowered by a factor of $(0.65 + 0.35/4)$, and so the power at the lower speed is $(28.0 \text{ kW})(0.50)(0.65 + 0.35/4) = 10.3 \text{ kW} = 13.8$ hp.

(c) Similarly, at the higher speed, $(28.0 \text{ kW})(2.0)(0.65 + 0.35 \times 4) = 114.8 \text{ kW} = 154$ hp.

Reflect: At low speeds rolling friction dominates the power requirement but at high speeds air resistance dominates.

*7.105. Set Up: Hooke's law states $U_{\text{el}} = \frac{1}{2}kx^2$ and $F = kx$.

Solve: (a) Yes, since U is a linear function of x^2.

(b) The graph shows $U_{\text{el}} = 5.0$ J when $x^2 = 4.0 \times 10^{-2}$ m^2; thus $k = \dfrac{2U_{\text{el}}}{x^2} = \dfrac{2(5.0 \text{ J})}{4.0 \times 10^{-2} \text{ m}^2} = 250$ N/m.

(c) $x = \sqrt{\dfrac{2U_{\text{el}}}{k}} = \sqrt{\dfrac{2(10.0 \text{ J})}{250 \text{ N/m}}} = 0.283 \text{ m} = 28.3$ cm

(d) $F = kx = (250 \text{ N/m})(0.283 \text{ m}) = 70.8$ N

(e) $U_{\text{el}} = \frac{1}{2}kx^2$; the graph is a parabola as shown below.

MOMENTUM

Solutions to Problems

*8.3. Set Up: The signs of the velocity components indicate their directions.

Solve: (a) $P_x = p_{Ax} + p_{Cx} = 0 + (10.0\text{ kg})(-3.0\text{ m/s}) = -30\text{ kg}\cdot\text{m/s}$

$$P_y = p_{Ay} + p_{Cy} = (5.0\text{ kg})(-11.0\text{ m/s}) + 0 = -55\text{ kg}\cdot\text{m/s}$$

(b) $P_x = p_{Bx} + p_{Cx} = (6.0\text{ kg})(10.0\text{ m/s}\cos 60°) + (10.0\text{ kg})(-3.0\text{ m/s}) = 0$

$$P_y = p_{By} + p_{Cy} = (6.0\text{ kg})(10.0\text{ m/s}\sin 60°) + 0 = 52\text{ kg}\cdot\text{m/s}$$

(c) $P_x = p_{Ax} + p_{Bx} + p_{Cx} = 0 + (6.0\text{ kg})(10.0\text{ m/s}\cos 60°) + (10.0\text{ kg})(-3.0\text{ m/s}) = 0$

$P_y = p_{Ay} + p_{By} + p_{Cy} = (5.0\text{ kg})(-11.0\text{ m/s}) + (6.0\text{ kg})(10.0\text{ m/s}\sin 60°) + 0 = -3.0\text{ kg}\cdot\text{m/s}$

Reflect: A has no x-component of momentum so P_x is the same in (b) and (c). C has no y-component of momentum so P_y in (c) is the sum of P_y in (a) and (b).

*8.5. Set Up: 145 g = 0.145 kg; 57 g = 0.057 kg

Solve: (a) $p = mv = (0.145\text{ kg})(45\text{ m/s}) = 6.5\text{ kg}\cdot\text{m/s}$

$$K = \tfrac{1}{2}mv^2 = \tfrac{1}{2}(0.145\text{ kg})(45\text{ m/s})^2 = 150\text{ J}$$

(b) (i) $K = \tfrac{1}{2}mv^2$ so $v = \sqrt{\dfrac{2K}{m}} = \sqrt{\dfrac{2(150\text{ J})}{0.057\text{ kg}}} = 73\text{ m/s}$

(ii) $p = mv$ so $v = \dfrac{p}{m} = \dfrac{6.5\text{ kg}\cdot\text{m/s}}{0.057\text{ kg}} = 110\text{ m/s}$

Reflect: K depends on v^2 and p depends on v so a smaller increase in v is required for the lighter ball to have the same kinetic energy.

*8.9. Set Up: $m_E = 5.98 \times 10^{24}\text{ kg}$. Consider the person and the earth to be an isolated system. Use coordinates where $+y$ is upward, in the direction the person jumps.

Solve: $P_{i,y} = P_{f,y}$. $P_{i,y} = 0$. The earth recoils in the $-y$ direction with speed v_E, so $0 = m_{person}v_{person} - m_E v_E$.

$$v_E = \left(\frac{m_{person}}{m_E} \right) v_{person} = \left(\frac{75 \text{ kg}}{5.98 \times 10^{24} \text{ kg}} \right)(2.0 \text{ m/s}) = 2.5 \times 10^{-23} \text{ m/s}$$

*8.13. Set Up: Let $+x$ be to the right.

Solve: (a) $P_{i,x} = P_{f,x}$ says $(0.250)v_{A,i} = (0.250 \text{ kg})(-0.120 \text{ m/s}) + (0.350 \text{ kg})(0.650 \text{ m/s})$ and $v_{A,i} = 0.790$ m/s

(b) $K_i = \frac{1}{2}(0.250 \text{ kg})(0.790 \text{ m/s})^2 = 0.0780$ J

$K_f = \frac{1}{2}(0.250 \text{ kg})(0.120 \text{ m/s})^2 + \frac{1}{2}(0.350 \text{ kg})(0.650 \text{ m/s})^2 = 0.0757$ J and $\Delta K = K_f - K_i = -0.0023$ J.

Reflect: The total momentum of the system is conserved but the total kinetic energy decreases.

*8.15. Set Up: Each horizontal component of momentum is conserved. $K = \frac{1}{2}mv^2$. Let $+x$ be the direction of Rebecca's initial velocity and let the $+y$ axis make an angle of $36.9°$ with respect to the direction of her final velocity. $v_{D1x} = v_{D1y} = 0$. $v_{R1x} = 13.0$ m/s; $v_{R1y} = 0$. $v_{R2x} = (8.00 \text{ m/s})\cos 53.1° = 4.80$ m/s; $v_{R2y} = (8.00 \text{ m/s})\sin 53.1° = 6.40$ m/s. Solve for v_{D2x} and v_{D2y}.

Solve: (a) $P_{1x} = P_{2x}$ gives $m_R v_{R1x} = m_R v_{R2x} + m_D v_{D2x}$.

$$v_{D2x} = \frac{m_R(v_{R1x} - v_{R2x})}{m_D} = \frac{(45.0 \text{ kg})(13.0 \text{ m/s} - 4.80 \text{ m/s})}{65.0 \text{ kg}} = 5.68 \text{ m/s}.$$

$P_{1y} = P_{2y}$ gives $0 = m_R v_{R2y} + m_D v_{D2y}$.

$$v_{D2y} = -\frac{m_R}{m_D} v_{R2y} = -\left(\frac{45.0 \text{ kg}}{65.0 \text{ kg}} \right)(6.40 \text{ m/s}) = -4.43 \text{ m/s}.$$

The directions of $\vec{v}_{R1}$, $\vec{v}_{R2}$, and $\vec{v}_{D2}$ are sketched in the figure below. We have $\tan \theta = \left| \frac{v_{D2y}}{v_{D2x}} \right| = \frac{4.43 \text{ m/s}}{5.68 \text{ m/s}}$ and

$\theta = 38.0°$. $v_D = \sqrt{v_{D2x}^2 + v_{D2y}^2} = 7.20$ m/s.

(b) $K_1 = \frac{1}{2}m_R v_{R1}^2 = \frac{1}{2}(45.0 \text{ kg})(13.0 \text{ m/s})^2 = 3.80 \times 10^3$ J.

$K_2 = \frac{1}{2}m_R v_{R2}^2 + \frac{1}{2}m_D v_{D2}^2 = \frac{1}{2}(45.0 \text{ kg})(8.00 \text{ m/s})^2 + \frac{1}{2}(65.0 \text{ kg})(7.20 \text{ m/s})^2 = 3.12 \times 10^3$ J.

$\Delta K = K_2 - K_1 = -680$ J.

Reflect: Each component of momentum is separately conserved. The kinetic energy of the system decreases.

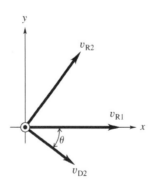

*8.19. Set Up: Use coordinates where $+x$ is to the right and $+y$ is upward.

Solve: *Collision*: There is no external horizontal force during the collision so $P_{i,x} = P_{f,x}$. This gives $(5.00 \text{ kg})(12.0 \text{ m/s}) = (10.0 \text{ kg})v_f$ and $v_f = 6.0$ m/s.

Motion after the collision: Only gravity does work and the initial kinetic energy of the combined chunks is converted entirely to gravitational potential energy when the chunk reaches its maximum height h above the valley floor. Conservation of energy gives

$$\tfrac{1}{2}m_{\text{tot}}v^2 = m_{\text{tot}}gh \ \text{ and } \ h = \frac{v^2}{2g} = \frac{(6.0 \text{ m/s})^2}{2(9.8 \text{ m/s}^2)} = 1.8 \text{ m}$$

Reflect: After the collision the energy of the system is $\tfrac{1}{2}m_{\text{tot}}v^2 = \tfrac{1}{2}(10.0 \text{ kg})(6.0 \text{ m/s})^2 = 180 \text{ J}$ when it is all kinetic energy and it is $m_{\text{tot}}gh = (10.0 \text{ kg})(9.8 \text{ m/s}^2)(1.8 \text{ m}) = 180 \text{ J}$ when it is all gravitational potential energy. Mechanical energy is conserved during the motion after the collision. But before the collision the total energy of the system is $\tfrac{1}{2}(5.0 \text{ kg})(12.0 \text{ m/s})^2 = 360 \text{ J}$; 50% of the mechanical energy is dissipated during the inelastic collision of the two chunks.

*8.21. Set Up: Let x be the direction of motion. Let each boxcar have mass m.
Solve: (a) $P_{i,x} = P_{f,x}$ says $(3m)(20.0 \text{ m/s}) = (4m)v_{f,x}$ and $v_{f,x} = 15.0 \text{ m/s}$.

(b) $K_i = \tfrac{1}{2}(3m)(20.0 \text{ m/s})^2 = 600m \text{ J/kg}$; $K_f = \tfrac{1}{2}(4m)(15.0 \text{ m/s})^2 = 450m \text{ J/kg}$

$$\Delta K = -150m \text{ J/kg} \ \text{ and } \ \frac{\Delta K}{K_i} = \frac{-150m \text{ J/kg}}{600m \text{ J/kg}} = -0.250;$$

25% of the original kinetic energy is dissipated. Kinetic energy is converted to other forms by work done by the forces during the collision.

*8.25. Set Up: Apply conservation of momentum to the collision and the work-energy equation to the motion after the collision. Let $+x$ be the direction the bullet was traveling before the collision.
Solve: *Motion after the collision*: $W_{\text{other}} + U_i + K_i = K_f + U_f$. There is no change in height so $U_i = U_f$. The block stops so $K_f = 0$. $n = mg$, $f_k = \mu_k mg$ and $W_{\text{other}} = W_f = -\mu_k mgs$. Let V be the speed of the block immediately after the collision, so $K_i = \tfrac{1}{2}mV^2$. Therefore, $-\mu_k mgs + \tfrac{1}{2}mV^2 = 0$ and

$$V = \sqrt{2\mu_k gs} = \sqrt{2(0.20)(9.80 \text{ m/s}^2)(0.230 \text{ m})} = 0.950 \text{ m/s}.$$

Collision: $m_{\text{bullet}}v_{\text{bullet, i}} = mV$, where $m = 1.20 \text{ kg} + 5.00 \times 10^{-3} \text{ kg} = 1.205 \text{ kg}$.

$$v_{\text{bullet,i}} = \frac{(1.205 \text{ kg})(0.950 \text{ m/s})}{5.00 \times 10^{-3} \text{ kg}} = 229 \text{ m/s}$$

Reflect: Our analysis neglects the friction force exerted by the surface on the block during the collision, since we assume momentum is conserved in the collision. The collision forces between the bullet and block are much larger than the friction force, so this approximation is justified.

*8.29. Set Up: Use coordinates where $+x$ is east and $+y$ is south. The system of two cars before and after the collision is sketched in the figure below. Neglect friction from the road during the collision. The enmeshed cars have mass $2000 \text{ kg} + 1500 \text{ kg} = 3500 \text{ kg}$.

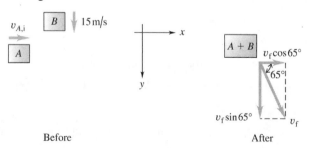

Before   After

Solve: There are no external horizontal forces during the collision, so $P_{i,x} = P_{f,x}$ and $P_{i,y} = P_{f,y}$.

(a) $P_{i,x} = P_{f,x}$ gives $(1500 \text{ kg})(15 \text{ m/s}) = (3500 \text{ kg})v_f \sin 65°$ and $v_f = 7.1 \text{ m/s}$

(b) $P_{i,y} = P_{f,y}$ gives $(2000 \text{ kg})v_{A,i} = (3500 \text{ kg})v_f \cos 65°.$ And then with $v_f = 7.1$ m/s, $v_{A,i} = 5.2$ m/s.

Reflect: Momentum is a vector and we must treat each component separately.

*8.31. Set Up: For an elastic collision with B initially stationary, the final velocities are
$$v_A = \left(\frac{m_A - m_B}{m_A + m_B}\right)v \text{ and } v_B = \left(\frac{2m_A}{m_A + m_B}\right)v.$$
Apply these equations with $m_A = 0.300$ kg, $m_B = 0.150$ kg, and $v = 0.80$ m/s.

Solve: (a) $v_A = \left(\dfrac{0.300 \text{ kg} - 0.150 \text{ kg}}{0.300 \text{ kg} + 0.150 \text{ kg}}\right)(0.80 \text{ m/s}) = 0.27$ m/s,

$$v_B = \left(\frac{2[0.300 \text{ kg}]}{0.300 \text{ kg} + 0.150 \text{ kg}}\right)(0.80 \text{ m/s}) = 1.07 \text{ m/s}.$$

The 0.300 kg glider moves to the right at 0.27 m/s and the 0.150 kg glider moves to the right at 1.07 m/s.

(b) $K_{A,f} = \frac{1}{2}m_A v_A^2 = \frac{1}{2}(0.300 \text{ kg})(0.27 \text{ m/s})^2 = 0.011$ J;

$$K_{B,f} = \frac{1}{2}m_B v_B^2 = \frac{1}{2}(0.150 \text{ kg})(1.07 \text{ m/s})^2 = 0.086 \text{ J}$$

Reflect: The relative velocity after the collision has magnitude 1.07 m/s $- 0.27$ m/s $= 0.80$ m/s, the same as before the collision. The initial kinetic energy of the system is $K_{A,i} = \frac{1}{2}(0.300 \text{ kg})(0.80 \text{ m/s})^2 = 0.096$ J. The final kinetic energy is $K_{A,f} + K_{B,f} = 0.097$ J, the same as the initial kinetic energy, apart from a slight difference due to rounding.

*8.33. Set Up: For an elastic collision with B initially stationary, the final velocities are
$$v_A = \left(\frac{m_A - m_B}{m_A + m_B}\right)v \text{ and } v_B = \left(\frac{2m_A}{m_A + m_B}\right)v.$$
Apply these equations with $m_A = 1.67 \times 10^{-27}$ kg, $m_B = 6.65 \times 10^{-27}$ kg, and $v = 258$ km/s.

Solve: (a) $v_A = \left(\dfrac{1.67 \times 10^{-27} \text{ kg} - 6.65 \times 10^{-27} \text{ kg}}{1.67 \times 10^{-27} \text{ kg} + 6.65 \times 10^{-27} \text{ kg}}\right)(258 \text{ km/s}) = -154$ km/s

$$v_B = \left(\frac{2[1.67 \times 10^{-27} \text{ kg}]}{1.67 \times 10^{-27} \text{ kg} + 6.65 \times 10^{-27} \text{ kg}}\right)(258 \text{ km/s}) = 104 \text{ km/s}$$

The proton recoils to the left at 154 km/s and the alpha particle travels to the right at 104 km/s.

(b) The proton has initial kinetic energy
$$K_{A,i} = \frac{1}{2}m_A v^2 = \frac{1}{2}(1.67 \times 10^{-27} \text{ kg})(2.58 \times 10^5 \text{ m/s})^2 = 5.56 \times 10^{-17} \text{ J}$$
and final kinetic energy
$$K_{A,f} = \frac{1}{2}m_A v_A^2 = \frac{1}{2}(1.67 \times 10^{-27} \text{ kg})(1.54 \times 10^5 \text{ m/s})^2 = 1.98 \times 10^{-17} \text{ J}.$$
The kinetic energy lost is $K_{A,i} - K_{A,f} = 3.58 \times 10^{-17}$ J.

(c) The kinetic energy gained by the alpha particle is
$$\frac{1}{2}(6.65 \times 10^{-27} \text{ kg})(1.04 \times 10^5 \text{ m/s})^2 = 3.60 \times 10^{-17} \text{ J}.$$

The energy lost by the proton is gained by the alpha particle. The total kinetic energy of the system is constant and the collision is elastic.

*8.37. Set Up: Assume the ball is initially moving to the right, and let this be the $+x$ direction. The ball stops, so its final velocity is zero.

Solve: (a) $J_x = mv_{f,x} - mv_{i,x} = 0 - (0.145 \text{ kg})(36.0 \text{ m/s}) = -5.22$ kg · m/s. The magnitude of the impulse applied to the ball is 5.22 kg · m/s.

(b) $J_x = F_x \Delta t$ so $F_x = \dfrac{J_x}{\Delta t} = \dfrac{-5.22 \text{ kg} \cdot \text{m/s}}{20.0 \times 10^{-3} \text{ s}} = -261 \text{ N}$

Reflect: The signs of J_x and F_x show that both these quantities are to the left.

*8.39. Set Up: (a) Take the $+x$ direction to be along the final direction of motion of the ball. The initial speed of the ball is zero.
(b) Take the $+x$ direction to be in the direction the ball is traveling before it is hit by the opponent.
Solve: (a) $J_x = mv_{f,x} - mv_{i,x} = (57 \times 10^{-3} \text{ kg})(73.14 \text{ m/s} - 0) = 4.2 \text{ kg} \cdot \text{m/s}$

$$F_x = \frac{J_x}{\Delta t} = \frac{4.2 \text{ kg} \cdot \text{m/s}}{30.0 \times 10^{-3} \text{ s}} = 140 \text{ N}$$

(b) $J_x = mv_{f,x} - mv_{i,x} = (57 \times 10^{-3} \text{ kg})(-55 \text{ m/s} - 73.14 \text{ m/s}) = -7.3 \text{ kg} \cdot \text{m/s}$

$$F_x = \frac{J_x}{\Delta t} = \frac{-7.3 \text{ kg} \cdot \text{m/s}}{30.0 \times 10^{-3} \text{ s}} = -240 \text{ N}$$

Reflect: The signs of J_x and F_x show their direction. $140 \text{ N} = 31 \text{ lb}$. This very attainable force has a large effect on the light ball. 140 N is 250 times the weight of the ball.

*8.43. Set Up: With $A = 1.5 \times 10^{-4} \text{ m}^2$ the maximum force without breaking the bone is
$$(1.5 \times 10^{-4} \text{ m}^2)(1 \times 10^8 \text{ N/m}^2) = 1.5 \times 10^4 \text{ N}.$$

Set the magnitude of the average force F_{av} during the collision equal to this value. Use coordinates where $+x$ is in his initial direction of motion. F_x is opposite to this direction, so $F_x = -1.5 \times 10^4 \text{ N}$.

Solve: $J_x = F_x \Delta t = (-1.5 \times 10^4 \text{ N})(10.0 \times 10^{-3} \text{ s}) = -150.0 \text{ N} \cdot \text{s}$

$$J_x = mv_{f,x} - mv_{i,x} \text{ and } v_{f,x} = 0.$$

$$v_{i,x} = -\frac{J_x}{m} = -\frac{-150 \text{ N} \cdot \text{s}}{70 \text{ kg}} = 2.1 \text{ m/s}$$

*8.45. Set Up: From the appendix we have $m_E = 5.97 \times 10^{24} \text{ kg}$, $m_M = 7.35 \times 10^{22} \text{ kg}$, and the distance between the earth and the moon is $3.84 \times 10^8 \text{ m}$. Represent the earth and the moon as point masses located at their respective centers. Use coordinates with the origin at the center of the earth and with the moon on the $+x$ axis; thus, $x_E = 0$ and $x_M = 3.84 \times 10^8 \text{ m}$.

Solve: $x_{\text{cm}} = \dfrac{m_E x_E + m_M x_M}{m_E + m_M} = \dfrac{(5.97 \times 10^{24} \text{ kg})(0) + (7.35 \times 10^{22} \text{ kg})(3.84 \times 10^8 \text{ m})}{(5.97 \times 10^{24} \text{ kg} + 7.35 \times 10^{22} \text{ kg})} = 4.67 \times 10^6 \text{ m}.$

The center of mass is $4.67 \times 10^6 \text{ m}$ from the center of the earth along the line that connects the center of the earth to that of the moon.

Reflect: Since the radius of the earth is $6.38 \times 10^6 \text{ m}$, the center of mass of the earth-moon system is roughly 1700 km below the earth's surface.

*8.49. Set Up: Use Eq. 8.18 to find the x and y coordinates of the center of mass of the machine part for each configuration of the part. In calculating the center of mass of the machine part, each uniform bar can be represented by a point mass at its geometrical center. Use coordinates with the axis at the hinge and the $+x$ and $+y$ axes along the horizontal and vertical bars in the figure in the problem. Let (x_i, y_i) and (x_f, y_f) be the coordinates of the bar before and after the vertical bar is pivoted. Let object 1 be the horizontal bar, object 2 be the vertical bar, and 3 be the ball.

Solve: $x_i = \dfrac{m_1 x_1 + m_2 x_2 + m_3 x_3}{m_1 + m_2 + m_3} = \dfrac{(4.00\ \text{kg})(0.750\ \text{m}) + 0 + 0}{4.00\ \text{kg} + 3.00\ \text{kg} + 2.00\ \text{kg}} = 0.333\ \text{m}.$

$$y_i = \frac{m_1 y_1 + m_2 y_2 + m_3 y_3}{m_1 + m_2 + m_3} = \frac{0 + (3.00\ \text{kg})(0.900\ \text{m}) + (2.00\ \text{kg})(1.80\ \text{m})}{9.00\ \text{kg}} = 0.700\ \text{m}.$$

$$x_f = \frac{(4.00\ \text{kg})(0.750\ \text{m}) + (3.00\ \text{kg})(-0.900\ \text{m}) + (2.00\ \text{kg})(-1.80\ \text{m})}{9.00\ \text{kg}} = -0.366\ \text{m}.$$

$y_f = 0.$ $x_f - x_i = -0.700\ \text{m}$ and $y_f - y_i = -0.700\ \text{m}.$ The center of mass moves 0.700 m to the right and 0.700 m upward.

Reflect: The vertical bar moves upward and to the right so it is sensible for the center of mass of the machine part to move in these directions.

*8.51. Set Up: $m_A = 1200\ \text{kg}$, $m_B = 1800\ \text{kg}.$ $M = m_A + m_B = 3000\ \text{kg}.$ Let $+x$ be to the right and let the origin be at the center of mass of the station wagon.

Solve: (a) $x_{cm} = \dfrac{m_A x_A + m_B x_B}{m_A + m_B} = \dfrac{0 + (1800\ \text{kg})(40.0\ \text{m})}{1200\ \text{kg} + 1800\ \text{kg}} = 24.0\ \text{m}$

The center of mass is between the two cars, 24.0 m to the right of the station wagon and 16.0 m behind the lead car.

(b) $P_x = m_A v_{A,i} + m_B v_{B,i} = (1200\ \text{kg})(12.0\ \text{m/s}) + (1800\ \text{kg})(20.0\ \text{m/s}) = 5.04 \times 10^4\ \text{kg} \cdot \text{m/s}$

(c) $v_{cm,x} = \dfrac{m_A v_{A,x} + m_B v_{B,x}}{m_A + m_B} = \dfrac{(1200\ \text{kg})(12.0\ \text{m/s}) + (1800\ \text{kg})(20.0\ \text{m/s})}{1200\ \text{kg} + 1800\ \text{kg}} = 16.8\ \text{m/s}$

(d) $P_x = M v_{cm,x} = (3000\ \text{kg})(16.8\ \text{m/s}) = 5.04 \times 10^4\ \text{kg} \cdot \text{m/s},$ the same as in part (b).

*8.53. Set Up: Use Eq. 8.25.

Solve: (a) $F = -v_{ex} \dfrac{\Delta m}{\Delta t} = -(1600\ \text{m/s}) \dfrac{-0.0500\ \text{kg}}{1.00\ \text{s}} = +80.0\ \text{N}$

(b) The absence of atmosphere would not prevent the rocket from operating. The rocket could be steered by ejecting the gas in a direction with a component perpendicular to the rocket's velocity and braked by ejecting it in a direction parallel (as opposed to antiparallel) to the rocket's velocity.

*8.57. Set Up: Use equation 8.24: $ma = -v_{ex} \dfrac{\Delta m}{\Delta t}.$ Assume that $\dfrac{\Delta m}{\Delta t}$ is constant over the 5.0 s interval and that m

doesn't change much during that interval. The thrust is $F = -v_{ex} \dfrac{\Delta m}{\Delta t}.$ Take m to have the constant value

$110\ \text{kg} + 70\ \text{kg} = 180\ \text{kg}.$ Note that the rate of change in mass, $\dfrac{\Delta m}{\Delta t}$, is negative since the mass of the MMU

decreases as gas is ejected.

Solve: (a) $\dfrac{\Delta m}{\Delta t} = -\dfrac{m}{v_{ex}} a = -\left(\dfrac{180\ \text{kg}}{490\ \text{m/s}}\right)(0.029\ \text{m/s}^2) = -0.0106\ \text{kg/s}.$ In 5.0 s the mass that is ejected is $(0.0106\ \text{kg/s})$

$(5.0\ \text{s}) = 0.053\ \text{kg}.$

(b) $F = -v_{ex} \dfrac{\Delta m}{\Delta t} = -(490\ \text{m/s})(-0.0106\ \text{kg/s}) = 5.19\ \text{N}.$

Reflect: The mass change in the 5.0 s is a very small fraction of the total mass m, so it is accurate to take m to be constant.

*8.59. Set Up: Momentum is conserved in the explosion. At the highest point the velocity of the boulder is zero. Since one fragment moves horizontally the other fragment also moves horizontally. Use projectile motion to relate the initial horizontal velocity of each fragment to its horizontal displacement. Use coordinates where $+x$ is north. Since both fragments start at the same height with zero vertical component of velocity, the time in the air, t, is the same for both. Call the fragments A and B, with A being the one that lands to the north. Therefore, we have $m_B = 3m_A.$

Solve: Apply $P_{1x} = P_{2x}$ to the collision: $0 = m_A v_{Ax} + m_B v_{Bx}$. $v_{Bx} = -\dfrac{m_A}{m_B} v_{Ax} = -v_{Ax}/3$. Apply projectile motion to

the motion after the collision: $x - x_0 = v_{0x}t$. Since t is the same, $\dfrac{(x - x_0)_A}{v_{Ax}} = \dfrac{(x - x_0)_B}{v_{Bx}}$ and

$(x - x_0)_B = \left(\dfrac{v_{Bx}}{v_{Ax}}\right)(x - x_0)_A = \left(\dfrac{-v_{Ax}/3}{v_{Ax}}\right)(x - x_0)_A = -(274 \text{ m})/3 = -91.3 \text{ m}$. The other fragment lands 91.3 m directly

south of the point of explosion.

Reflect: The fragment that has three times the mass travels one-third as far.

*8.63. Set Up: Apply conservation of momentum to the collision between the two people. Apply conservation of energy to the motion of the stuntman before the collision and to the entwined people after the collision. For the motion of the stuntman, we have $y_1 - y_2 = 5.0$ m. Let v_S be the magnitude of his horizontal velocity just before the collision. Let V be the speed of the entwined people just after the collision. Let d be the distance they slide along the floor.

Solve: (a) Motion before the collision: $K_1 + U_1 = K_2 + U_2$. $K_1 = 0$ and $\frac{1}{2}mv_S^2 = mg(y_1 - y_2)$. $v_S = \sqrt{2g(y_1 - y_2)} =$

$\sqrt{2(9.80 \text{ m/s}^2)(5.0 \text{ m})} = 9.90 \text{ m/s}$.

$$\text{Collision: } m_S v_S = m_{\text{tot}} V. \ V = \dfrac{m_S}{m_{\text{tot}}} v_S = \left(\dfrac{80.0 \text{ kg}}{150.0 \text{ kg}}\right)(9.90 \text{ m/s}) = 5.28 \text{ m/s}.$$

(b) Motion after the collision: $K_1 + U_1 + W_{\text{other}} = K_2 + U_2$ gives $\frac{1}{2}m_{\text{tot}}V^2 - \mu_k m_{\text{tot}} g d = 0$.

$$d = \dfrac{V^2}{2\mu_k g} = \dfrac{(5.28 \text{ m/s})^2}{2(0.250)(9.80 \text{ m/s}^2)} = 5.7 \text{ m}.$$

Reflect: Mechanical energy is dissipated in the inelastic collision, so the kinetic energy just after the collision is less than the initial potential energy of the stuntman.

*8.65. Set Up: Apply conservation of energy to the motion before and after the collision and apply conservation of momentum to the collision. Let v be the speed of the mass released at the rim just before it strikes the second mass. Let each object have mass m.

Solve: Conservation of energy says $\frac{1}{2}mv^2 = mgR$; $v = \sqrt{2gR}$.

Set Up: This is speed v_1 for the collision. Let v_2 be the speed of the combined object just after the collision.

Solve: Conservation of momentum applied to the collision gives $mv_1 = 2mv_2$ so $v_2 = v_1/2 = \sqrt{gR/2}$

Set Up: Apply conservation of energy to the motion of the combined object after the collision. Let y_3 be the final height above the bottom of the bowl.

Solve: $\frac{1}{2}(2m)v_2^2 = (2m)gy_3$.

$y_3 = \dfrac{v_2^2}{2g} = \dfrac{1}{2g}\left(\dfrac{gR}{2}\right) = R/4.$

Reflect: Mechanical energy is lost in the collision, so the final gravitational potential energy is less than the initial gravitational potential energy.

*8.69. Set Up: Let $+x$ be to the right. $w_A = 800$ N, $w_B = 600$ N and $w_C = 1000$ N.

Solve: $P_{i,x} = P_{f,x}$ gives $0 = m_A v_{A,f,x} + m_B v_{B,f,x} + m_C v_{C,f,x}$.

$$v_{C,f,x} = \dfrac{m_A v_{A,f,x} + m_B v_{B,f,x}}{m_C} = \dfrac{w_A v_{A,f,x} + w_B v_{B,f,x}}{w_C}.$$

$$v_{C,f,x} = \frac{(800 \text{ N})(-[5.00 \text{ m/s}]\cos 30.0°) + (600 \text{ N})(+[7.00 \text{ m/s}]\cos 36.9°)}{1000 \text{ N}} = -0.105 \text{ m/s}$$

The sleigh's velocity is 0.105 m/s, to the left.

Reflect: The vertical component of the momentum of the system consisting of the two people and the sleigh is not conserved, because of the net force exerted on the sleigh by the ice while they jump.

*8.71. Set Up: The center of mass of each piece of length L is at its center.

Solve: (a) From symmetry, the center of mass is on the vertical axis, a distance $(L/2)\cos(\alpha/2)$ below the apex.

(b) The center of mass is on the vertical axis of symmetry, a distance $2(L/2)/3 = L/3$ above the center of the horizontal segment.

(c) Using the wire frame as a coordinate system, the coordinates of the center of mass are equal and each is equal to $(L/2)/2 = L/4$. The center of mass is along the bisector of the angle, a distance $L/\sqrt{8}$ from the corner.

(d) By symmetry, the center of mass is at the center of the equilateral triangle, a distance $(L/2)\tan 60° = L/\sqrt{12}$ above the center of the horizontal segment.

*8.75. Set Up: Apply conservation of energy to the motion before and after the collision. Apply conservation of momentum to the collision. First consider the motion after the collision. The combined object has mass $m_{tot} = 25.0$ kg. Apply $\Sigma\vec{F} = m\vec{a}$ to the object at the top of the circular loop, where the object has speed v_3. The acceleration is $a_{rad} = v_3^2/R$, downward.

Solve: $T + mg = m\dfrac{v_3^2}{R}$.

The minimum speed v_3 for the object not to fall out of the circle is given by setting $T = 0$. This gives $v_3 = \sqrt{Rg}$, where $R = 3.50$ m.

Set Up: Next, use conservation of energy with point 2 at the bottom of the loop and point 3 at the top of the loop. Take $y = 0$ at point 2. Only gravity does work, so $K_2 + U_2 = K_3 + U_3$.

Solve: $\frac{1}{2}m_{tot}v_2^2 = \frac{1}{2}m_{tot}v_3^2 + m_{tot}g(2R)$.

Use $v_3 = \sqrt{Rg}$ and solve for v_2: $v_2 = \sqrt{5gR} = 13.1$ m/s.

Set Up: Now apply conservation of momentum to the collision between the dart and the sphere. Let v_1 be the speed of the dart before the collision.

Solve: $(5.00 \text{ kg})v_1 = (25.0 \text{ kg})(13.1 \text{ m/s})$.

$v_1 = 65.5$ m/s.

Reflect: The collision is inelastic and mechanical energy is removed from the system by the negative work done by the forces between the dart and the sphere.

*8.79. Set Up: Apply conservation of momentum. Take $+x$ to be to the right. For (a) the collision is elastic. Let A be the cart that is initially moving to the right. For (b) the two carts stick together and the combined object has final speed v_f.

Solve: (a) $P_{i,x} = P_{f,x}$ gives $Mv_0 + M(-v_0) = Mv_A + Mv_B$ and $v_A = -v_B$. For an elastic collision $v_{B,f} - v_{A,f} = -(v_{B,i} - v_{A,i})$. This gives $v_B - v_A = -(-v_0 - v_0) = 2v_0$. $v_A = -v_B$ so $v_B = v_0$ and $v_A = -v_0$.

A rebounds with speed v_0 and B rebounds with speed v_0.

(b) $P_{i,x} = P_{f,x}$ gives $Mv_0 + M(-v_0) = (2M)v_f$ and $v_f = 0$. Both carts are rest after the collision.

Reflect: Collision (b) is totally inelastic. All the initial kinetic energy is converted to other forms. Any collision in which the objects stick together must be inelastic.

*8.81. Set Up: $m_n = 1.675 \times 10^{-27}$ kg, $m_p = 1.673 \times 10^{-27}$ kg and $m_e = 9.109 \times 10^{-31}$ kg. After the decay the electron moves in the $+x$ direction with speed v_e and the proton moves in the $-x$ direction with speed v_p. Apply conservation of momentum.

Solve: (a) $P_{i,x} = P_{f,x}$ gives $0 = m_e v_e + m_p(-v_p)$. $\dfrac{v_e}{v_p} = \dfrac{m_p}{m_e} = 1837$

(b) $\dfrac{K_e}{K_p} = \dfrac{\frac{1}{2}m_e v_e^2}{\frac{1}{2}m_p v_p^2} = \dfrac{\frac{1}{2}m_e(1837 v_p)^2}{\frac{1}{2}m_p v_p^2} = \left(\dfrac{m_e}{m_p}\right)(1837)^2 = 1837.$

(c) The final speed of the electron is much greater than that of the proton.

Solutions to Passage Problems

*8.85. Set Up: Choose the squid and the ejected seawater (squirt) as the "system." Assume that the initial momentum of the system is zero.
Solve: Assuming that the external forces are negligible during the squirt, the total momentum of the squid/squirt system must be conserved. The momentum of the squid must be equal and opposite to that of the squirt. Thus, the correct answer is E.

*8.87. Set Up: Assume three identical squirts, each contributing -1.5 kg·m/s to the squid's total momentum.
Solve: The total momentum of the squid after three squirts is -4.5 kg·m/s, thus the squid's velocity is
$$v_x = \frac{p_x}{m} = \frac{-4.5 \text{ kg·m/s}}{2.5 \text{ kg}} = -1.8 \text{ m/s} = -180 \text{ cm/s. Thus, the correct answer is C.}$$

ROTATIONAL MOTION

Problems 3, 5, 7, 11, 13, 17, 21, 25, 29, 31, 33, 37, 39, 43, 47, 49, 51, 55, 57, 59, 63, 65, 69, 73, 75, 77

Solutions to Problems

*9.3. Set Up: For one revolution, $\Delta\theta = 2\pi$ rad. Assume constant angular velocity, so

$$\omega = \frac{\Delta\theta}{\Delta t}.$$

The second hand makes 1 revolution in 1 minute = 60.0 s. The minute hand makes 1 revolution in 1 h = 3600 s, and the hour hand makes 1 revolution in 12 h = 43,200 s.

Solve: (a) *second hand* $\omega = \dfrac{2\pi \text{ rad}}{60 \text{ s}} = 0.105$ rad/s;

minute hand $\omega = \dfrac{2\pi \text{ rad}}{3600 \text{ s}} = 1.75 \times 10^{-3}$ rad/s;

hour hand $\omega = \dfrac{2\pi \text{ rad}}{43,200 \text{ s}} = 1.45 \times 10^{-4}$ rad/s

(b) The period is the time for 1 revolution. Second hand, 1 min; minute hand, 1 h; hour hand, 12 h.
Reflect: When the angular velocity is constant, $\omega = \omega_{\text{av}}$.

*9.5. Set Up: 1 rev = 2π rad

Solve: (a) The period is the time for 1 rev, so period = 2.25 s.

(b) $\omega = \dfrac{\theta}{t} = \dfrac{2\pi \text{ rad}}{2.25 \text{ s}} = 2.79$ rad/s

Reflect: The angular speed and period don't depend on the radius of the wheel.

*9.7. Set Up: According the appendix, the distance between the earth and the moon is 3.84×10^8 m. Since the radius of both the earth and the moon are small in comparison, we will assume that this is roughly the distance, r, that the laser beam travels. The relation between the angular velocity of the beam and its linear velocity on the lunar surface is $v = r\omega$.

Solve: (a) The speed of the laser beam across the moon's surface is

$$v = r\omega = (3.84 \times 10^8 \text{ m})(1.50 \times 10^{-3} \text{ rad/s}) = 5.76 \times 10^5 \text{ m/s}.$$

(b) Assume that the laser light spreads out from a point source on the earth to a 6.00 km diameter spot on the moon. The angle of divergence (in radians) is the angle that subtends a 6.00 km arc on the lunar surface. Thus,

$$\theta = \frac{s}{r} = \frac{6.00 \times 10^3 \text{ m}}{3.84 \times 10^8 \text{ m}} = 1.56 \times 10^{-5} \text{ rad}$$

Reflect: The actual distance that the beam travels depends on the exact location where the beam leaves the earth and where it hits the lunar surface. Since the radius of the earth is roughly 0.06×10^8 m and the radius of the moon is 0.02×10^8 m, the distance traveled by the beam is known to an accuracy of only two significant figures. The answers should be rounded accordingly.

*9.11. Set Up: $\omega_0 = 0$. $\omega = (78.0 \text{ rpm})\left(\frac{2\pi \text{ rad}}{1 \text{ rev}}\right)\left(\frac{1 \text{ min}}{60 \text{ s}}\right) = 8.17 \text{ rad/s}$

Solve: **(a)** $\omega = \omega_0 + \alpha t$ gives $\alpha = \dfrac{\omega - \omega_0}{t} = \dfrac{8.17 \text{ rad/s} - 0}{3.50 \text{ s}} = 2.33 \text{ rad/s}^2$

(b) $\theta - \theta_0 = \omega_0 t + \frac{1}{2}\alpha t^2 = \frac{1}{2}(2.33 \text{ rad/s}^2)(3.50 \text{ s})^2 = (14.27 \text{ rad})\left(\dfrac{360°}{2\pi \text{ rad}}\right) = 818°$

*9.13. Set Up: 570 rpm = 59.7 rad/s; 1600 rpm = 167.6 rad/s

Solve: $\omega = \omega_0 + \alpha t$ so $\alpha = \dfrac{\omega - \omega_0}{t} = \dfrac{167.6 \text{ rad/s} - 59.7 \text{ rad/s}}{(133 \text{ min})(60 \text{ s/1 min})} = 0.0135 \text{ rad/s}^2$

*9.17. Set Up: 500.0 rpm = 8.33 rev/s. Let the direction of rotation of the flywheel be positive.

Solve: **(a)** $(\theta - \theta_0) = \left(\dfrac{\omega_0 + \omega}{2}\right)t$ gives

$$\omega = \frac{2(\theta - \theta_0)}{t} - \omega_0 = \frac{2(200.0 \text{ rev})}{30.0 \text{ s}} - 8.33 \text{ rev/s} = 5.00 \text{ rev/s} = 300 \text{ rpm}$$

(b) Use information in part (a) to find α: $\omega = \omega_0 + \omega_0 t$ gives

$$\alpha = \frac{\omega - \omega_0}{t} = \frac{5.00 \text{ rev/s} - 8.33 \text{ rev/s}}{30.0 \text{ s}} = -0.111 \text{ rev/s}^2.$$

Then, with $\omega = 0$ and $\omega_0 = 8.33$ rev/s the equation $\omega = \omega_0 + \omega_0 t$ gives

$$t = \frac{\omega - \omega_0}{\alpha} = \frac{0 - 8.33 \text{ rev/s}}{-0.111 \text{ rev/s}^2} = 75.0 \text{ s.}$$

$$\theta - \theta_0 = \left(\frac{\omega_0 + \omega}{2}\right)t = \left(\frac{8.33 \text{ rev/s} + 0}{2}\right)(75.0 \text{ s}) = 312 \text{ rev.}$$

Reflect: The angular acceleration is negative because the wheel is slowing down.

*9.21. Set Up: First, convert all distance measurements into feet. Use $v = r\omega$ to find v. $r = 12.0 \text{ in} = 1.00 \text{ ft}$. From the appendix, we have 1 mi/h = 1.466 ft/s.

Solve: Convert the speed of the car: $v = \left(63 \dfrac{\text{mi}}{\text{h}}\right)\left(\dfrac{1.466 \text{ ft/s}}{1 \text{ mi/h}}\right) = 92 \text{ ft/s}$. Find the angular velocity of the tires:

$$\omega = \frac{v}{r} = \frac{92 \text{ ft/s}}{1.00 \text{ ft}} = 92 \text{ rad/s.}$$

Reflect: $v = r\omega$ requires that ω be in rad/s.

*9.25. Set Up: $r_1 = 4.0$ in., $r_2 = 1.5$ in. and $\omega_2 = 75$ rpm. The 3 in. sprocket and 24 in. wheel are mounted on the same axle and turn at the same rate.

Solve: All points on the chain have the same speed so $v_1 = v_2$; points on the rim of each sprocket move at the same tangential speed. $v = r\omega$ gives $r_1\omega_1 = r_2\omega_2$.

$$\omega_1 = \left(\frac{r_2}{r_1}\right)\omega_2 = \left(\frac{1.5 \text{ in.}}{4.0 \text{ in.}}\right)(75 \text{ rpm}) = 28 \text{ rpm}$$

Reflect: The large sprocket turns at a slower rate than the small sprocket.

*9.29. Set Up: Use Eq. (9.15) to relate ω to a_{rad} and use $\sum \vec{F} = m\vec{a}$ to relate a_{rad} to F_{rad}. Use Eq. (9.13) to relate ω to v, where v is the tangential speed.

Solve: (a) $a_{rad} = r\omega^2$ and $F_{rad} = ma_{rad} = mr\omega^2$

$$\frac{F_{rad,2}}{F_{rad,1}} = \left(\frac{\omega_2}{\omega_1}\right)^2 = \left(\frac{640 \text{ rev/min}}{423 \text{ rev/min}}\right)^2 = 2.29$$

(b) Using $v = r\omega$ we have $\dfrac{v_2}{v_1} = \dfrac{\omega_2}{\omega_1} = \dfrac{640 \text{ rev/min}}{423 \text{ rev/min}} = 1.51$

(c) Using $v = r\omega$ and $\omega = (640 \text{ rev/min})\left(\dfrac{1 \text{ min}}{60 \text{ s}}\right)\left(\dfrac{2\pi \text{ rad}}{1 \text{ rev}}\right) = 67.0 \text{ rad/s}$ we have

$v = r\omega = (0.235 \text{ m})(67.0 \text{ rad/s}) = 15.7 \text{ m/s}$.

$$a_{rad} = r\omega^2 = (0.235 \text{ m})(67.0 \text{ rad/s})^2 = 1060 \text{ m/s}^2 \text{ so } \frac{a_{rad}}{g} = \frac{1060 \text{ m/s}^2}{9.80 \text{ m/s}^2} = 108; \ a = 108g.$$

Reflect: In parts (a) and (b), since a ratio is used the units cancel and there is no need to convert ω to rad/s. In part (c), v and a_{rad} are calculated from ω, and ω must be in rad/s.

*9.31. Set Up: I for the object is the sum of the values of I for each part. For the bar, for an axis perpendicular to the bar, use the appropriate expression from Table 9.2. For a point mass, we have $I = mr^2$, where r is the distance of the mass from the axis.

Solve: (a) $I = I_{bar} + I_{balls} = \dfrac{1}{12}M_{bar}L^2 + 2m_{balls}\left(\dfrac{L}{2}\right)^2$.

$$I = \frac{1}{12}(4.00 \text{ kg})(2.00 \text{ m})^2 + 2(0.500 \text{ kg})(1.00 \text{ m})^2 = 2.33 \text{ kg} \cdot \text{m}^2$$

(b) $I = \dfrac{1}{3}m_{bar}L^2 + m_{ball}L^2 = \dfrac{1}{3}(4.00 \text{ kg})(2.00 \text{ m})^2 + (0.500 \text{ kg})(2.00 \text{ m})^2 = 7.33 \text{ kg} \cdot \text{m}^2$

(c) We have $I = 0$ because all the masses are on the axis, where $r = 0$.

Reflect: The moment of inertia for an object depends on both the location and orientation of the axis of rotation in relation to the object. The axis of rotation with the lowest moment inertia is always one of those that pass through the object's center of mass.

*9.33. Set Up: (a) Each mass is a distance

$$\frac{\sqrt{2(0.400 \text{ m})^2}}{2} = \frac{0.400 \text{ m}}{\sqrt{2}}$$

from the axis. (b) each mass is 0.200 m from the axis. (c) two masses are on the axis and two are $\dfrac{0.400 \text{ m}}{\sqrt{2}}$ from the axis.

Solve: (a) $I = \sum mr^2 = 4(0.200 \text{ kg})\left(\dfrac{0.400 \text{ m}}{\sqrt{2}}\right)^2 = 0.0640 \text{ kg} \cdot \text{m}^2$

(b) $I = 4(0.200 \text{ kg})(0.200 \text{ m})^2 = 0.0320 \text{ kg} \cdot \text{m}^2$

(c) $I = 2(0.200 \text{ kg})\left(\dfrac{0.400 \text{ m}}{\sqrt{2}}\right)^2 = 0.0320 \text{ kg} \cdot \text{m}^2$

Reflect: The value of I depends on the location of the axis.

*9.37. Set Up: $K = \frac{1}{2}I\omega^2$. Use Table 9.2 to relate I to the mass M of the disk. 45.0 rpm = 4.71 rad/s. For a uniform solid disk we have $I = \frac{1}{2}MR^2$.

Solve: (a) $I = \dfrac{2K}{\omega^2} = \dfrac{2(0.250 \text{ J})}{(4.71 \text{ rad/s})^2} = 0.0225 \text{ kg} \cdot \text{m}^2$.

(b) $I = \frac{1}{2}MR^2$ and $M = \dfrac{2I}{R^2} = \dfrac{2(0.0225 \text{ kg} \cdot \text{m}^2)}{(0.300 \text{ m})^2} = 0.500 \text{ kg}$.

Reflect: No matter what the shape is, the rotational kinetic energy is proportional to the mass of the object.

*9.39. Set Up: $K = \frac{1}{2}I\omega^2$, with ω in rad/s. 1 rev/min = $(2\pi/60)$ rad/s. $\Delta K = -500$ J

Solve: $\omega_i = 650$ rev/min = 68.1 rad/s. $\omega_f = 520$ rev/min = 54.5 rad/s.

$\Delta K = K_f - K_i = \frac{1}{2}I(\omega_f^2 - \omega_i^2)$ and

$I = \dfrac{2(\Delta K)}{\omega_f^2 - \omega_i^2} = \dfrac{2(-500 \text{ J})}{(54.5 \text{ rad/s})^2 - (68.1 \text{ rad/s})^2} = 0.600 \text{ kg} \cdot \text{m}^2$

*9.43. Set Up: The speed v of the weight is related to ω of the cylinder by $v = R\omega$, where $R = 0.325$ m. Use coordinates where $+y$ is upward and $y_i = 0$ for the weight. $y_f = -h$, where h is the unknown distance the weight descends. Let $m = 1.50$ kg and $M = 3.25$ kg. For the cylinder $I = \frac{1}{2}MR^2$.

Solve: (a) Conservation of energy says $K_i + U_i = K_f + U_f$. $K_i = 0$ and $U_i = 0$. $U_f = mgy_f = -mgh$.

$$K_f = \frac{1}{2}mv^2 + \frac{1}{2}I\omega^2 = \frac{1}{2}mv^2 + \frac{1}{2}\left(\frac{1}{2}MR^2\right)\left(\frac{v}{R}\right)^2 = \left(\frac{1}{2}m + \frac{1}{4}M\right)v^2$$

$$\left(\frac{1}{2}m + \frac{1}{4}M\right)v^2 - mgh = 0$$

$$h = \frac{\left(\frac{1}{2}m + \frac{1}{4}M\right)v^2}{mg} = \frac{\left[\frac{1}{2}(1.50 \text{ kg}) + \frac{1}{4}(3.25 \text{ kg})\right](2.50 \text{ m/s})^2}{(1.50 \text{ kg})(9.80 \text{ m/s}^2)} = 0.664 \text{ m}$$

(b) $\omega = \dfrac{v}{R} = \dfrac{2.50 \text{ m/s}}{0.325 \text{ m}} = 7.69$ rad/s

Reflect: The net work done by the rope that connects the cylinder and weight is zero. The speed v of the weight equals the tangential speed at the outer surface of the cylinder, and this gives $v = R\omega$.

*9.47. Set Up: Since there is rolling without slipping, $v_{cm} = R\omega$. The kinetic energy is given by Eq. (9.19). We have $\omega = 3.00$ rad/s and $R = 0.600$ m. For a hoop rotating about an axis at its center we have $I = MR^2$.

Solve: (a) $v_{cm} = R\omega = (0.600 \text{ m})(3.00 \text{ rad/s}) = 1.80$ m/s.

(b) $K = \frac{1}{2}Mv_{cm}^2 + \frac{1}{2}I\omega^2 = \frac{1}{2}Mv_{cm}^2 + \frac{1}{2}(MR^2)(v_{cm}/R^2) = Mv_{cm}^2 = (2.20 \text{ kg})(1.80 \text{ m/s})^2 = 7.13$ J

Reflect: For the special case of a hoop, the total kinetic energy is equally divided between the motion of the center of mass and the rotation about the axis through the center of mass.

*9.49. Set Up: The ball has moment of inertia $I_{cm} = \frac{2}{3}mR^2$. Rolling without slipping means $v_{cm} = R\omega$. Use coordinates where $+y$ is upward and $y = 0$ at the bottom of the hill, so $y_i = 0$ and $y_f = h = 5.00$ m.

Solve: (a) Conservation of energy gives $K_i + U_i = K_f + U_f$. $U_i = 0$, $K_f = 0$ (stops). Therefore $K_i = U_f$ and $\frac{1}{2}mv_{cm}{}^2 + \frac{1}{2}I_{cm}\omega^2 = mgh$.

$$\tfrac{1}{2}I_{cm}\omega^2 = \tfrac{1}{2}\left(\tfrac{2}{3}mR^2\right)\left(\frac{v_{cm}}{R}\right)^2 = \tfrac{1}{3}mv_{cm}{}^2 \text{ so } \tfrac{5}{6}mv_{cm}{}^2 = mgh.$$

$$v_{cm} = \sqrt{\frac{6gh}{5}} = \sqrt{\frac{6(9.80 \text{ m/s}^2)(5.00 \text{ m})}{5}} = 7.67 \text{ m/s}$$

and

$$\omega = \frac{v_{cm}}{R} = \frac{7.67 \text{ m/s}}{0.113 \text{ m}} = 67.9 \text{ rad/s}.$$

(b) $K_{rot} = \frac{1}{2}I\omega^2 = \frac{1}{3}mv_{cm}{}^2 = \frac{1}{3}(0.426 \text{ kg})(7.67 \text{ m/s})^2 = 8.35$ J

Reflect: Its translational kinetic energy at the base of the hill is $\frac{1}{2}mv_{cm}{}^2 = \frac{3}{2}K_{rot} = 12.52$ J. Its total kinetic energy is 20.9 J. This equals its final potential energy: $mgh = (0.426 \text{ kg})(9.80 \text{ m/s}^2)(5.00 \text{ m}) = 20.9$ J

*9.51. Set Up: Apply Eq. (9.19). For an object that is rolling without slipping we have $v_{cm} = R\omega$.

Solve: The fraction of the total kinetic energy that is rotational is

$$\frac{(1/2)I_{cm}\omega^2}{(1/2)Mv_{cm}{}^2 + (1/2)I_{cm}\omega^2} = \frac{1}{1 + (M/I_{cm})v_{cm}{}^2/\omega^2} = \frac{1}{1 + (MR^2/I_{cm})}$$

(a) $I_{cm} = (1/2)MR^2$, so the above ratio is 1/3.

(b) $I_{cm} = (2/5)MR^2$ so the above ratio is 2/7.

(c) $I_{cm} = (2/3)MR^2$ so the ratio is 2/5.

(d) $I_{cm} = (5/8)MR^2$ so the ratio is 5/13.

Reflect: The moment of inertia of each object takes the form $I = \beta MR^2$. The ratio of rotational kinetic energy to total kinetic energy can be written as $\dfrac{1}{1 + 1/\beta} = \dfrac{\beta}{1 + \beta}$. The ratio increases as β increases.

*9.55. Set Up: The linear acceleration a of the elevator equals the tangential acceleration of a point on the rim of the shaft. $a = 0.150g = 1.47$ m/s. For the shaft, $R = 0.0800$ m.

Solve: $a_{tan} = R\alpha$ so $a = R\alpha$ and $\alpha = \dfrac{1.47 \text{ m/s}^2}{0.0800 \text{ m}} = 18.4 \text{ rad/s}^2$

Reflect: In $a_{tan} = R\alpha$, α is in units of rad/s^2.

*9.57. Set Up: The distance d the car travels equals the arc length s traveled by a point on the rim of the tire, so $d = r\theta$. The odometer reading depends on the angle through which the wheels have turned.

Solve: (a) $d = 0.10$ mi $= 528$ ft. $r = 12$ in. $= 1$ ft. $\theta = \dfrac{d}{r} = \dfrac{528 \text{ ft}}{1 \text{ ft}} = 528 \text{ rad} = 84.0 \text{ rev}.$

(b) $d = r\theta$. 5000 rev $= 3.14 \times 10^4$ rad. $d = (1 \text{ ft})(3.14 \times 10^4 \text{ rad}) = 3.14 \times 10^4$ ft $= 5.95$ mi

(c) With proper 24 in. diameter tires the angular displacement for $d = 500$ mi $= 2.64 \times 10^6$ ft is

$$\theta = \frac{d}{r} = \frac{2.64 \times 10^6 \text{ ft}}{1 \text{ ft}} = 2.64 \times 10^6 \text{ rad}.$$

With 28 in. tires this θ corresponds to $d = r\theta = \left(\frac{14}{12} \text{ ft}\right)(2.64 \times 10^6 \text{ rad}) = 3.08 \times 10^6 \text{ ft} = 583 \text{ mi}.$

Reflect: In $s = r\theta$ the angle θ must be expressed in radians.

*9.59. Set Up: The translational kinetic energy is $K = \frac{1}{2}mv^2$ and the kinetic energy of the rotating flywheel is $K = \frac{1}{2}I\omega^2$. Use the scale speed to calculate the actual speed v. From that calculate K for the car and then solve for ω that gives this K for the flywheel.

Solve: (a) $\dfrac{v_{\text{toy}}}{v_{\text{scale}}} = \dfrac{L_{\text{toy}}}{L_{\text{real}}}$

$v_{\text{toy}} = v_{\text{scale}}\left(\dfrac{L_{\text{toy}}}{L_{\text{real}}}\right) = (700 \text{ km/h})\left(\dfrac{0.150 \text{ m}}{3.0 \text{ m}}\right) = 35.0 \text{ km/h}$

$v_{\text{toy}} = (35.0 \text{ km/h})(1000 \text{ m/1 km})(1 \text{ h/3600 s}) = 9.72 \text{ m/s}$

(b) $K = \frac{1}{2}mv^2 = \frac{1}{2}(0.180 \text{ kg})(9.72 \text{ m/s})^2 = 8.50 \text{ J}$

(c) $K = \frac{1}{2}I\omega^2$ gives that $\omega = \sqrt{\dfrac{2K}{I}} = \sqrt{\dfrac{2(8.50 \text{ J})}{4.00 \times 10^{-5} \text{ kg} \cdot \text{m}^2}} = 652 \text{ rad/s}$

Reflect: $K = \frac{1}{2}I\omega^2$ gives ω in rad/s. 652 rad/s = 6200 rev/min so the rotation rate of the flywheel is very large.

*9.63. Set Up: All points on the belt (which is shown in the figure below) move with the same speed. Since the belt doesn't slip, the speed of the belt is the same as the speed of a point on the rim of the shaft and on the rim of the wheel, and these speeds are related to the angular speed of each circular object by $v = r\omega$.

Solve: (a) $v_1 = r_1\omega_1$

$\omega_1 = (60.0 \text{ rev/s})(2\pi \text{ rad/1 rev}) = 377 \text{ rad/s}$

$v_1 = r_1\omega_1 = (0.45 \times 10^{-2} \text{ m})(377 \text{ rad/s}) = 1.70 \text{ m/s}$

(b) $v_1 = v_2$

$r_1\omega_1 = r_2\omega_2$

$\omega_2 = (r_1/r_2)\omega_1 = (0.45 \text{ cm/2.00 cm})(377 \text{ rad/s}) = 84.8 \text{ rad/s}$

Reflect: The wheel has a larger radius than the shaft so it turns slower to have the same tangential speed for points on the rim.

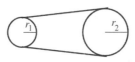

*9.65. Set Up: My total mass is $m = 90 \text{ kg}$. I model my head as a uniform sphere of radius 8 cm. I model my trunk and legs as a uniform solid cylinder of radius 12 cm. I model my arms as slender rods of length 60 cm. $\omega = 72 \text{ rev/min} = 7.5 \text{ rad/s}$.

Solve: (a) $I_{\text{tot}} = \frac{2}{5}(0.070m)(0.080 \text{ m})^2 + \frac{1}{2}(0.80m)(0.12 \text{ m})^2 + 2\left(\frac{1}{3}\right)(0.13m)(0.60 \text{ m})^2 = 3.3 \text{ kg} \cdot \text{m}^2$

(b) $K_{\text{rot}} = \frac{1}{2}I\omega^2 = \frac{1}{2}(3.3 \text{ kg} \cdot \text{m}^2)(7.5 \text{ rad/s})^2 = 93 \text{ J}$

Reflect: According to these estimates about 85% of the total I is due to the outstretched arms. If the initial translational kinetic energy $\frac{1}{2}mv^2$ of the skater is converted to this rotational kinetic energy as he goes into a spin, his initial speed must be 1.4 m/s.

*9.69. Set Up: $I_{cm} = \frac{2}{5}mR^2$. If the stone rolls without slipping, $v_{cm} = R\omega$. If there is no friction the stone slides without rolling and has no rotational kinetic energy. Use coordinates where $+y$ is upward and $y = 0$ at the bottom of the hill.

Solve: After the stone is launched into the air its translational kinetic energy is converted to potential energy. At its maximum height h, $mgh = \frac{1}{2}mv_{cm}^2$, where v_{cm} is its translational speed at the bottom of the hill.

(a) Apply conservation of energy to the motion down the hill: $mgH = \frac{1}{2}mv_{cm}^2$ and $v_{cm}^2 = 2gH$. Then $mgh = \frac{1}{2}mv_{cm}^2 = \frac{1}{2}m(2gH)$ and $h = H$.

(b) Now the initial potential energy is converted to both translational and rotational kinetic energy as the stone rolls down the hill:

$$mgH = \frac{1}{2}mv_{cm}^2 + \frac{1}{2}I_{cm}\omega^2$$

$$\frac{1}{2}I_{cm}\omega^2 = \frac{1}{2}\left(\frac{2}{5}mR^2\right)\left(\frac{v_{cm}}{R}\right)^2 = \frac{1}{5}mv_{cm}^2 \text{ so } mgH = \frac{7}{10}mv_{cm}^2 \text{ and } v_{cm}^2 = \frac{10}{7}gH$$

This gives $mgh = \frac{1}{2}m\left(\frac{10}{7}gH\right)$ and $h = \frac{5}{7}H$.

(c) In (a) all the initial gravitational potential energy is converted back to final gravitational potential energy; the kinetic energy at the final maximum height is zero. In (b) the stone gains rotational kinetic energy as it rolls down the hill and it still has this rotational kinetic energy at its maximum height; not all the initial potential energy is converted into the final potential energy.

Reflect: Both answers do not depend on the mass or radius of the stone. But the answer to (b) depends on how the mass is distributed; the answer would be different for a hollow rolling sphere.

*9.73. Set Up: Apply conservation of energy to the system consisting of blocks A and B and the pulley. The system at points 1 and 2 of its motion is sketched in the figure below.

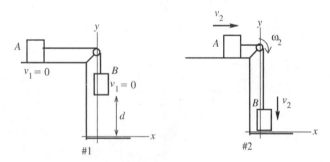

#1   #2

Use the work-energy relation $K_1 + U_1 + W_{other} = K_2 + U_2$. Use coordinates where $+y$ is upward and where the origin is at the position of block B after it has descended. The tension in the rope does positive work on block A and negative work of the same magnitude on block B, so the net work done by the tension in the rope is zero. Both blocks have the same speed.

Solve: Gravity does work on block B and kinetic friction does work on block A. Therefore $W_{other} = W_f = -\mu_k m_A gd$.

$K_1 = 0$ (system is released from rest)

$U_1 = m_B gy_{B1} = m_B gd$; $\ U_2 = m_B gy_{B2} = 0$

$K_2 = \frac{1}{2}m_A v_2^2 + \frac{1}{2}m_B v_2^2 + \frac{1}{2}I\omega_2^2$.

But $v(\text{blocks}) = R\omega(\text{pulley})$, so $\omega_2 = v_2/R$ and

$K_2 = \frac{1}{2}(m_A + m_B)v_2^2 + \frac{1}{2}I(v_2/R)^2 = \frac{1}{2}(m_A + m_B + I/R^2)v_2^2$

Putting all this into the work-energy relation gives

$m_B gd - \mu_k m_A gd = \frac{1}{2}(m_A + m_B + I/R^2)v_2^2$

$$(m_A + m_B + I/R^2)v_2{}^2 = 2gd(m_B - \mu_k m_A)$$

$$v_2 = \sqrt{\frac{2gd(m_B - \mu_k m_A)}{m_A + m_B + I/R^2}}$$

Reflect: If $m_B \gg m_A$ and I/R^2, then $v_2 = \sqrt{2gd}$; block B falls freely. If I is very large, v_2 is very small. Must have $m_B > \mu_k m_A$ for motion, so the weight of B will be larger than the friction force on A. I/R^2 has units of mass and is in a sense the "effective mass" of the pulley.

Solutions to Passage Problems

*9.75. Set Up: Let us consider counterclockwise rotations to be positive and clockwise rotations to be negative. Thus, $\omega_0 = -30$ rad/s and $\omega = +30$ rad/s. Since the angular acceleration is constant, we may use equation 9.7: $\omega = \omega_0 + \alpha t$.

Solve: $\alpha = \dfrac{\omega - \omega_0}{t} = \dfrac{(+30 \text{ rad/s}) - (-30 \text{ rad/s})}{10 \text{ s}} = 6 \text{ rad/s}^2$. The correct answer is D.

*9.77. Set Up: Set $\omega_0 = -30$ rad/s. Since the angular acceleration is constant, we may use equation 9.11: $\Delta\theta = \omega_0 t + \dfrac{1}{2}\alpha t^2$. We know that $\alpha = 6 \text{ rad/s}^2$ from Problem 9.75.

Solve: The displacement from 0 s to 20 s is: $\Delta\theta = \omega_0 t + \dfrac{1}{2}\alpha t^2 = (-30 \text{ rad/s})(20 \text{ s}) + \dfrac{1}{2}(6 \text{ rad/s}^2)(20 \text{ s})^2 = 600$ rad. Since the displacement from 0 s to 10 s is zero, this is also the displacement from 10 s to 20 s. Alternatively, we may set $\omega_0 = +30$ rad/s and $t = 10$ s to obtain $\Delta\theta = \omega_0 t + \dfrac{1}{2}\alpha t^2 = (+30 \text{ rad/s})(10 \text{ s}) + \dfrac{1}{2}(6 \text{ rad/s}^2)(10 \text{ s})^2 = 600$ rad. In either case, the correct answer is D.

10

DYNAMICS OF ROTATIONAL MOTION

Problems 1, 3, 9, 11, 13, 17, 19, 21, 25, 27, 29, 33, 37, 41, 43, 47, 49, 51, 55, 57, 61, 63, 65, 69, 71, 75, 79, 81

Solutions to Problems

*10.1. Set Up: Let counterclockwise torques be positive. $\tau = Fl$ with $l = r\sin\phi$.
Solve: (a) $\tau = + (10.0 \text{ N})(4.00 \text{ m})\sin 90.0° = 40.0 \text{ N} \cdot \text{m}$, counterclockwise.

(b) $\tau = + (10.0 \text{ N})(4.00 \text{ m})\sin 60.0° = 34.6 \text{ N} \cdot \text{m}$, counterclockwise.

(c) $\tau = + (10.0 \text{ N})(4.00 \text{ m})\sin 30.0° = 20.0 \text{ N} \cdot \text{m}$, counterclockwise.

(d) $\tau = - (10.0 \text{ N})(2.00 \text{ m})\sin 60.0° = -17.3 \text{ N} \cdot \text{m}$, clockwise.

(e) $\tau = 0$ since the force acts on the axis and $l = 0$.

(f) $\tau = 0$ since the line of action of the force passes through the location of the axis and $l = 0$.

Reflect: The torque of a force depends on the direction of the force and where it is applied to the object.

*10.3. Set Up: Use $\tau = Fl = rF\sin\phi$ to calculate the magnitude of each torque and use the right-hand rule to determine the direction of each torque. Add the torques to find the net torque. Let counterclockwise torques be positive. For the 11.9 N force (F_1) we have $r = 0$. For the 14.6 N force (F_2), $r = 0.350$ m and $\phi = 40.0°$. For the 8.50 N force (F_3), $r = 0.350$ m and $\phi = 90.0°$.

Solve: $\tau_1 = 0$. $\tau_2 = -(14.6 \text{ N})(0.350 \text{ m})\sin 40.0° = -3.285 \text{ N} \cdot \text{m}$.

$\tau_3 = +(8.50 \text{ N})(0.350 \text{ m})\sin 90.0° = +2.975 \text{ N} \cdot \text{m}$. $\sum \tau = -3.285 \text{ N} \cdot \text{m} + 2.975 \text{ N} \cdot \text{m} = -0.31 \text{ N} \cdot \text{m}$. The net torque is 0.31 N·m and is clockwise.
Reflect: If we treat the torques as vectors, $\vec{\tau}_2$ is into the page and $\vec{\tau}_3$ is out of the page.

*10.9. Set Up: Let the direction the grindstone is rotating be positive. From Table 9.2, $I = \frac{1}{2}mR^2$. Use the information about the motion to calculate α. In $\sum \tau = I\alpha$, α must be in rad/s^2.
Solve: $\omega_0 = 850 \text{ rev/min} = 89.0 \text{ rad/s}$. $\omega = 0$. $t = 7.50$ s. $\omega = \omega_0 + \alpha t$ gives

$$\alpha = \frac{\omega - \omega_0}{t} = \frac{0 - 89.0 \text{ rad/s}}{7.50 \text{ s}} = -11.9 \text{ rad/s}^2.$$

The net torque is due to the kinetic friction force f_k between the ax and the grindstone. $\sum \tau = I\alpha$ gives $-f_k R = \left(\frac{1}{2}mR^2\right)\alpha$ and

$$f_k = \frac{-mR\alpha}{2} = -\frac{(50.0 \text{ kg})(0.260 \text{ m})(-11.9 \text{ rad/s}^2)}{2} = 77.4 \text{ N}.$$

$f_k = \mu_k n$ so $\mu_k = \dfrac{f_k}{n} = \dfrac{77.4 \text{ N}}{160 \text{ N}} = 0.484$

Reflect: $\sum \tau = I\alpha$ relates the forces on the rotating object to the motion of the object.

*10.11. Set Up: For the pulley $I = \frac{1}{2}M(R_1^2 + R_2^2)$, where $R_1 = 30.0$ cm and $R_2 = 50.0$ cm. The free-body diagrams for the stone and for the pulley are given in the figure below. Apply $\sum \vec{F} = m\vec{a}$ to the stone. The stone accelerates downward, so use coordinates with $+y$ downward. Apply $\sum \tau = I\alpha$ to the rotation of the pulley, with clockwise as the positive sense of rotation. n is the normal force applied to the pulley by the axle. $a = R_2\alpha$.

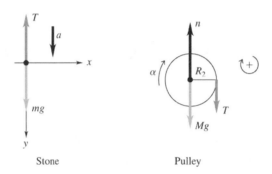

Stone   Pulley

Solve: **(a)** $\sum F_y = ma_y$ applied to the stone gives $mg - T = ma$. $\sum \tau = I\alpha$ applied to the pulley gives $TR_2 = I\alpha$.

$$I = \frac{1}{2}M(R_1^2 + R_2^2) \text{ and } \alpha = a/R_2, \text{ so}$$

$$T = \frac{1}{2}M(1 + [R_1/R_2]^2)a$$

Combining the two equations to eliminate T gives

$$a = \frac{g}{1 + \frac{1}{2}(M/m)(1 + [R_1/R_2]^2)} = \frac{9.80 \text{ m/s}^2}{1 + \frac{1}{2}(10.0 \text{ kg}/2.00 \text{ kg})(1 + [30 \text{ cm}/50 \text{ cm}]^2)}$$

$$a = 2.23 \text{ m/s}^2$$

(b) $T = \frac{1}{2}(10.0 \text{ kg})(1 + [30 \text{ cm}/50 \text{ cm}]^2)(2.23 \text{ m/s}^2) = 15.2 \text{ N}$

(c) $\alpha = a/R_2 = (2.23 \text{ m/s}^2)/(0.500 \text{ m}) = 4.46 \text{ rad/s}^2$

Reflect: The tension in the wire is less than the weight of the stone (19.6 N). For the stone, the downward force is greater than the upward force and the stone accelerates downward.

*10.13. Set Up: For the pulley $I = \frac{1}{2}MR^2$. The elevator has

$$m_1 = \frac{22,500 \text{ N}}{9.80 \text{ m/s}^2} = 2300 \text{ kg}.$$

The free-body diagrams for the elevator, the pulley, and the counterweight are given in the figure below. Apply $\sum \vec{F} = m\vec{a}$ to the elevator and to the counterweight. For the elevator take $+y$ upward and for the counterweight take $+y$ downward, in each case in the direction of the acceleration of the object. Apply $\sum \tau = I\alpha$ to the pulley, with clockwise as the positive sense of rotation. n is the normal force applied to the pulley by the axle. The elevator and counterweight each have acceleration a. $a = R\alpha$.

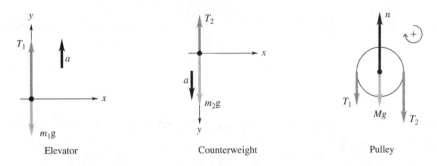

Elevator   Counterweight   Pulley

Solve: **(a)** and **(b)** Calculate the acceleration of the elevator: $y - y_0 = v_{0y}t + \frac{1}{2}a_y t^2$ gives

$$a = \frac{2(y - y_0)}{t^2} = \frac{2(6.75 \text{ m})}{(3.00 \text{ s})^2} = 1.50 \text{ m/s}^2$$

$\sum F_y = ma_y$ for the elevator gives $T_1 - m_1 g = m_1 a$ and

$$T_1 = m_1(a + g) = (2300 \text{ kg})(1.50 \text{ m/s}^2 + 9.80 \text{ m/s}^2) = 2.60 \times 10^4 \text{ N}$$

$\sum \tau = I\alpha$ for the pulley gives $(T_2 - T_1) = \left(\frac{1}{2}MR^2\right)\alpha$. With $\alpha = a/R$ this becomes $T_2 - T_1 = \frac{1}{2}Ma$.

$$T_2 = T_1 + \frac{1}{2}Ma = 2.60 \times 10^4 \text{ N} + \frac{1}{2}(875 \text{ kg})(1.50 \text{ m/s}^2) = 2.67 \times 10^4 \text{ N}$$

$\sum F_y = ma_y$ for the counterweight gives $m_2 g - T_2 = m_2 a$ and

$$m_2 = \frac{T_2}{g - a} = \frac{2.67 \times 10^4 \text{ N}}{9.80 \text{ m/s}^2 - 1.50 \text{ m/s}^2} = 3.22 \times 10^3 \text{ kg}$$

and $w = 3.16 \times 10^4 \text{ N}$.

Reflect: The tension in the cable must be different on either side of the pulley in order to produce the net torque on the pulley required to give it an angular acceleration. The tension in the cable attached to the elevator is greater than the weight of the elevator and the elevator accelerates upward. The tension in the cable attached to the counterweight is less than the weight of the counterweight and the counterweight accelerates downward.

*10.17. Set Up: Follow the method of Example 10.5 except use $I = \frac{1}{2}MR^2$ for the rolling cylinder rather than

$I = \frac{2}{5}MR^2$ for the rolling sphere. Choose the positive x-axis pointing up the ramp, so the x-component of gravity and the acceleration of the disk are both negative. Since the disk is slowing down, both τ and α are negative. If the disk does not slip, its rotational rate must be decreasing as its speed decreases. This implies that the force of friction at the point where the disk contacts the ramp must act *up* the ramp to oppose the rotation of the disk.

Solve:
Translation: $\sum F_x = -Mg \sin\theta + f_s = Ma_{cm,x}$

Rotation:

$$\sum \tau = -f_s R = I_{cm}\alpha = \left(\frac{1}{2}MR^2\right)\alpha$$

Start with the rotation equations. Eliminate α by using the equation $a_{cm} = R\alpha$, which is valid since the disk is not slipping.

Thus, we obtain $f_s R = -\frac{1}{2}MR^2\left(\frac{a_{cm,x}}{R}\right)$ which simplifies to $f_s = -\frac{1}{2}Ma_{cm,x}$. Next, substitute this into the translation

equation to obtain $-Mg\sin\theta - \frac{1}{2}Ma_{cm,x} = Ma_{cm,x}$. Finally, solve for $a_{cm,x}$ to obtain: solve for $a_{cm,x} = -\frac{2}{3}g\sin\theta =$

$-\frac{2}{3}(9.80 \text{ m/s}^2)\sin 10.0° = -1.135 \text{ m/s}^2$.

To determine the time that it takes for the disk to stop: set $v_{0,x} = +2.50$ m/s, $v_x = 0$, and use $v_x = v_{0,x} + a_x t$. Thus,

we have $t = \dfrac{v_x - v_{0,x}}{a_x} = \dfrac{0 - 2.50 \text{ m/s}}{-1.135 \text{ m/s}^2} = 2.20$ s.

Reflect: It is surprising that—in order to create the necessary change in rotational speed—the frictional force acts up the slope regardless of whether the disk is rolling up or down the slope. Note that you can also solve this problem using energy conservation methods.

*10.19. Set Up: (a) Let the direction of rotation of the merry-go-round be positive. Apply $\Sigma\tau = I\alpha$ to the merry-go-round to calculate α and then use a constant acceleration equation to calculate ω after 20.0 s. **(b)** $W = \tau\Delta\theta$.

Calculate $\Delta\theta = \theta - \theta_0$ from a constant acceleration equation. **(c)** $P_{av} = \dfrac{W}{\Delta t}$.

Solve: (a) $\Sigma\tau = I\alpha$ gives

$$\alpha = \frac{\Sigma\tau}{I} = \frac{FR}{I} = \frac{(25.0 \text{ N})(4.40 \text{ m})}{24.5 \text{ kg}\cdot\text{m}^2} = 4.49 \text{ rad/s}^2.$$

$$\omega = \omega_0 + \alpha t = 0 + (4.49 \text{ rad/s}^2)(20.0 \text{ s}) = 89.8 \text{ rad/s}$$

(b) $\theta - \theta_0 = \omega_0 t + \frac{1}{2}\alpha t^2 = \frac{1}{2}(4.49 \text{ rad/s}^2)(20.0 \text{ s})^2 = 898$ rad

$$W = \tau\Delta\theta = (25.0 \text{ N})(4.40 \text{ m})(898 \text{ rad}) = 9.88 \times 10^4 \text{ J}$$

(c) $P_{av} = \dfrac{W}{\Delta t} = \dfrac{9.88 \times 10^4 \text{ J}}{20.0 \text{ s}} = 4.94 \times 10^3 \text{ W}$

Reflect: The power applied when $\omega = 89.8$ rad/s, at the end of the 20.0 s time interval, is

$$P = \tau\omega = (25.0 \text{ N})(4.40 \text{ m})(89.8 \text{ rad/s}) = 9.88 \times 10^3 \text{ W}.$$

At $t = 0$, $\omega = 0$ and $P = 0$. ω and hence P increase linearly in time, so

$$P_{av} = \frac{P_f - P_i}{2} = \frac{9.88 \times 10^3 \text{ W}}{2} = 4.94 \times 10^3 \text{ W}.$$

This agrees with our answer in part (c).

*10.21. Set Up: $P = \tau\omega$, with ω in rad/s. $\omega = 4000.0$ rev/min $= 419$ rad/s. The speed v of the weight is $v = R\omega$. Since ω and v are constant, the tension in the rope equals the weight of the object.

Solve: (a) $\tau = \dfrac{P}{\omega} = \dfrac{150 \times 10^3 \text{ W}}{419 \text{ rad/s}} = 358 \text{ N}\cdot\text{m}$

(b) $\tau = wR$ so $w = \dfrac{\tau}{R} = \dfrac{358 \text{ N}\cdot\text{m}}{0.200 \text{ m}} = 1790$ N

(c) $v = R\omega = (0.200 \text{ m})(419 \text{ rad/s}) = 83.8$ m/s

Reflect: The rate at which the weight gains gravitational potential energy, wv, equals the power output of the motor, so

$$v = \frac{P}{w} = \frac{150 \times 10^3 \text{ W}}{1790 \text{ N}} = 83.8 \text{ m/s},$$

which agrees with the result we calculated for part **(c)**.

*10.25. Set Up: $l = 1.6$ m

Solve: (a) $v = 0$ so $L = 0$

(b) Find v: $v_y^2 = v_{0y}^2 + 2a_y(y - y_0)$ gives $v = \sqrt{2g(y - y_0)} = \sqrt{2(9.80 \text{ m/s}^2)(2.5 \text{ m})} = 7.0$ m/s

$$L = mvl = (4.0 \text{ kg})(7.0 \text{ m/s})(1.6 \text{ m}) = 44.8 \text{ kg} \cdot \text{m}^2/\text{s, counterclockwise}$$

Reflect: The angular momentum increases as the brick falls because of the external torque of the gravity force.

*10.27. Set Up: The figure below shows the force on the drawbridge. The weight acts at the center of the drawbridge and this force has moment arm l. Take clockwise rotation to be positive.

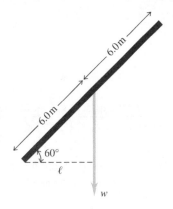

Solve: (a) $\tau = wl = (15,000 \text{ N})(6.0 \text{ m})(\cos 60.0°) = 45,000 \text{ N} \cdot \text{m}$

(b) $\dfrac{\Delta L}{\Delta t} = \Sigma \tau = 45,000 \text{ kg} \cdot \text{m}^2/\text{s}^2$

Reflect: The torque changes as the angle of the bridge above the horizontal changes.

*10.29. Set Up: For a thin-walled hollow cylinder $I = mR^2$. For a slender rod rotating about an axis through its center, $I = \frac{1}{12}ml^2$.

Solve: $L_i = L_f$ so $I_i\omega_i = I_f\omega_f$. $I_i = 0.40 \text{ kg} \cdot \text{m}^2 + \frac{1}{12}(8.0 \text{ kg})(1.8 \text{ m})^2 = 2.56 \text{ kg} \cdot \text{m}^2$. $I_f = 0.40 \text{ kg} \cdot \text{m}^2 + (8.0 \text{ kg})$ $(0.25 \text{ m})^2 = 0.90 \text{ kg} \cdot \text{m}^2$.

$$\omega_f = \left(\frac{I_i}{I_f}\right)\omega_i = \left(\frac{2.56 \text{ kg} \cdot \text{m}^2}{0.90 \text{ kg} \cdot \text{m}^2}\right)(0.40 \text{ rev/s}) = 1.14 \text{ rev/s}$$

*10.33. Set Up: $L_i = L_f$. $I_{\text{child}} = mr^2$, where r is his distance from the axis. $T_i = 6.00$ s so

$$\omega_i = \frac{2\pi \text{ rad}}{6.00 \text{ s}} = 1.047 \text{ rad/s}.$$

Solve: $I_i = 1200 \text{ kg} \cdot \text{m}^2$, since the child is at $r = 0$.

$$I_f = 1200 \text{ kg} \cdot \text{m}^2 + (40.0 \text{ kg})(2.00 \text{ m})^2 = 1360 \text{ kg} \cdot \text{m}^2$$

$I_i\omega_i = I_f\omega_f$ gives

$$\omega_f = \left(\frac{I_i}{I_f}\right)\omega_i = \left(\frac{1200 \text{ kg} \cdot \text{m}^2}{1360 \text{ kg} \cdot \text{m}^2}\right)(1.047 \text{ rad/s}) = 0.924 \text{ rad/s}.$$

Reflect: When some of the mass moves farther from the axis, the moment of inertia increases and the angular velocity decreases.

*10.37. Set Up: The force diagram for the leg is given in the figure below. The weight of each piece acts at the center of mass of that piece. The mass of the upper leg is $m_{ul} = (0.215)(37 \text{ kg}) = 7.955 \text{ kg}$. The mass of the lower leg is $m_{ll} = (0.140)(37 \text{ kg}) = 5.18 \text{ kg}$. Use the coordinates shown, with the origin at the hip and the x axis along the leg.

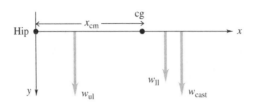

Solve: $x_{cm} = \dfrac{x_{ul}m_{ul} + x_{ll}m_{ll} + x_{cast}m_{cast}}{m_{ul} + m_{ll} + m_{cast}}$

$$x_{cm} = \frac{(18.0 \text{ cm})(7.955 \text{ kg}) + (69.0 \text{ cm})(5.18 \text{ kg}) + (78.0 \text{ cm})(5.50 \text{ kg})}{7.955 \text{ kg} + 5.18 \text{ kg} + 5.50 \text{ kg}} = 49.9 \text{ cm}$$

*10.41. Set Up: Apply the first and second conditions of equilibrium to the shelf. The free-body diagram for the shelf is given in the following figure. Take the axis at the left-hand end of the shelf and let counterclockwise torque be positive. The center of gravity of the uniform shelf is at its center.

Solve: **(a)** $\Sigma\tau = 0$ gives $-w_t(0.200 \text{ m}) - w_s(0.300 \text{ m}) + T_2(0.400 \text{ m}) = 0$.

$$T_2 = \frac{(25.0 \text{ N})(0.200 \text{ m}) + (50.0 \text{ N})(0.300 \text{ m})}{0.400 \text{ m}} = 50.0 \text{ N}$$

$\Sigma F_y = 0$ gives $T_1 + T_2 - w_t - w_s = 0$ and $T_1 = 25.0 \text{ N}$. The tension in the left-hand wire is 25.0 N and the tension in the right-hand wire is 50.0 N.

Reflect: We can verify that $\Sigma\tau = 0$ is zero for any axis, for example for an axis at the right-hand end of the shelf.

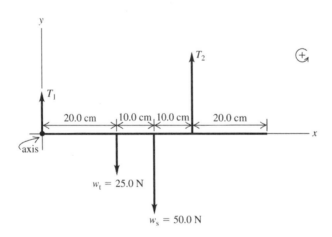

*10.43. Set Up: The free-body diagram for the boom is given in the figure below. Let the length of the boom be L.

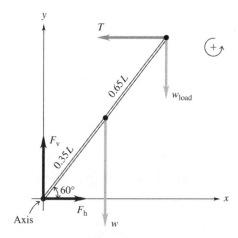

Solve: (a) $\sum \tau = 0$ gives $T(L\sin 60.0°) - w_{\text{load}}(L\cos 60.0°) - w(0.35L\cos 60.0°) = 0$ and

$$T = \frac{w_{\text{load}}\cos 60.0° + w(0.35\cos 60.0°)}{\sin 60.0°} = \frac{(5000\text{ N})\cos 60.0° + (2600\text{ N})(0.35\cos 60.0°)}{\sin 60.0°} = 3.41 \times 10^3\text{ N}$$

(b) $\sum F_x = 0$ gives $F_h - T = 0$ and $F_h = 3410$ N. $\sum F_y = 0$ gives $F_v - w - w_{\text{load}} = 0$ and $F_v = 5000\text{ N} + 2600\text{ N} = 7600\text{ N}$

*10.47. Set Up: The center of gravity of the beam is 2.0 m from the hinge. Use coordinates with $+y$ upward and $+x$ to the right. Take the pivot at the hinge and let counterclockwise torque be positive.
Solve: (a) The free-body diagram for the beam is shown in the figure below. The tension T in the wire has been replaced by its x and y components.

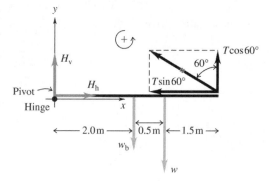

(b) $\sum \tau = 0$ gives $(T\cos 60.0°)(4.0\text{ m}) - w_b(2.0\text{ m}) - w(2.5\text{ m}) = 0$

$$T = \frac{(2500\text{ N})(2.00\text{ m}) + (3500\text{ N})(2.50\text{ m})}{2.00\text{ m}} = 6880\text{ N}$$

(c) $\sum F_x = 0$ gives $H_h - T\sin 60.0° = 0$ and $H_h = T\sin 60.0° = 5960$ N
$\sum F_y = 0$ gives $H_v + T\cos 60.0° - w_b - w = 0$ and $H_v = w_b + w - T\cos 60.0° = 2560$ N

*10.49. Set Up: The system of the person and diving board is at rest so the two conditions of equilibrium apply. The free-body diagram for the diving board is given in the figure below. Take the origin of coordinates at the left-hand end of the board (point A). $\vec{F}_1$ is the force applied at the support point and $\vec{F}_2$ is the force at the end that is held down.

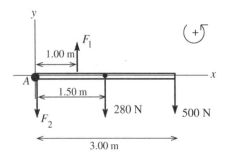

Solve: (a) $\sum \tau_A = 0$ gives $+F_1(1.0 \text{ m}) - (500 \text{ N})(3.00 \text{ m}) - (280 \text{ N})(1.50 \text{ m}) = 0$

$$F_1 = \frac{(500 \text{ N})(3.00 \text{ m}) + (280 \text{ N})(1.50 \text{ m})}{1.00 \text{ m}} = 1920 \text{ N}$$

(b) $\sum F_y = ma_y$

$F_1 - F_2 - 280 \text{ N} - 500 \text{ N} = 0$

$F_2 = F_1 - 280 \text{ N} - 500 \text{ N} = 1920 \text{ N} - 280 \text{ N} - 500 \text{ N} = 1140 \text{ N}$

Reflect: We can check our answers by calculating the net torque about some point and checking that $\tau_z = 0$ for that point also. Net torque about the right-hand end of the board:

$(1140 \text{ N})(3.00 \text{ m}) + (280 \text{ N})(1.50 \text{ m}) - (1920 \text{ N})(2.00 \text{ m}) = 3420 \text{ N} \cdot \text{m} + 420 \text{ N} \cdot \text{m} - 3840 \text{ N} \cdot \text{m} = 0,$ which checks.

*10.51. Set Up: Let the forearm be at an angle ϕ below the horizontal. Take the pivot at the elbow joint and let counterclockwise torques be positive. Let $+y$ be upward and let $+x$ be to the right. Each forearm has mass $m_{\text{arm}} = \frac{1}{2}(0.0600)(72 \text{ kg}) = 2.16 \text{ kg}$. The weight held in each hand is $w = mg$, with $m = 7.50 \text{ kg}$. $\vec{T}$ is the force the biceps muscle exerts on the forearm. $\vec{E}$ is the force exerted by the elbow and has components E_v and E_h.

Solve: (a) The free-body diagram for the forearm is given in the figure below.

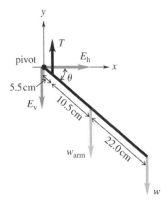

(b) $\sum \tau = 0$ gives $T(5.5 \text{ cm})(\cos \theta) - w_{\text{arm}}(16.0 \text{ cm})(\cos \theta) - w(38.0 \text{ cm})(\cos \theta) = 0$

$$T = \frac{16.0 w_{\text{arm}} + 38.0 w}{5.5} = \frac{16.0(2.16 \text{ kg})(9.80 \text{ m/s}^2) + 38.0(7.50 \text{ kg})(9.80 \text{ m/s}^2)}{5.5} = 569 \text{ N}$$

(c) $\sum F_x = 0$ gives $E_h = 0$. $\sum F_y = 0$ gives $T - E_v - w_{\text{arm}} - w = 0$, so

$$E_v = T - w_{\text{arm}} - w = 569 \text{ N} - (2.16 \text{ kg})(9.80 \text{ m/s}^2) - (7.50 \text{ kg})(9.80 \text{ m/s}^2) = 474 \text{ N}$$

Since we calculate E_v to be positive, we correctly assumed that it was downward when we drew the free-body diagram.

(d) The factor $\cos \theta$ divides out of the $\sum \tau = 0$ equation in part (b), so the force T stays the same as she raises her arm.

*10.55. Set Up: According to section 10.7 the direction of the angular momentum vector is given by the right-hand rule. For counterclockwise rotations in the plane of this page (positive rotations), the angular momentum vector is perpendicular to the page and points toward you. For clockwise rotations in the plane of this page (negative rotations), the rotation vector is perpendicular to the page and points away from you.

Solve: (a) As you see it, the minute hand of the clock appears to move clockwise, so its angular momentum vector is perpendicular to the face of the clock and directed away from you (i.e., inward).
(b) If you stand facing the right side of the backward moving car, the car will be moving to your left and its right front tire will appear to rotate counterclockwise. Thus, its angular momentum vector points toward you (i.e., perpendicular to the wheel and pointed outwards from the right side of the car).
(c) Assuming that the ice skater is spinning clockwise (as seen from above) about an axis that is perpendicular to the ice, the angular momentum vector of the ice skater will point downward.
(d) As viewed from above its north pole, the earth appears to rotate counterclockwise. Thus, its angular momentum vector points toward you (i.e., aligned with the earth's axis and pointing upward from the north pole). Note: If you switched your view to above the south pole of the earth, the earth would appear to rotate clockwise. Thus, its angular momentum vector would point away from you—however, this still means that it points out of the earth's north pole.

Reflect: For simple problems it is sufficient to follow the convention that angular momentum is taken to be positive for counterclockwise rotations and negative for clockwise rotations; however, in more complex problems it is often better to treat angular momentum as a vector.

*10.57. Set Up: Apply $\sum F_y = 0$ with $+y$ upward and apply $\sum \tau = 0$ with the pivot at the point of suspension for each rod.

Solve: (a) The free-body diagram for each rod is given in the figure below.

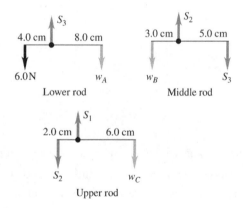

(b) $\sum \tau = 0$ for the lower rod: $(6.0 \text{ N})(4.0 \text{ cm}) = w_A(8.0 \text{ cm})$ and $w_A = 3.0 \text{ N}$.
$\sum F_y = 0$ for the lower rod: $S_3 = 6.0 \text{ N} + w_A = 9.0 \text{ N}$

$\sum \tau = 0$ for the middle rod: $w_B(3.0 \text{ cm}) = (5.0 \text{ cm})S_3$ and $w_B = \left(\dfrac{5.0}{3.0}\right)(9.0 \text{ N}) = 15.0 \text{ N}$.

$\sum F_y = 0$ for the middle rod: $S_2 = 9.0 \text{ N} + S_3 = 24.0 \text{ N}$

$\sum \tau = 0$ for the upper rod: $S_2(2.0 \text{ cm}) = w_C(6.0 \text{ cm})$ and $w_C = \left(\dfrac{2.0}{6.0}\right)(24.0 \text{ N}) = 8.0 \text{ N}$.

$\sum F_y = 0$ for the upper rod: $S_1 = S_2 + w_C = 32.0 \text{ N}$

*10.61. Set Up: Apply $\sum \tau = 0$ to the slab. The free-body diagram is given in Figure (a) below. We have $\tan \beta = \dfrac{3.75 \text{ m}}{1.75 \text{ m}}$ so $\beta = 65.0°$. Also, $20.0° + \beta + \alpha = 90.0°$ so we have $\alpha = 5.0°$. The distance from the axis to the center of the block is $\sqrt{\left(\dfrac{3.75 \text{ m}}{2}\right)^2 + \left(\dfrac{1.75 \text{ m}}{2}\right)^2} = 2.07 \text{ m}$.

Solve: **(a)** $w(2.07 \text{ m})\sin 5.0° - T(3.75 \text{ m})\sin 52.0° = 0$. $T = 0.061w$. Each worker must exert a force of $0.012w$, where w is the weight of the slab.

(b) As θ increases, the moment arm for w decreases and the moment arm for T increases, so the worker needs to exert less force.

(c) Note that $T \to 0$ when w passes through the support point. This situation is sketched in Figure (b) below. $\tan \theta = \dfrac{(1.75 \text{ m})/2}{(3.75 \text{ m})/2}$ and $\theta = 25.0°$. If θ exceeds this value the gravity torque causes the slab to tip over.

Reflect: The moment arm for T is much greater than the moment arm for w, so the force the workers apply is much less than the weight of the slab.

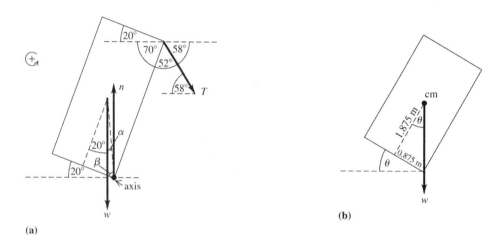

(a)

(b)

*10.63. Set Up: A accelerates downward, B accelerates upward and the wheel turns clockwise. Apply $\sum F_y = ma_y$ to blocks A and B. Let $+y$ be downward for A and y be upward for B. Apply $\sum \tau = I\alpha$ to the wheel, with the clockwise sense of rotation positive. Each block has the same magnitude of acceleration, a, and $a = R\alpha$. Call the tension in the cord between C and A T_A and the tension between C and B T_B.

Solve: For A, $\sum F_y = ma_y$ gives $m_A g - T_A = m_A a$. For B, $\sum F_y = ma_y$ gives $T_B - m_B g = m_B a$. For the wheel, $T_A R - T_B R = I\alpha = I(a/R)w$ and $T_A - T_B = \left(\dfrac{I}{R^2}\right) a$. Adding these three equations gives

$$(m_A - m_B)g = \left(m_A + m_B + \dfrac{I}{R^2}\right) a.$$

$$a = \left(\dfrac{m_A - m_B}{m_A + m_B + I/R^2}\right) g = \left(\dfrac{4.00 \text{ kg} - 2.00 \text{ kg}}{4.00 \text{ kg} + 2.00 \text{ kg} + (0.300 \text{ kg})/(0.120 \text{ m})^2}\right)(9.80 \text{ m/s}^2) = 0.730 \text{ m/s}^2$$

$$\alpha = \dfrac{a}{R} = \dfrac{0.730 \text{ m/s}^2}{0.120 \text{ m}} = 6.08 \text{ rad/s}^2$$

$$T_A = m_A(g - a) = (4.00 \text{ kg})(9.80 \text{ m/s}^2 - 0.730 \text{ m/s}^2) = 36.3 \text{ N}$$

$$T_B = m_B(g + a) = (2.00 \text{ kg})(9.80 \text{ m/s}^2 + 0.730 \text{ m/s}^2) = 21.1 \text{ N}$$

Reflect: The tensions must be different in order to produce a torque that accelerates the wheel when the blocks accelerate.

*10.65. Set Up: The free-body diagram for the leg is given in Figure (a) below. Let the pivot be at the hip joint and take counterclockwise torques to be positive. There is a force $\vec{F}$ of unknown magnitude and direction that the hip joint exerts on the leg, but this force produces zero torque.

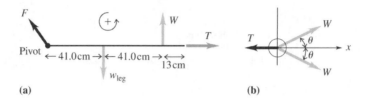

(a)   (b)

Solve: (a) $\sum \tau = 0$ gives $w_{leg}(41.0 \text{ cm}) - W(82.0 \text{ cm}) = 0$ and

$W = \frac{1}{2}w_{leg} = \frac{1}{2}(14.2 \text{ kg})(9.80 \text{ m/s}^2) = 69.6 \text{ N}$

(b) The free-body diagram for the pulley attached to the foot is sketched in Figure (b) above. $\sum F_x = 0$ gives $2W\cos\theta - T = 0$, so $\theta = 85.1°$.

(c) From (b), $T = 2W\cos\theta$ and the maximum traction force T is $T = 2W = 139 \text{ N}$, when $\theta = 0°$.

*10.69. Set Up: Apply the conditions of equilibrium to the cylinder. The free-body diagram for the cylinder is given in the figure below. The center of gravity of the cylinder is at its geometrical center. The cylinder has radius R.

Solve: (a) T produces a clockwise torque about the center of gravity so there must be a friction force that produces a counterclockwise torque about this axis.

(b) Applying $\sum \tau = 0$ to an axis at the center of gravity gives $-TR + fR = 0$ and $T = f$. The condition $\sum \tau = 0$ applied to an axis at the point of contact between the cylinder and the ramp gives $-T(2R) + MgR\sin\theta = 0$. $T = (Mg/2)\sin\theta$.

Reflect: We can show that $\sum F_x = 0$ and $\sum F_y = 0$, for the x- and y-axes that are parallel and perpendicular to the ramp (respectively), or for x- and y-axes that are horizontal and vertical (respectively).

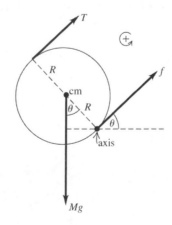

*10.71. Set Up: Apply $\sum \tau_z = 0$ to the wheel. Take torques about the upper corner of the curb.

Solve: The force $\vec{F}$ acts at a perpendicular distance $R - h$ and the weight acts at a perpendicular distance $\sqrt{R^2 - (R-h)^2} = \sqrt{2Rh - h^2}$. Setting the torques equal for the minimum necessary force, $F = mg \dfrac{\sqrt{2Rh - h^2}}{R - h}$.

(b) The torque due to gravity is the same, but the force $\vec{F}$ acts at a perpendicular distance $2R - h$, so the minimum force is $(mg)\sqrt{2Rh - hv}/(2R - h)$.

Reflect: (c) Less force is required when the force is applied at the top of the wheel, since in this case $\vec{F}$ has a larger moment arm.

*10.75. Set Up: For an axis about one edge, the door has

$$I = \tfrac{1}{3}ml^2 = \tfrac{1}{3}\left(\frac{750\ \text{N}}{9.80\ \text{m/s}^2}\right)(1.25\ \text{m})^2 = 39.9\ \text{kg} \cdot \text{m}^2.$$

Exena applies a torque $Fl = (220\ \text{N})(1.25\ \text{m}) = 275\ \text{N} \cdot \text{m}$.

Solve: $\sum \tau = I\alpha$ gives

$$\alpha = \frac{275\ \text{N} \cdot \text{m}}{39.0\ \text{kg} \cdot \text{m}^2} = 6.89\ \text{rad/s}^2.$$

The door turns through $\pi/2$ rad and $\omega_0 = 0$. $\theta - \theta_0 = \omega_0 t + \tfrac{1}{2}\alpha t^2$ gives

$$t = \sqrt{\frac{2(\theta - \theta_0)}{\alpha}} = \sqrt{\frac{2(\pi/2)\ \text{rad}}{6.89\ \text{rad/s}^2}} = 0.675\ \text{s}$$

Solutions to Passage Problems

*10.79. Set Up: Let F be the thrust of a single rocket located at a distance r from the axis of the space station; its torque is given by $\tau = rF \sin \theta$. In this case the rockets are located on the outside surface where $r = 250$ m and maximum torque occurs when $\theta = 90°$ (i.e., when the thrust is perpendicular to the radius r. If all the rockets are oriented to produce the same sense of rotation (i.e., clockwise or counterclockwise) their torques will add, so that the net torque will be 40 times the torque of a single rocket.

Solve: We have the net torque of all 40 rockets to be $\tau_{net} = 40rF \sin \theta = 40(250\ \text{m})(1000\ \text{N}) = 1.00 \times 10^7\ \text{N} \cdot \text{m}$. Thus, the correct answer is C.

*10.81. Set Up: We know that $\omega_0 = 0$ and $\omega = 0.20$ rad/s. From the previous problem we know that $\alpha = 10^{-5}$ rad/s^2. Since α is constant, we can use $\omega = \omega_0 + \alpha t$ to find t.

Solve: We have $t = \dfrac{\omega - \omega_0}{\alpha} = \dfrac{0.20\ \text{rad/s} - 0}{10^{-5}\ \text{rad/s}^2} = 2 \times 10^4$ s, which is about 5.6 hours. Thus, the correct answer is A.

11

ELASTICITY AND PERIODIC MOTION

Problems 1 , 5, 7, 9, 13, 15, 19, 23, 25, 29, 33, 35, 37, 41, 43, 45, 49, 51, 55, 59, 61, 65, 67

Solutions to Problems

*11.1. Set Up: $A = \pi r^2$, with $r = 2.75 \times 10^{-4}$ m. The force F_T applied to the end of the wire is $mg = (25.0$ kg$)$ $(9.80$ m/s$^2) = 245$ N. stress $= F_T/A$ and strain $= \Delta l/l_0$. $Y =$ stress/strain

Solve: (a) stress $= \dfrac{245\ \text{N}}{\pi(2.75 \times 10^{-4}\ \text{m})^2} = 1.03 \times 10^9$ Pa

(b) strain $= \Delta l/l_0 = \dfrac{1.10 \times 10^{-3}\ \text{m}}{0.750\ \text{m}} = 1.47 \times 10^{-3}$

(c) $Y = \dfrac{\text{stress}}{\text{strain}} = \dfrac{1.03 \times 10^9\ \text{Pa}}{1.47 \times 10^{-3}} = 7.01 \times 10^{11}$ Pa

Reflect: Our result for Y is about the same size as the values in Table 11.1. In SI units the stress is very large. The strain is dimensionless and small, so Y is very large and has the same units (Pa) as stress.

*11.5. Set Up: $A = 50.0$ cm$^2 = 50.0 \times 10^{-4}$ m^2. $Y = \dfrac{l_0 F_\perp}{A\,\Delta l}$

Solve: relaxed: $Y = \dfrac{(0.200\ \text{m})(25.0\ \text{N})}{(50.0 \times 10^{-4}\ \text{m}^2)(3.0 \times 10^{-2}\ \text{m})} = 3.33 \times 10^4$ Pa

maximum tension: $Y = \dfrac{(0.200\ \text{m})(500\ \text{N})}{(50.0 \times 10^{-4}\ \text{m}^2)(3.0 \times 10^{-2}\ \text{m})} = 6.67 \times 10^5$ Pa

*11.7. Set Up: For steel, $Y = 2.0 \times 10^{11}$ Pa. $A = \pi d^2/4$.

Solve: $Y = \dfrac{l_0 F_\perp}{A\,\Delta l}$ gives $A = \dfrac{l_0 F_\perp}{Y\Delta l} = \dfrac{(2.00\ \text{m})(400.0\ \text{N})}{(2.0 \times 10^{11}\ \text{Pa})(0.25 \times 10^{-2}\ \text{m})} = 1.6 \times 10^{-6}$ m^2

$$d = \sqrt{\dfrac{4A}{\pi}} = \sqrt{\dfrac{4(1.6 \times 10^{-6}\ \text{m}^2)}{\pi}} = 1.43\ \text{mm}$$

Reflect: The thinner the wire the more it stretches for the same applied force. The length of this wire changes by 0.12%.

*11.9. Set Up: $Y = \dfrac{F_T/A}{\Delta l/l_0}$ so $F_T = \left(\dfrac{YA}{l_0}\right)\Delta l$ and $k = \dfrac{YA}{l_0}$. From Problem 11.8, $k = 4.6 \times 10^5 \,\mathrm{N/m}$ for the natural

Achilles tendon. $A = \pi r^2$

Solve: (a) $k = \dfrac{YA}{l_0}$ so $A = \dfrac{kl_0}{Y} = \dfrac{(4.6 \times 10^5 \,\mathrm{N/m})(0.25 \,\mathrm{m})}{30 \times 10^9 \,\mathrm{Pa}} = 3.8 \times 10^{-6} \,\mathrm{m}^2$

$A = \pi r^2$ so $r = \sqrt{A/\pi} = 1.1 \,\mathrm{mm}$ and the diameter is 2.2 mm.

(b) The natural tendon has $r = \sqrt{(78.1 \,\mathrm{mm}^2)/\pi} = 4.99 \,\mathrm{mm}$ and diameter 10.0 mm. The artificial tendon's diameter is much smaller.

Reflect: The artificial tendon has a larger Y and therefore a smaller diameter.

*11.13. Set Up: At the surface the pressure is $1.0 \times 10^5 \,\mathrm{Pa}$, so $\Delta p = 1.16 \times 10^8 \,\mathrm{Pa}$. $V_0 = 1.00 \,\mathrm{m}^3$. Density $= m/V$.

At the surface $1.00 \,\mathrm{m}^3$ of water has mass $1.03 \times 10^3 \,\mathrm{kg}$.

Solve: (a) $B = -\dfrac{(\Delta p)V_0}{\Delta V}$ gives $\Delta V = -\dfrac{(\Delta p)V_0}{B} = -\dfrac{(1.16 \times 10^8 \,\mathrm{Pa})(1.00 \,\mathrm{m}^3)}{2.2 \times 10^9 \,\mathrm{Pa}} = -0.0527 \,\mathrm{m}^3$

(b) At this depth $1.03 \times 10^3 \,\mathrm{kg}$ of seawater has volume $V_0 + \Delta V = 0.9473 \,\mathrm{m}^3$. The density is

$$\frac{1.03 \times 10^3 \,\mathrm{kg}}{0.9473 \,\mathrm{m}^3} = 1.09 \times 10^3 \,\mathrm{kg/m}^3.$$

*11.15. Set Up: Use Eq. (11.10). Same material implies same S

Solve: $S = \dfrac{\text{stress}}{\text{strain}}$ so $\text{strain} = \dfrac{\text{stress}}{S} = \dfrac{F_\parallel/A}{S}$ and same forces implies same $F_\parallel$.

For the smaller object, $(\text{strain})_1 = F_\parallel/A_1 S$

For the larger object, $(\text{strain})_2 = F_\parallel/A_2 S$

$$\frac{(\text{strain})_2}{(\text{strain})_1} = \left(\frac{F_\parallel}{A_2 S}\right)\left(\frac{A_1 S}{F_\parallel}\right) = \frac{A_1}{A_2}$$

Larger solid has triple each edge length, so $A_2 = 9A_1$, and $\dfrac{(\text{strain})_2}{(\text{strain})_1} = \dfrac{1}{9}$

Reflect: The larger object has a smaller deformation.

*11.19. Set Up: $S = \dfrac{F_\parallel}{A\phi}$. $F_\parallel = F \sin 12°$. ϕ is in radians. $F = 8mg$, with $m = 10 \,\mathrm{kg}$. $\pi \,\mathrm{rad} = 180°$

Solve: $\phi = \dfrac{F_\parallel}{AS} = \dfrac{8mg \sin 12°}{(10 \times 10^{-4} \,\mathrm{m}^2)(12 \times 10^6 \,\mathrm{Pa})} = 0.1494 \,\mathrm{rad} = 8.6°$

Reflect: The shear modulus of cartilage is much less than the values for metals given in Table 11.1.

*11.23. Set Up: The frequency f in Hz is the number of cycles per second. The angular frequency ω is $\omega = 2\pi f$

and has units of radians. The period T and frequency f are related by $T = \dfrac{1}{f}$.

Solve: (a) $T = \dfrac{1}{f} = \dfrac{1}{466 \,\mathrm{Hz}} = 2.15 \times 10^{-3} \,\mathrm{s}$. $\omega = 2\pi f = 2\pi(466 \,\mathrm{Hz}) = 2.93 \times 10^3 \,\mathrm{rad/s}$

(b) $f = \dfrac{1}{T} = \dfrac{1}{50.0 \times 10^{-6} \,\mathrm{s}} = 2.00 \times 10^4 \,\mathrm{Hz}$. $\omega = 2\pi f = 1.26 \times 10^5 \,\mathrm{rad/s}$

(c) $f = \dfrac{\omega}{2\pi}$ so f ranges from $\dfrac{2.7 \times 10^{15}\,\text{Hz}}{2\pi\,\text{rad}} = 4.3 \times 10^{14}\,\text{Hz}$ to $\dfrac{4.7 \times 10^{15}\,\text{Hz}}{2\pi\,\text{rad}} = 7.5 \times 10^{14}\,\text{Hz}$

$T = \dfrac{1}{f}$ so T ranges from $\dfrac{1}{7.5 \times 10^{14}\,\text{Hz}} = 1.3 \times 10^{-15}\,\text{s}$ to $\dfrac{1}{4.3 \times 10^{14}\,\text{Hz}} = 2.3 \times 10^{-15}\,\text{s}$

(d) $T = \dfrac{1}{f} = \dfrac{1}{5.0 \times 10^6\,\text{Hz}} = 2.0 \times 10^{-7}\,\text{s}$ and $\omega = 2\pi f = 2\pi(5.0 \times 10^6\,\text{Hz}) = 3.1 \times 10^7\,\text{rad/s}$

Reflect: Visible light has much higher frequency than either sounds we can hear or ultrasound. Ultrasound is sound with frequencies higher than what the ear can hear. Large f corresponds to small T.

*11.25. Set Up: The amplitude is the maximum displacement from equilibrium. In one period the object goes from $x = +A$ to $x = -A$ and returns.

Solve: (a) $A = 0.120\,\text{m}$

(b) $0.800\,\text{s} = T/2$ so the period is $1.60\,\text{s}$

(c) $f = \dfrac{1}{T} = 0.625\,\text{Hz}$

*11.29. Set Up: Velocity, position, and total energy are related by $E = \frac{1}{2}kA^2 = \frac{1}{2}mv_x^2 + \frac{1}{2}kx^2$. The maximum speed is at $x = 0$.

Solve: (a) $E = \frac{1}{2}mv_x^2 + \frac{1}{2}kx^2 = \frac{1}{2}(0.150\,\text{kg})(0.30\,\text{m/s})^2 + \frac{1}{2}(300.0\,\text{N/m})(0.012\,\text{m})^2$

$$E = 6.75 \times 10^{-3}\,\text{J} + 2.16 \times 10^{-2}\,\text{J} = 2.84 \times 10^{-2}\,\text{J}.$$

(b) $E = \frac{1}{2}kA^2$ so $A = \sqrt{\dfrac{2E}{k}} = \sqrt{\dfrac{2(2.84 \times 10^{-2}\,\text{J})}{300.0\,\text{N/m}}} = 0.0138\,\text{m}.$

(c) For $x = 0$, $\frac{1}{2}mv_{max}^2 = E$ and $v_{max} = \sqrt{\dfrac{2E}{m}} = \sqrt{\dfrac{2(2.84 \times 10^{-2}\,\text{J})}{0.150\,\text{kg}}} = 0.615\,\text{m/s}.$

Reflect: When $x = 0.0120\,\text{m}$ the system has both kinetic and elastic potential energy. When $x = \pm A$ the energy is all elastic potential energy and when $x = 0$ the energy is all kinetic energy.

*11.33. Set Up: $K + U = E$, with $E = \frac{1}{2}kA^2$ and $U = \frac{1}{2}kx^2$

Solve: $U = K$ says $2U = E$. This gives $2\left(\frac{1}{2}kx^2\right) = \frac{1}{2}kA^2$, so $x = A/\sqrt{2}$.

Reflect: When $x = A/2$ the kinetic energy is three times the elastic potential energy.

*11.35. Set Up: The period is the time for one cycle. A is the maximum value of x.
Solve: (a) From the figure, $T = 0.80\,\text{s}$.

(b) $f = \dfrac{1}{T} = 1.25\,\text{Hz}$

(c) $\omega = 2\pi f = 7.85\,\text{rad/s}$

(d) From the figure, $A = 3.0\,\text{cm}$

(e) $T = 2\pi\sqrt{\dfrac{m}{k}}$ so $k = m\left(\dfrac{2\pi}{T}\right)^2 = (2.40\,\text{kg})\left(\dfrac{2\pi}{0.80\,\text{s}}\right)^2 = 148\,\text{N/m}$

*11.37. Set Up: Equation 11.26 gives $T = 2\pi\sqrt{\dfrac{m}{k}}$, so we find $k = m\left(\dfrac{2\pi}{T}\right)^2$. Equation 11.18 gives $v_{max} = \sqrt{\dfrac{k}{m}}\,A$.

From the analysis of Example 11.15 we know that $a_{max} = \dfrac{k}{m}A$.

Solve: **(a)** $k = m\left(\dfrac{2\pi}{T}\right)^2 = (0.355 \text{ kg})\left(\dfrac{2\pi}{2.15 \text{ s}}\right)^2 = 3.03 \text{ N/m}.$

(b) As x varies from $-A$ to $+A$ the glider moves a distance of $2A = 1.80 \text{ m} - 1.06 \text{ m} = 0.74 \text{ m},$ thus $A = \dfrac{0.74 \text{ m}}{2} = 0.37 \text{ m}.$ Finally, we calculate $v_{max} = \sqrt{\dfrac{k}{m}} A = \sqrt{\dfrac{3.03 \text{ N/m}}{0.355 \text{ kg}}} (0.37 \text{ m}) = 1.1 \text{ m/s}$

(c) $a_{max} = \dfrac{k}{m} A = \dfrac{3.03 \text{ N/m}}{0.355 \text{ kg}} \cdot (0.37 \text{ m}) = 3.2 \text{ m/s}^2$

Reflect: We measure the position of the mass in a harmonic oscillator from its equilibrium position. For an ideal spring, the equilibrium position is halfway between the two extreme positions of the vibrating mass.

*11.41. Set Up: The period is the time for 1 cycle; after time T the motion repeats. The graph shows $v_{max} = 20.0 \text{ cm/s.}$ $\frac{1}{2} m v_x{}^2 + \frac{1}{2} k x^2 = \frac{1}{2} k A^2$

Solve: **(a)** $T = 1.60 \text{ s}$

(b) $f = \dfrac{1}{T} = 0.625 \text{ Hz}$

(c) $\omega = 2\pi f = 3.93 \text{ rad/s}$

(d) $v_x = v_{max}$ when $x = 0$ so $\frac{1}{2} k A^2 = \frac{1}{2} m v_{max}{}^2.$ $A = v_{max} \sqrt{\dfrac{m}{k}}.$ $f = \dfrac{1}{2\pi}\sqrt{\dfrac{k}{m}}$ so $A = v_{max}/(2\pi f).$ From the graph in the problem, $v_{max} = 0.20 \text{ m/s}.$

$$A = (0.20 \text{ m/s})/(2\pi)(0.625 \text{ Hz}) = 0.0.051 \text{ m} = 5.1 \text{ cm}$$

The mass is at $x = \pm A$ when $v_x = 0,$ and this occurs at $t = 0.4 \text{ s}, \; 1.2 \text{ s}, \text{ and } 1.8 \text{ s.}$

(e) $-kx = m a_x$ gives

$$a_{max} = \dfrac{kA}{m} = (2\pi f)^2 A = (4\pi^2)(0.625 \text{ Hz})^2 (0.051 \text{ m}) = 0.79 \text{ m/s}^2 = 79 \text{ cm/s}^2.$$

The acceleration is maximum when $x = \pm A$ and this occurs at the times given in (d).

(f) $T = 2\pi \sqrt{\dfrac{m}{k}}$ so $m = k\left(\dfrac{T}{2\pi}\right)^2 = (75 \text{ N/m})\left(\dfrac{1.60 \text{ s}}{2\pi}\right)^2 = 4.9 \text{ kg}$

Reflect: The speed is maximum at $x = 0,$ when $a_x = 0.$ The magnitude of the acceleration is maximum at $x = \pm A,$ where $v_x = 0.$

*11.43. Set Up: $f = \dfrac{1}{2\pi}\sqrt{\dfrac{k}{m}}$

Solve: **(a)** $f_s = \dfrac{1}{2\pi}\sqrt{\dfrac{k}{m_s}},$ $f_{s+v} = \dfrac{1}{2\pi}\sqrt{\dfrac{k}{m_s + m_v}}$

$$\dfrac{f_{s+v}}{f_s} = \left(\dfrac{1}{2\pi}\sqrt{\dfrac{k}{m_s + m_v}}\right)\left(2\pi\sqrt{\dfrac{m_s}{k}}\right) = \sqrt{\dfrac{m_s}{m_s + m_v}} = \dfrac{1}{\sqrt{1 + m_v/m_s}}$$

(b) $\left(\dfrac{f_{s+v}}{f_s}\right)^2 = \dfrac{1}{1 + m_v/m_s}.$ Solving for m_v gives

$$m_v = m_s\left(\left[\frac{f_s}{f_{s+v}}\right]^2 - 1\right) = (2.10\times10^{-16}\,\text{g})\left(\left[\frac{2.00\times10^{15}\,\text{Hz}}{2.87\times10^{14}\,\text{Hz}}\right]^2 - 1\right) = 9.99\times10^{-15}\,\text{g}$$

$m_V = 9.99$ femtograms

Reflect: When the mass increases, the frequency of oscillation increases.

*11.45. Set Up: $T = 2\pi\sqrt{\dfrac{L}{g}}$. The motion specified is one-half of a cycle so the time is $t = T/2$.

Solve: $T = 2\pi\sqrt{\dfrac{3.50\,\text{m}}{9.80\,\text{m/s}^2}} = 3.75$ s, so $t = 1.88$ s

Reflect: The period of a simple pendulum does not depend on its amplitude, as long as the amplitude is small, and does not depend on the mass of the bob. The period depends only on the length of the pendulum and the value of g.

*11.49. Set Up: $T = 2\pi\sqrt{\dfrac{L}{g}}$ and $f = \dfrac{1}{2\pi}\sqrt{\dfrac{g}{L}}$.

Solve: (a) $L \to 2L$, so $T_{\text{new}} = \sqrt{2}T$.

(b) $L = \dfrac{g}{(2\pi f)^2}$. $f \to 3f$, so $L_{\text{new}} = L/9$

(c) $g \to 10g$ and T unchanged, so $L_{\text{new}} = 10L$

(d) $g \to 10g$, so $T_{\text{new}} = T/\sqrt{10}$

(e) The period does not depend on m, so $T_{\text{new}} = T$.

*11.51. Set Up: As shown in the figure below, the height h above the lowest point of the swing is $h = L - L\cos\theta = L(1 - \cos\theta)$.

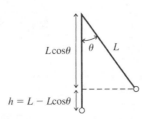

Solve: (a) At the maximum angle of swing, $K = 0$ and $E = mgh$.

$$E_i = mgL(1 - \cos\theta_i) = (2.50\,\text{kg})(9.80\,\text{m/s}^2)(1.45\,\text{m})(1 - \cos11°) = 0.653\,\text{J}$$

$$E_f = mgL(1 - \cos\theta_f) = (2.50\,\text{kg})(9.80\,\text{m/s}^2)(1.45\,\text{m})(1 - \cos4.5°) = 0.110\,\text{J}$$

The mechanical energy lost is $E_i - E_f = 0.543$ J.

(b) The mechanical energy has been converted to other forms by air resistance and by dissipative forces within the rope.

Reflect: After a while the rock will have come to rest and then all its initial mechanical energy has been "lost," has been converted to other forms.

*11.55. Set Up: $T = 2\pi\sqrt{\dfrac{m}{k}}$. M is the mass of the empty car and the mass of the loaded car is $M + 250$ kg.

Solve: The period of the empty car is $T_E = 2\pi\sqrt{\dfrac{M}{k}}$. The period of the loaded car is $T_L = 2\pi\sqrt{\dfrac{M + 250\,\text{kg}}{k}}$.

$$k = \frac{(250 \text{ kg})(9.80 \text{ m/s}^2)}{4.00 \times 10^{-2} \text{ m}} = 6.125 \times 10^4 \text{ N/m}$$

$$M = \left(\frac{T_L}{2\pi}\right)^2 k - 250 \text{ kg} = \left(\frac{1.08 \text{ s}}{2\pi}\right)^2 (6.125 \times 10^4 \text{ N/m}) - 250 \text{ kg} = 1.56 \times 10^3 \text{ kg}.$$

$$T_E = 2\pi \sqrt{\frac{1.56 \times 10^3 \text{ kg}}{6.125 \times 10^4 \text{ N/m}}} = 1.00 \text{ s}$$

*11.59. Set Up: The bounce frequency is given by Eq. (11.25) and the pendulum frequency by Eq. (11.31). Use the relation between these two frequencies that is specified in the problem to calculate the equilibrium length L of the spring, when the apple hangs at rest on the end of the spring.

Solve: For vertical SHM: $f_b = \frac{1}{2\pi}\sqrt{\frac{k}{m}}$. For pendulum motion (small amplitude): $f_p = \frac{1}{2\pi}\sqrt{\frac{g}{L}}$. The problem specifies

that $f_p = \frac{1}{2}f_b$ so $\frac{1}{2\pi}\sqrt{\frac{g}{L}} = \frac{1}{2}\frac{1}{2\pi}\sqrt{\frac{k}{m}}$. Solve for L: $g/L = k/4m$ so $L = 4gm/k = 4w/k = 4(1.00 \text{ N})/1.50 \text{ N/m} = 2.67 \text{ m}$.

Reflect: This is the *stretched* length of the spring, its length when the apple is hanging from it. (Note: Small angle of swing means v is small as the apple passes through the lowest point, so a_{rad} is small and the component of mg perpendicular to the spring is small. Thus the amount the spring is stretched changes very little as the apple swings back and forth.)

Set Up: Use Newton's second law to calculate the distance that the spring is stretched from its unstretched length when the apple hangs from it. The free-body diagram for the apple hanging at rest on the end of the spring is given in the figure below.

Solve: $\sum F_y = ma_y$

$k\Delta L - mg = 0$

$\Delta L = mg/k = w/k = 1.00 \text{ N}/1.50 \text{ N/m} = 0.667 \text{ m}$

Thus the unstretched length of the spring is $2.67 \text{ m} - 0.67 \text{ m} = 2.00 \text{ m}$.

Reflect: The spring shortens to its unstretched length when the apple is removed.

*11.61. Set Up: Apply conservation of energy to the motion before and after the collision. Apply conservation of linear momentum to the collision. After the collision the system moves as a simple pendulum. If the maximum

angular displacement is small we have $f = \frac{1}{2\pi}\sqrt{\frac{g}{L}}$. In the motion before and after the collision there is energy

conversion between gravitational potential energy mgh, where h is the height above the lowest point in the motion, and kinetic energy.

Solve: Energy conservation during downward swing: $m_2 g h_0 = \frac{1}{2}m_2 v^2$ and $v = \sqrt{2gh_0} = \sqrt{2(9.8 \text{ m/s}^2)(0.100 \text{ m})} = 1.40 \text{ m/s}$.

Momentum conservation during collision: $m_2 v = (m_2 + m_3)V$ and $V = \frac{m_2 v}{m_2 + m_3} = \frac{(2.00 \text{ kg})(1.40 \text{ m/s})}{5.00 \text{ kg}} = 0.560 \text{ m/s}$.

Energy conservation during upward swing: $Mgh_f = \frac{1}{2}MV^2$ and $h_f = V^2/2g = \frac{(0.560 \text{ m/s})^2}{2(9.80 \text{ m/s}^2)} = 0.0160 \text{ m} = 1.60 \text{ cm}.$

The figure below shows how the maximum angular displacement is calculated from h_f. We have $\cos\theta = \frac{48.4 \text{ cm}}{50.0 \text{ cm}}$,

so $\theta = 14.5°$. $f = \frac{1}{2\pi}\sqrt{\frac{g}{l}} = \frac{1}{2\pi}\sqrt{\frac{9.80 \text{ m/s}^2}{0.500 \text{ m}}} = 0.705 \text{ Hz}.$

Reflect: Note that $\sin(14.5°) = 0.250$. Since the maximum angle, $14.5° = 0.253 \text{ rad}$, is small, the approximation $\sin\theta \approx \theta_{rad}$ is valid. Thus, the condition for Eq. (11.32) to be valid is met.

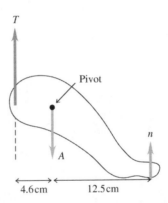

*11.65. Set Up: $Y = \dfrac{F_T/A}{\Delta l/l_0}$. The foot is in rotational equilibrium. Let counterclockwise torques be positive.

Solve: (a) The free-body diagram for the foot is given in the figure below, T is the tension in the tendon and A is the force exerted on the foot by the ankle. $n = (75 \text{ kg})g$, the weight of the person.

T

Pivot

n

A

4.6 cm   12.5 cm

(b) Apply $\Sigma\tau = 0$, with the pivot at the ankle: $T(4.6 \text{ cm}) - n(12.5 \text{ cm}) = 0$

$$T = \left(\frac{12.5 \text{ cm}}{4.6 \text{ cm}}\right)(75 \text{ kg})(9.80 \text{ m/s}^2) = 2000 \text{ N, } 2.72 \text{ times his weight.}$$

(c) The foot pulls downward on the tendon with a force of 2000 N.

$$\Delta l = \left(\frac{F_T}{YA}\right)l_0 = \frac{2000 \text{ N}}{(1470 \times 10^6 \text{ Pa})(78 \times 10^{-6} \text{ m}^2)}(25 \text{ cm}) = 4.4 \text{ mm}$$

*11.67. Set Up: The tension is the same at all points along the composite rod. Apply Eqs. (11.2) and (11.4) to relate the elongations, stresses, and strains for each rod in the compound.

Solve: Each piece of the composite rod is subjected to a tensile force of 4.00×10^4 N.

(a) $Y = \dfrac{F_\perp l_0}{A\,\Delta l}$ so $\Delta l = \dfrac{F_\perp l_0}{YA}$

$\Delta l_b = \Delta l_n$ gives that $\dfrac{F_\perp l_{0,b}}{Y_b A_b} = \dfrac{F_\perp l_{0,n}}{Y_n A_n}$ (b for brass and n for nickel); $l_{0,n} = L$

But the $F_\perp$ is the same for both, so

$$l_{0,n} = \frac{Y_n}{Y_b} \frac{A_n}{A_b} l_{0,b}$$

$$L = \left(\frac{21 \times 10^{10} \text{ Pa}}{9.0 \times 10^{10} \text{ Pa}}\right)\left(\frac{1.00 \text{ cm}^2}{2.00 \text{ cm}^2}\right)(1.40 \text{ m}) = 1.63 \text{ m}$$

(b) stress $= F_\perp / A = T/A$

brass: stress $= T/A = (4.00 \times 10^4 \text{ N})/(2.00 \times 10^{-4} \text{ m}^2) = 2.00 \times 10^8 \text{ Pa}$

nickel: stress $= T/A = (4.00 \times 10^4 \text{ N})/(1.00 \times 10^{-4} \text{ m}^2) = 4.00 \times 10^8 \text{ Pa}$

(c) $Y = $ stress/strain and strain $= $ stress/Y

brass: strain $= (2.00 \times 10^8 \text{ Pa})/(9.0 \times 10^{10} \text{ Pa}) = 2.22 \times 10^{-3}$

nickel: strain $= (4.00 \times 10^8 \text{ Pa})/(21 \times 10^{10} \text{ Pa}) = 1.90 \times 10^{-3}$

Reflect: Larger Y means less Δl and smaller A means greater Δl, so the two effects largely cancel and the lengths don't differ greatly. Equal Δl and nearly equal l means the strains are nearly the same. But equal tensions and A differing by a factor of 2 means the stresses differ by a factor of 2.

MECHANICAL WAVES AND SOUND

Problems 1, 5, 7, 11, 15, 17, 19, 23, 25, 29, 33, 35, 39, 41, 45, 47, 49, 53, 55, 59, 61, 65, 67, 69, 73, 75, 79

Solutions to Problems

*12.1. Set Up: $f_1 = \dfrac{v}{2L} = \dfrac{326 \text{ m/s}}{2(0.330 \text{ m})} = 494$ Hz.

Solve: (a) $v = \dfrac{6.0 \text{ m}}{5.0 \text{ s}} = 1.2$ m/s. For $f = 20{,}000$ Hz, $\lambda = \dfrac{v}{f} = \dfrac{344 \text{ m/s}}{20{,}000 \text{ Hz}} = 1.7$ cm. For $f = 20$ Hz,

$$\lambda = \frac{v}{f} = \frac{344 \text{ m/s}}{20 \text{ Hz}} = 17 \text{ m}.$$

The range of wavelengths is 1.7 cm to 17 m.

(b) $v = c = 3.00 \times 10^8$ m/s. For $\lambda = 700$ nm,

$$f = \frac{c}{\lambda} = \frac{3.00 \times 10^8 \text{ m/s}}{700 \times 10^{-9} \text{ m}} = 4.3 \times 10^{14} \text{ Hz}.$$

For $\lambda = 400$ nm,

$$f = \frac{c}{\lambda} = \frac{3.00 \times 10^8 \text{ m/s}}{400 \times 10^{-9} \text{ m}} = 7.5 \times 10^{14} \text{ Hz}.$$

The range of frequencies for visible light is 4.3×10^{14} Hz to 7.5×10^{14} Hz.

(c) $v = 344$ m/s. $\lambda = \dfrac{v}{f} = \dfrac{344 \text{ m/s}}{23 \times 10^3 \text{ Hz}} = 1.5$ cm

(d) $v = 1480$ m/s. $\lambda = \dfrac{v}{f} = \dfrac{1480 \text{ m/s}}{23 \times 10^3 \text{ Hz}} = 6.4$ cm

Reflect: For a given v, larger f corresponds to smaller λ. For the same f, λ increases when v increases.

*12.5. Set Up: $v = \sqrt{\dfrac{F_\perp}{\mu}}$. The mass per unit length of the wire is

$$\mu = \frac{m}{L} = \frac{0.0600 \text{ kg}}{4.00 \text{ m}} = 0.015 \text{ kg/m}.$$

Solve: $v = \sqrt{\dfrac{1000 \text{ N}}{0.0150 \text{ kg/m}}} = 258$ m/s

*12.7. Set Up: $\mu = 0.0550$ kg/m. $F_{\perp} = mg$, the weight of the hanging mass. The frequency of the waves is 120 Hz. $v = f\lambda$.

Solve: (a) $v = \sqrt{\dfrac{F_{\perp}}{\mu}} = \sqrt{\dfrac{(1.50 \text{ kg})(9.80 \text{ m/s}^2)}{0.0550 \text{ kg/m}}} = 16.3$ m/s

(b) $\lambda = \dfrac{v}{f} = \dfrac{16.3 \text{ m/s}}{120 \text{ Hz}} = 0.136$ m

(c) Doubling the mass increases both the wave speed and the wavelength by a factor of $\sqrt{2}$; $v = 23.1$ m/s and $\lambda = 0.192$ m.

Reflect: The speed of the waves on the rope is much less than the speed of sound in air.

*12.11. Set Up: $v = f\lambda$. $T = 1/f$. Wave number $k = 2\pi/\lambda$. The general form for $y(x,t)$ is

$$y(x,t) = A\sin 2\pi \left(\frac{t}{T} - \frac{x}{\lambda} \right).$$

Solve: (a) $f = \dfrac{v}{\lambda} = \dfrac{8.00 \text{ m/s}}{0.320 \text{ m}} = 25.0$ Hz. $T = 1/f = 0.0400$ s. $k = \dfrac{2\pi}{\lambda} = 19.6$ rad/m.

(b) $y(x,t) = -(0.0700 \text{ m})\sin 2\pi \left(\dfrac{t}{0.0400 \text{ s}} - \dfrac{x}{0.320 \text{ m}} \right)$

(c) $y(0.360 \text{ m}, 0.150 \text{ s}) = -(0.0700 \text{ m})\sin 2\pi \left(\dfrac{0.150 \text{ s}}{0.0400 \text{ s}} - \dfrac{0.360 \text{ m}}{0.320 \text{ m}} \right)$

$y = -(0.0700 \text{ m})\sin(16.5 \text{ rad}) = 0.0498$ m

*12.15. Set Up: The waves obey the principle of superposition: the resultant displacement of the string is the algebraic sum of the displacements due to each wave. Each wave travels 1 cm each second.

Solve: The locations of each wave pulse and the resultant displacement of the string is shown at each time in Figures (a)–(d) below.

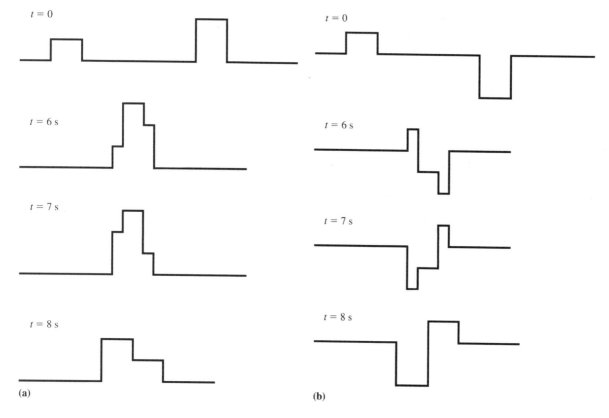

(a)   (b)

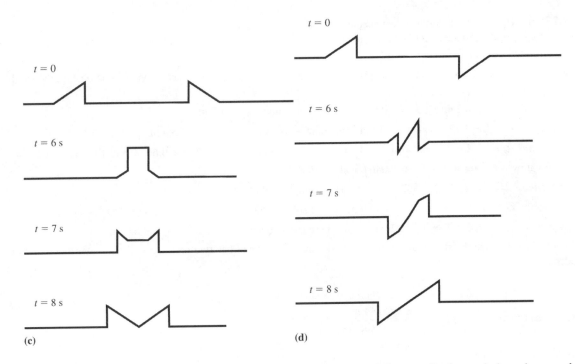

(c)   (d)

Reflect: When the pulses displace the string in the same direction they reinforce each other and give a larger pulse and when the pulses displace the string in opposite directions they tend to cancel and give a smaller pulse.

*12.17. Set Up: $v = \sqrt{\dfrac{F_{\perp}}{\mu}}$ where $\mu = m/L$. $f_1 = \dfrac{v}{2L}$. The nth harmonic has frequency $f_n = nf_1$.

Solve: (a) $v = \sqrt{\dfrac{F_{\perp}}{m/L}} = \sqrt{\dfrac{F_{\perp}L}{m}} = \sqrt{\dfrac{(800 \text{ N})(0.400 \text{ m})}{3.00 \times 10^{-3} \text{ kg}}} = 327 \text{ m/s}.$

$$f_1 = \frac{v}{2L} = \frac{327 \text{ m/s}}{2(0.400 \text{ m})} = 409 \text{ Hz.}$$

(b) $n = \dfrac{10,000 \text{ Hz}}{f_1} = 24.4.$ The 24th harmonic is the highest that could be heard.

*12.19. Set Up: For the fundamental, $f_1 = \dfrac{v}{2x}$, where x is the length of the portion of the string that is free to vibrate. The wave speed v depends on the tension and linear mass density of the string, so it is the same no matter where the finger is placed.

(a) For $x = L = 0.600$ m, $f_1 = 440$ Hz. $v = 2xf_1 = 2(0.600 \text{ m})(440 \text{ Hz}) = 528$ m/s. Then for $f_1 = 587$ Hz,

$$x = \frac{v}{2f_1} = \frac{528 \text{ m/s}}{2(587 \text{ Hz})} = 0.450 \text{ m} = 45.0 \text{ cm}$$

(b) For $f_1 = 392$ Hz,

$$\lambda = \frac{v}{2f_1} = 0.673 \text{ m.}$$

The maximum length of the vibrating string, the distance between the bridge and the upper end of the fingerboard, is 0.600 m, so this is not possible. A G_4 note cannot be played without retuning.

Reflect: The speed of the waves on the string is larger than the speed of sound in air.

*12.23. Set Up: An open end is a displacement antinode. A closed end is a displacement node. Adjacent nodes are a distance $\lambda/2$ apart. Adjacent antinodes are also a distance $\lambda/2$ apart, and the node to antinode distance is $\lambda/4$. Let x be the distance from the left-hand end of the pipe.

Solve: (a) There is an antinode at $x = 0$. The first node is at $x = \lambda/4$ and each successive node is $\lambda/2$ farther to the right. Each successive overtone adds one node.

fundamental: $\lambda_1 = 2L = 2.40$ m. $\lambda_1/4 = 0.60$ m. There is a node at $x = 0.60$ m.

1st overtone: $\lambda_2 = L = 2.40$ m. $\lambda_2/4 = 0.30$ m. There are a nodes at $x = 0.30$ m, 0.90 m.

2nd overtone: $\lambda_3 = 2L/3 = 0.80$ m. $\lambda_3/4 = 0.20$ m. There are a nodes at $x = 0.20$ m, 0.60 m, and 1.00 m.

(b) There now is a node at $x = 0$. Successive nodes are $\lambda/2$ farther to the right.

fundamental: $\lambda_1 = 4L = 4.80$ m. $\lambda_1/2 = 2.40$ m. There only a node is at $x = 0$.

1st overtone: $\lambda_3 = 4L/3 = 1.60$ m. $\lambda_3/2 = 0.80$ m. There are a nodes at $x = 0$ and $x = 0.80$ m.

2nd overtone: $\lambda_5 = 4L/5 = 0.96$ m. $\lambda_5/2 = 0.48$ m. There are a nodes at $x = 0$, 0.48 m, and 0.96 m.

Reflect: In each case the fundamental has one node, the 1st overtone has two nodes, and the 2nd overtone has three nodes.

*12.25. Set Up: A pipe open at one end and closed at the other is a stopped pipe.

Solve: (a) For an open pipe, $f_1 = \dfrac{v}{2L} = \dfrac{344 \text{ m/s}}{2(4.88 \text{ m})} = 35.2$ Hz.

(b) For a stopped pipe, $f_1 = \dfrac{v}{4L} = \dfrac{35.2 \text{ Hz}}{2} = 17.6$ Hz.

*12.29. Set Up: For a stopped pipe, $f_1 = \dfrac{v}{4L}$. $f_1 = 220$ Hz

Solve: $L = \dfrac{v}{4f_1} = \dfrac{344 \text{ m/s}}{4(220 \text{ Hz})} = 39.1$ cm. This result is a reasonable value for the mouth to diaphragm distance for a typical adult.

Reflect: 1244 Hz is not an integer multiple of the fundamental frequency of 220 Hz, it is 5.65 times the fundamental. The production of sung notes is more complicated than harmonics of an air column of fixed length.

*12.33. Set Up: The path difference for the two sources is d. For destructive interference, the path difference is a half-integer number of wavelengths. For constructive interference, the path difference is an integer number of wavelengths. $\lambda = v/f$

Solve: $\lambda = \dfrac{v}{f} = \dfrac{344 \text{ m/s}}{725 \text{ Hz}} = 0.474$ m

(a) Will first produce destructive interference when $d = \lambda/2 = 0.237$ m.

(b) Will next produce destructive interference when $d = 3\lambda/2 = 0.711$ m.

(c) Will first produce constructive interference again when $d = \lambda = 0.474$ m.

*12.35. Set Up: One speaker is 4.50 m from the microphone and the other is 4.03 m from the microphone, so the path difference is 0.42 m. For constructive interference, the path difference is an integer number of wavelengths. For destructive interference, the path difference is a half-integer number of wavelengths. $f = v/\lambda$

Solve: (a) $\lambda = 0.42$ m gives $f = \dfrac{v}{\lambda} = 820$ Hz;

$2\lambda = 0.42$ m gives $\lambda = 0.21$ m and $f = \dfrac{v}{\lambda} = 1640$ Hz;

$3\lambda = 0.42$ m gives $\lambda = 0.14$ m and $f = v = 2460$ Hz, and so on.

The frequencies for constructive interference are $n(820 \text{ Hz})$, $n = 1, 2, 3, \ldots$.

(b) $\lambda/2 = 0.42$ m gives $\lambda = 0.84$ m and $f = \dfrac{v}{\lambda} = 410$ Hz;

$3\lambda/2 = 0.42$ m gives $\lambda = 0.28$ m and $f = \dfrac{v}{\lambda} = 1230$ Hz;

$5\lambda/2 = 0.42$ m gives $\lambda = 0.168$ m and $f = \dfrac{v}{\lambda} = 2050$ Hz, and so on.

The frequencies for destructive interference are $(2n + 1)(410 \text{ Hz})$, $n = 0, 1, 2, \ldots$.

Reflect: The frequencies for constructive interference lie midway between the frequencies for destructive interference.

*12.39. Set Up: $P = I(4\pi r^2)$ and $\dfrac{I_1}{I_2} = \dfrac{r_2^2}{r_1^2}$

Solve: (a) $P = I(4\pi r^2) = (6.50 \text{ W/m}^2)(4\pi)(2.50 \text{ m})^2 = 511$ W

(b) $I_2 = I_1\left(\dfrac{r_1}{r_2}\right)^2 = (6.50 \text{ W/m}^2)\left(\dfrac{2.50 \text{ m}}{7.00 \text{ m}}\right)^2 = 0.829 \text{ W/m}^2$

(c) All the power radiated in one second is received by the walls of the room in one second. Each second the walls receive 511 J.

*12.41. Set Up: $\beta = 10\log\left(\dfrac{I}{I_0}\right)$, with $I_0 = 10^{-12} \text{ W/m}^2$.

(a) $\beta = 10\log\left(\dfrac{1.27 \times 10^{-4} \text{ W/m}^2}{1 \times 10^{-12} \text{ W/m}^2}\right) = 81.0$ dB

(b) $\beta = 10\log\left(\dfrac{6.53 \times 10^{-10} \text{ W/m}^2}{1 \times 10^{-12} \text{ W/m}^2}\right) = 28.1$ dB

(c) $\beta = 10\log\left(\dfrac{1.5 \times 10^{-14} \text{ W/m}^2}{1 \times 10^{-12} \text{ W/m}^2}\right) = -18.2$ dB. β is negative when $I < I_0$.

*12.45. Set Up: Intensity is energy per unit time per unit area. $\beta = 10\log\left(\dfrac{I}{I_0}\right)$, with $I_0 = 1 \times 10^{-12} \text{ W/m}^2$. The area of the eardrum is $A = \pi r^2$, with $r = 4.2 \times 10^{-3}$ m. Part (b) of Problem 12.42 gave $v \doteq 0.074$ mm/s.

Solve: (a) $\beta = 110$ dB gives $11.0 = \log\left(\dfrac{I}{I_0}\right)$ and $I = (10^{11})I_0 = 0.100 \text{ W/m}^2$.

$$E = IAt = (0.100 \text{ W/m}^2)\pi(4.2 \times 10^{-3} \text{ m})^2(1 \text{ s}) = 5.5 \text{ } \mu\text{J}$$

(b) $K = \frac{1}{2}mv^2$ so

$$v = \sqrt{\dfrac{2K}{m}} = \sqrt{\dfrac{2(5.5 \times 10^{-6} \text{ J})}{2.0 \times 10^{-6} \text{ kg}}} = 2.3 \text{ m/s}.$$

This is about 31,000 times faster than the speed in Problem 12.46b.

Reflect: Even though the sound wave intensity level is very high, the rate at which energy is delivered to the eardrum is very small, because the area of the eardrum is very small.

*12.47. Set Up: Apply $\beta = (10 \text{ dB})\log(I/I_0)$, where $I_0 = 10^{-12} \text{ W/m}^2$. For part (b) use $I = I_0 10^{(\beta/10 \text{ dB})}$, which is given in Example 12.8.

Solve: (a) $\beta = (10 \text{ dB})\log\left(\dfrac{0.500 \text{ } \mu\text{W/m}^2}{10^{-12} \text{ W/m}^2}\right) = 57$ dB.

(b) $I = I_0 10^{(\beta/10 \text{ dB})} = (10^{-12} \text{ W/m}^2)10^{10.3} = 2.00 \times 10^{-2} \text{ W/m}^2$.

Reflect: As expected, the sound intensity is larger for the jackhammer.

*12.49. Set Up: $f_{\text{beat}} = f_1 - f_2$ and $f = \dfrac{v}{\lambda}$

Solve: $f_1 = \dfrac{v}{\lambda_1} = \dfrac{344 \text{ m/s}}{0.3394 \text{ m}} = 1014 \text{ Hz}; \quad f_2 = \dfrac{v}{\lambda_2} = \dfrac{344 \text{ m/s}}{0.3440 \text{ m}} = 1000 \text{ Hz}$

$f_{\text{beat}} = 1014 \text{ Hz} - 1000 \text{ Hz} = 14 \text{ Hz}$, so 14 beats each second.

*12.53. Set Up: The positive direction is from listener to source. $f_S = 392$ Hz.

(a) $v_S = 0$. $v_L = -15.0$ m/s.

$$f_L = \left(\frac{v + v_L}{v + v_S}\right)f_S = \left(\frac{344 \text{ m/s} - 15.0 \text{ m/s}}{344 \text{ m/s}}\right)(392 \text{ Hz}) = 375 \text{ Hz}$$

(b) $v_S = +35.0$ m/s. $v_L = +15.0$ m/s.

$$f_L = \left(\frac{v + v_L}{v + v_S}\right)f_S = \left(\frac{344 \text{ m/s} + 15.0 \text{ m/s}}{344 \text{ m/s} + 35.0 \text{ m/s}}\right)(392 \text{ Hz}) = 371 \text{ Hz}$$

(c) $f_{\text{beat}} = f_1 - f_2 = 4$ Hz

*12.55. Set Up: Choose the positive direction pointing from the motorcycle toward the car. The car is stationary, so $v_S = 0$. Use $f_L = \left(\dfrac{v + v_L}{v + v_S}\right)f_S$.

Solve: We have $f_L = \dfrac{v + v_L}{v + v_S}f_S = (1 + v_L/v)f_S$, which gives $v_L = v\left(\dfrac{f_L}{f_S} - 1\right) = (344 \text{ m/s})\left(\dfrac{490 \text{ Hz}}{520 \text{ Hz}} - 1\right) = -19.8$ m/s. You must be traveling at 19.8 m/s.

Reflect: $v_L < 0$ means that the listener is moving away from the source.

*12.59. Set Up: An observer on a *stationary* boat would measure the frequency of the waves as $f = \dfrac{v}{\lambda} = \dfrac{16.5 \text{ m/s}}{40.0 \text{ m}} = 0.4125$ Hz —so for simplicity imagine that the waves are generated by a stationary source ($v_s = 0$) with $f_s = f = 0.4125$ Hz. Thus, we will apply Eq. (12.19) for the special case of a stationary source. As usual, we choose the positive direction facing toward the source of the waves (i.e., westward), so that $v_L > 0$ in part (a) and $v_L < 0$ in part (b).

Solve: (a) $f_L = \left(\dfrac{v + v_L}{v}\right)f_S = \left(\dfrac{16.5 \text{ m/s} + 5.00 \text{ m/s}}{16.5 \text{ m/s}}\right)(0.4125 \text{ Hz}) = 0.5375$ Hz. Thus, the time interval between crests, as seen by an observer on the ship, is $T_L = \dfrac{1}{f_L} = \dfrac{1}{0.5375 \text{ Hz}} = 1.86$ s.

(b) The analysis is the same as in part (a)—except that $v_L = -5.00$ m/s :

$f_L = \left(\dfrac{v + v_L}{v}\right)f_S = \left(\dfrac{16.5 \text{ m/s} + -5.00 \text{ m/s}}{16.5 \text{ m/s}}\right)(0.4125 \text{ Hz}) = 0.2875$ Hz. Thus, in this case the time interval between

crests, as seen by an observer on the ship, is $T_L = \dfrac{1}{f_L} = \dfrac{1}{0.2875 \text{ Hz}} = 3.48$ s.

Reflect: In reality we do not know that the source of the waves is stationary. However, if the source of the waves were moving with a velocity v_S it would need to generate a frequency f_S so that a *stationary* boat would still observe the known frequency of $0.\underline{4}125$ Hz $=\left(\dfrac{v}{v+v_S}\right)f_S$. Thus, the frequency observed by a *moving* boat would be

$$f_L=\left(\dfrac{v+v_L}{v+v_S}\right)f_S=\left(\dfrac{v+v_L}{v}\right)\left(\dfrac{v}{v+v_S}\right)f_S=\left(\dfrac{v+v_L}{v}\right)(0.\underline{4}125\text{ Hz}),\text{ which is the same as we obtained by assuming a}$$

stationary source.

*12.61. Set Up: $f_S=1000$ Hz. The positive direction is from the listener to the source.

(a) $v_S=-(344\text{ m/s})/2=-172\text{ m/s},\ v_L=0.$

$$f_L=\left(\dfrac{v+v_L}{v+v_S}\right)f_S=\left(\dfrac{344\text{ m/s}}{344\text{ m/s}-172\text{ m/s}}\right)(1000\text{ Hz})=2000\text{ Hz}$$

(b) $v_S=0,\ v_L=+172\text{ m/s}.$

$$f_L=\left(\dfrac{v+v_L}{v+v_S}\right)f_S=\left(\dfrac{344\text{ m/s}+172\text{ m/s}}{344\text{ m/s}}\right)(1000\text{ Hz})=1500\text{ Hz}$$

(c) The answer in (b) is much less than the answer in (a). It is the velocity of the source and listener relative to the air that determines the effect, not the relative velocity of the source and listener relative to each other.

*12.65. Set Up: $\beta=10\log\dfrac{I}{I_0}$ gives I if β is specified. Then $I=\dfrac{P}{4\pi r^2}$ relates I, r and P.

Solve: (a) $\beta=100$ dB so $\log\dfrac{I}{I_0}=10$. $\dfrac{I}{I_0}=10^{10}$ and $I=(10^{-12}\text{ W/m}^2)(10^{10})=1\times10^{-2}\text{ W/m}^2.$

$$r=\sqrt{\dfrac{P}{4\pi I}}=\sqrt{\dfrac{20.0\text{ W}}{4\pi(1\times10^{-2}\text{ W/m}^2)}}=12.6\text{ m}$$

(b) $\beta=60$ dB so $\log\dfrac{I}{I_0}=6$. $\dfrac{I}{I_0}10^6$ and $I=(10^{-12}\text{ W/m}^2)(10^6)=1\times10^{-6}\text{ W/m}^2$

$$r=\sqrt{\dfrac{20.0\text{ W}}{4\pi(1\times10^{-6}\text{ W/m}^2)}}=1.26\times10^3\text{ m}=1.26\text{ km}$$

*12.67. Set Up: The flute acts as a stopped pipe and its harmonic frequencies are given by Eq. (12.12). The resonant frequencies of the string are $f_n=nf_1, n=1,\ 2,\ 3,\dots$ The string resonates when the string frequency equals the flute frequency. For the string we have $f_{1s}=600.0$ Hz. For the flute, the fundamental frequency is

$$f_{1f}=\dfrac{v}{4L}=\dfrac{344.0\text{ m/s}}{4(0.1075\text{ m})}=800.0\text{ Hz. Let }n_f\text{ label the harmonics of the flute and let }n_s\text{ label the harmonics of the}$$

string.

Solve: For the flute and string to be in resonance, $n_f f_{1f}=n_s f_{1s}$, where $f_{1s}=600.0$ Hz is the fundamental frequency for the string. $n_s=n_f(f_{1f}/f_{1s})=\tfrac{4}{3}n_f.$ Thus, n_s is an integer when $n_f=3N, N=1,\ 3,\ 5,\ \dots$ (the flute has only odd harmonics). $n_f=3N$ gives $n_s=4N$

Flute harmonic $3N$ resonates with string harmonic $4N$, $N=1,\ 3,\ 5,\dots.$

Reflect: We can check our results for some specific values of N. For $N=1$, $n_f=3$ and $f_{3f}=2400$ Hz. For this N, $n_s=4$ and $f_{4s}=2400$ Hz. For $N=3$, $n_f=9$ and $f_{9f}=7200$ Hz, and $n_s=12$, $f_{12s}=7200$ Hz. Our general results do give equal frequencies for the two objects.

*12.69. Set Up: According to convention the positive direction always points from the listener to the source. Thus, your velocity is positive when you walk toward a stationary source and negative when you walk away from a stationary source—and so we may write $v_L=\pm|v_L|$ to handle both cases. Since you are simultaneously moving toward one source and away from the other source, you will simultaneously hear each source Doppler shifted—one source will be shifted to a higher frequency, f_+, and the other to a lower frequency, f_-. These two frequencies will mix to produce beats with a beat frequency $f_{beat} = f_+ - f_-$.

Solve: (a) As you walk toward a speaker you have $v_L = +|v_L|$ so $f_+ = \left(\dfrac{v + |v_L|}{v}\right) f_S$. Similarly, as you walk away

from a speaker you have $v_L = -|v_L|$ so $f_- = \left(\dfrac{v - |v_L|}{v}\right) f_S$.

Also, we know $f_{beat} = f_+ - f_- = \left(\dfrac{v + |v_L|}{v}\right) f_S - \left(\dfrac{v - |v_L|}{v}\right) f_S = \dfrac{2|v_L|}{v} f_S$

This reduces to $|v_L| = \dfrac{1}{2} \dfrac{f_{beat}}{f_S} \cdot v = \dfrac{1}{2} \left(\dfrac{2.50 \text{ Hz}}{229 \text{ Hz}}\right) \cdot (344 \text{ m/s}) = \underline{1.878} \text{ m/s} = 1.88 \text{ m/s}.$

(b) According to the results of part (a) $f_{beat} = \dfrac{2(1.878 \text{ m/s})}{344 \text{ m/s}} \cdot (573 \text{ Hz}) = 6.26 \text{ Hz}.$

Reflect: Alternatively, consider the standing wave established between the two speakers with antinodes spaced by $\dfrac{\lambda}{2} = \dfrac{v}{2f_S}$. As you walk for a time t and a distance $|v_L|t$, you will cross $\dfrac{|v_L|t}{v/2f_S}$ antinodes. As a result, you will

perceive a beat frequency of $f_{beat} = \dfrac{\text{number of antinodes crossed}}{\text{time}} = \dfrac{|v_L|}{v/2f_S}$. Thus, we have $|v_L| = \dfrac{1}{2} \dfrac{f_{beat}}{f_S} \cdot v$, which is the result we obtained in part (a) by using the Doppler effect.

*12.73. Set Up: Apply the Doppler effect formula $f_L = \left(\dfrac{v + v_L}{v + v_S}\right) f_S$. In the SHM the source moves toward and away from the listener, with maximum speed $\omega_p A_p$. The direction from listener to source is positive.

Solve: (a) The maximum velocity of the siren is $\omega_p A_p = 2\pi f_p A_p$. You hear a sound with frequency $f_L = f_{siren} v/$ $(v + v_S)$, where v_S varies between $+2\pi f_p A_p$ and $-2\pi f_p A_p$. $f_{L-max} = f_{siren} v/(v - 2\pi f_p A_p)$ and $f_{L-min} = f_{siren} v/$ $(v + 2\pi f_p A_p)$.

(b) The maximum (minimum) frequency is heard when the platform is passing through equilibrium and moving up (down).

Reflect: When the platform is moving upward the frequency you hear is greater than f_{siren} and when it is moving downward the frequency you hear is less than f_{siren}. When the platform is at its maximum displacement from equilibrium its speed is zero and the frequency you hear is f_{siren}.

*12.75. Set Up: The greatest frequency shift from the Doppler effect occurs when one speaker is moving away and one is moving toward the person. The speakers have speed $v_0 = r\omega$, where $r = 0.75$ m. $f_L = \left(\dfrac{v + v_L}{v + v_S}\right) f_S$, with the positive direction from the listener to the source. $v = 344$ m/s

Solve: (a) $f = \dfrac{v}{\lambda} = \dfrac{344 \text{ m/s}}{0.313 \text{ m}} = 1100$ Hz.

$$\omega = (75 \text{ rpm})\left(\dfrac{2\pi \text{ rad}}{1 \text{ rev}}\right)\left(\dfrac{1 \text{ min}}{60 \text{ s}}\right) = 7.85 \text{ rad/s}$$

and $v_0 = (0.75 \text{ m})(7.85 \text{ rad/s}) = 5.89$ m

Speaker A, moving toward the listener:

$$f_{LA} = \left(\dfrac{v}{v - 5.89 \text{ m/s}}\right)(1100 \text{ Hz}) = 1119 \text{ Hz}.$$

Speaker B, moving toward the listener:

$$f_{LB} = \left(\frac{v}{v + 5.89 \text{ m/s}} \right)(1100 \text{ Hz}) = 1081 \text{ Hz}$$

$$f_{\text{beat}} = f_1 - f_2 = 1119 \text{ Hz} - 1081 \text{ Hz} = 38 \text{ Hz}$$

(b) A person can her individual beats only up to about 7 Hz and this beat frequency is much larger than that.

Reflect: As the turntable rotates faster the beat frequency at this position of the speakers increases.

Solutions to Passage Problems

*12.79. Set Up: For a string fixed at both ends we have $f_1 = \dfrac{v}{2L}$, where $v = \sqrt{\dfrac{T}{\mu}}$ is the speed of the wave on the string.

Solve: Assuming that L remains nearly fixed, both T and μ will remain nearly fixed. Thus, the fundamental frequency does not change significantly. The correct answer is A. Note that, although most objects do expand when heated, we will show in Chapter 14 that it would take a large change in temperature to change the length of a typical string or pipe by an amount comparable to the given 2% change in the speed of sound.

FLUID MECHANICS

Problems 1, 5, 7, 11, 15, 19, 23, 27, 29, 31, 35, 37, 39, 43, 45, 49, 51, 53, 57, 59, 63, 65, 69, 71, 75, 77

Solutions to Problems

*13.1. Set Up: The density of gold is $19.3 \times 10^3 \, \text{kg/m}^3$.

Solve: $V = (5.0 \times 10^{-3} \, \text{m})(15.0 \times 10^{-3} \, \text{m})(30.0 \times 10^{-3} \, \text{m}) = 2.25 \times 10^{-6} \, \text{m}^3$.

$$\rho = \frac{m}{V} = \frac{0.0158 \, \text{kg}}{2.25 \times 10^{-6} \, \text{m}^3} = 7.02 \times 10^3 \, \text{kg/m}^3.$$

The metal is not pure gold.

*13.5. Set Up: $\rho = m/V$. For a sphere, $V = \frac{4}{3}\pi r^3$.

Solve: (a) The mass of the nugget is $\dfrac{\$1.00 \times 10^6}{\$426.6/\text{ounce}} = (2344 \text{ ounces})\left(\dfrac{31.1035 \text{ g}}{1 \text{ ounce}}\right) = 7.29 \times 10^4 \text{ g}$

$$V = \frac{m}{\rho} = \frac{7.29 \times 10^4 \, \text{g}}{19.3 \, \text{g/cm}^3} = 3.78 \times 10^3 \, \text{cm}^3$$

$$r = \left(\frac{3V}{4\pi}\right)^{1/3} = \left[\frac{3(3.78 \times 10^3 \, \text{cm}^3)}{4\pi}\right]^{1/3} = 9.66 \, \text{cm}$$

and its diameter is $2r = 19.3$ cm.

(b) The platinum nugget would have mass

$$m = \rho V = (21.4 \, \text{g/cm}^3)(3.78 \times 10^3 \, \text{cm}^3) = 8.09 \times 10^4 \, \text{g} = 2.60 \times 10^3 \text{ troy ounces}$$

and would be worth 2.29 million dollars.

Reflect: The platinum nugget is worth more because a nugget of the same size has more mass, since platinum's density is greater, and because it is worth more per ounce. The value if platinum is larger than the value if gold by $(21.4/19.3)(879.00/426.60)$.

*13.7. Set Up: $\rho = m/V = m/L^3$

Solve: (a) $m = \rho L^3$ so $\rho_1 L_1^3 = \rho_2 L_2^3$ and $\rho_2 = \rho_1 \left(\dfrac{L_1}{L_2}\right)^3 = \rho \left(\dfrac{L_1}{L_1/2}\right)^3 = 8\rho$

(b) $L_2 = L_1 \left(\dfrac{\rho_1}{\rho_2}\right)^{1/3} = L \left(\dfrac{\rho_1}{3\rho_1}\right)^{1/3} = L/3^{1/3}$

Reflect: If the volume is decreased while the mass is kept the same, then the density increases.

*13.11. Set Up: $5\,\mathrm{L} = 5 \times 10^{-3}\,\mathrm{m}^3$. $\rho = m/V$. $F = pA$. $1\,\mathrm{cm}^2 = 10^{-4}\,\mathrm{m}^2$. A sphere of radius r has volume $V = \frac{4}{3}\pi r^3$.
Specific gravity $= 5.0$ means $\rho = 5.0 \times 10^3\,\mathrm{kg/m}^3$.
Solve: (a) $m = \rho V = (1050\,\mathrm{kg/m}^3)(5 \times 10^{-3}\,\mathrm{m}^3) = 5.25\,\mathrm{kg}$
(b) $F = pA = (13{,}000\,\mathrm{Pa})(1.0 \times 10^{-4}\,\mathrm{m}^2) = 1.3\,\mathrm{N}$
(c) $m = \rho V = (5.0 \times 10^3\,\mathrm{kg/m}^3)(\frac{4}{3})\pi(3.75 \times 10^{-6}\,\mathrm{m})^3 = 1.1 \times 10^{-12}\,\mathrm{kg} = 1.1 \times 10^{-9}\,\mathrm{g}$

*13.15. Set Up: (a) Gauge pressure, $\Delta p = p - p_{\mathrm{atm}}$, is related to depth according to $\Delta p = \rho g h$. According to Table 13.1 the density of seawater is $1.03 \times 10^3\,\mathrm{kg/m}^3$, which is the same as that of freshwater to two significant figures. $1\,\mathrm{atm} = 1.01 \times 10^5\,\mathrm{Pa}$. **(b)** A uniform pressure p applied perpendicular to a surface of area A produces a force of pA on the surface. The lateral (side) area of a cylinder of radius r and length l is $2\pi r l$, and the area of each end cap is πr^2.
Solve: (a) The gauge pressure at 400 meters is $\Delta p = \rho g h = (1.03 \times 10^3\,\mathrm{kg/m}^3)(9.80\,\mathrm{m/s}^2)(400\,\mathrm{m}) = 4 \times 10^6\,\mathrm{Pa}$, which is equal to $(4 \times 10^6\,\mathrm{Pa})\left(\dfrac{1\,\mathrm{atm}}{1.01 \times 10^5\,\mathrm{Pa}}\right) = 40\,\mathrm{atm}$. Similarly, at a depth of 3000 meters we have $\Delta p = (1.03 \times 10^3\,\mathrm{kg/m}^3)(9.80\,\mathrm{m/s}^2)(3000\,\mathrm{m}) = 3 \times 10^7\,\mathrm{Pa} = 300\,\mathrm{atm}$. Thus, the pressure ranges from 40 atm to 300 atm.
(b) We are asked to find the total *inward* pressure force acting on the surface of the whale at a depth of 3000 meters. Since the external pressure on all sides of the whale acts inward, we can simply multiply the external pressure times the total surface area of the whale. Modeling the whale as a cylinder, its total surface area is $A = 2\pi r^2 + 2\pi r l = 2\pi(2\,\mathrm{m})^2 + 2\pi(2\,\mathrm{m})(16\,\mathrm{m}) = 226\,\mathrm{m}^2$. Thus, the total inward force is $F = pA = (3 \times 10^7\,\mathrm{Pa})(226\,\mathrm{m}^2) = 7 \times 10^9\,\mathrm{N}$.
Reflect: According to Archimedes's principle, the total upward buoyant force on the whale is equal to the weight of water that would occupy the whale's volume. This is $\rho g V = (1.03 \times 10^3\,\mathrm{kg/m}^3)(9.80\,\mathrm{m/s}^2)\pi(2\,\mathrm{m})^2(16\,\mathrm{m}) = 2 \times 10^6\,\mathrm{N}$, which is much smaller that the result in part (b) as we would expect.

*13.19. Set Up: $p = p_0 + \rho g h$. $F = pA$. For seawater, $\rho = 1.03 \times 10^3\,\mathrm{kg/m}^3$
Solve: The force F that must be applied is the difference between the upward force of the water and the downward forces of the air and the weight of the hatch. The difference between the pressure inside and out is the gauge pressure, so

$$F = (\rho g h)\,A - w = (1.03 \times 10^3\,\mathrm{kg/m}^3)(9.80\,\mathrm{m/s}^2)(30\,\mathrm{m})(0.75\,\mathrm{m}^2) - 300\,\mathrm{N} = 2.27 \times 10^5\,\mathrm{N}.$$

Reflect: The force due to the gauge pressure of the water is much larger than the weight of the hatch and would be impossible for the crew to apply it just by pushing.

*13.23. Set Up: $p_0 = p_{\mathrm{surface}} + \rho g h$ where p_{surface} is the pressure at the surface of a liquid and p_0 is the pressure at a depth h below the surface.
Solve: (a) For the oil layer, $p_{\mathrm{surface}} = p_{\mathrm{atm}}$ and p_0 is the pressure at the oil-water interface.

$$p_0 - p_{\mathrm{atm}} = p_{\mathrm{gauge}} = \rho g h = (600\,\mathrm{kg/m}^3)(9.80\,\mathrm{m/s}^2)(0.120\,\mathrm{m}) = 706\,\mathrm{Pa}$$

(b) For the water layer, $p_{\mathrm{surface}} = 706\,\mathrm{Pa} + p_{\mathrm{atm}}$.

$$p_0 - p_{\mathrm{atm}} = p_{\mathrm{gauge}} = 706\,\mathrm{Pa} + \rho g h = 706\,\mathrm{Pa} + (1.00 \times 10^3\,\mathrm{kg/m}^3)(9.80\,\mathrm{m/s}^2)(0.250\,\mathrm{m}) = 3.16 \times 10^3\,\mathrm{Pa}$$

Reflect: The gauge pressure at the bottom of the barrel is due to the combined effects of the oil layer and water layer. The pressure at the bottom of the oil layer is the pressure at the top of the water layer.

*13.27. Set Up: Pascal's law says the pressure is the same everywhere in the hydraulic fluid, so $F_1/A_1 = F_2/A_2$.
$F_1 = 100$ N and $F = (3500 \text{ kg})(9.80 \text{ m/s}^2) = 3.43 \times 10^4$ N, the weight of the car and platform that is being lifted. The
volume of fluid displaced at each piston when the pistons move distances d_1 and d_2 is the same, so $d_1 A_1 = d_2 A_2$.

Solve: **(a)** $A = \pi r^2$ so $\dfrac{F_1}{r_1^2} = \dfrac{F_2}{r_2^2}$. $r_2 = r_1 \sqrt{\dfrac{F_2}{F_1}} = (0.125 \text{ m})\sqrt{\dfrac{3.43 \times 10^4 \text{ N}}{100 \text{ N}}} = 2.32$ m and the diameter is 4.64 m.

(b) $d_1 A_1 = d_2 A_2$.

$$d_2 = d_1 \frac{A_1}{A_2} = d_1 \frac{\pi r_1^2}{\pi r_2^2} = (50 \text{ cm})\left(\frac{0.125 \text{ m}}{2.32 \text{ m}}\right)^2 = 1.45 \text{ mm}$$

Reflect: The work done by the man is $F_1 d_1 = (100 \text{ N})(0.50 \text{ m}) = 50$ J. The work done on the car is $F_2 d_2 = (3.43 \times 10^4 \text{ N})(1.43 \times 10^{-3} \text{ m}) = 50$ J. Energy conservation requires that these two quantities be equal, and they are equal.

*13.29. Set Up: The density of aluminum is 2.7×10^3 kg/m³. The density of water is 1.00×10^3 kg/m³. $\rho = m/V$.
The buoyant force is $F_B = \rho_{\text{water}} V_{\text{obj}} g$.

Solve: **(a)** $T = mg = 89$ N so $m = 9.08$ kg. $V = \dfrac{m}{\rho} = \dfrac{9.08 \text{ kg}}{2.7 \times 10^3 \text{ kg/m}^3} = 3.36 \times 10^{-3}$ m³ = 3.4 L.

(b) When the ingot is totally immersed in the water while suspended, $T + F_B - mg = 0$.

$$F_B = \rho_{\text{water}} V_{\text{obj}} g = (1.00 \times 10^3 \text{ kg/m}^3)(3.36 \times 10^{-3} \text{ m}^3)(9.80 \text{ m/s}^2) = 32.9 \text{ N}.$$

$$T = mg - F_B = 89 \text{ N} - 32.9 \text{ N} = 56 \text{ N}.$$

Reflect: The buoyant force is equal to the difference between the apparent weight when the object is submerged in the fluid and the actual gravity force on the object.

*13.31. Set Up: $F_B = \rho_{\text{fluid}} g V_{\text{sub}}$, where V_{sub} is the volume of the object that is below the fluid's surface.
Solve: **(a)** Floats, so the buoyant force equals the weight of the object: $F_B = mg$. Using Archimedes's principle gives
$\rho_w g V = mg$ and

$$V = \frac{m}{\rho_w} = \frac{5750 \text{ kg}}{1.00 \times 10^3 \text{ kg/m}^3} = 5.75 \text{ m}^3.$$

(b) $F_B = mg$ and $\rho_w g V_{\text{sub}} = mg$. $V_{\text{sub}} = 0.80V = 4.60$ m³, so the mass of the floating object is

$$m = \rho_w V_{\text{sub}} = (1.00 \times 10^3 \text{ kg/m}^3)(4.60 \text{ m}^3) = 4600 \text{ kg}.$$

He must throw out $5750 \text{ kg} - 4600 \text{ kg} = 1150$ kg.
Reflect: He must throw out 20% of the boat's mass.

*13.35. Set Up: $F_B = \rho_{\text{water}} V_{\text{obj}} g$. The net force on the sphere is zero.
Solve: **(a)** $F_B = (1000 \text{ kg/m}^3)(0.650 \text{ m}^3)(9.80 \text{ m/s}^2) = 6.37 \times 10^3$ N
(b) $F_B = T + mg$ and

$$m = \frac{F_B - T}{g} = \frac{6.37 \times 10^3 \text{ N} - 900 \text{ N}}{9.80 \text{ m/s}^2} = 558 \text{ kg}.$$

(c) Now $F_B = \rho_{water} V_{sub} g$, where V_{sub} is the volume of the sphere that is submerged. $F_B = mg$. $\rho_{water} V_{sub} = mg$ and

$$V_{sub} = \frac{m}{\rho_{water}} = \frac{558 \text{ kg}}{1000 \text{ kg/m}^3} = 0.558 \text{ m}^3.$$

$$\frac{V_{sub}}{V_{obj}} = \frac{0.558 \text{ m}^3}{0.650 \text{ m}^3} = 0.858 = 85.8\%$$

Reflect: When the sphere is totally submerged, the buoyant force on it is greater than its weight. When it is floating, it needs to be only partially submerged in order to produce a buoyant force equal to its weight.

*13.37. Set Up: $F_B = \rho_{fluid} g V_{sub}$. For a floating object, $F_B = mg = \rho_{obj} V_{obj} g$. $\rho_{ice} = 0.92 \times 10^3 \text{ kg/m}^3$.

Solve: **(a)** $F_B = mg$ says $\rho_{sw} g V_{sub} = \rho_{ice} g V_{ice}$. $V_{sub} = 0.90 V_{ice}$, so $0.90 \rho_{sw} = \rho_{ice}$ and

$$\rho_{sw} = \frac{\rho_{ice}}{0.90} = 1.02 \times 10^3 \text{ kg/m}^3.$$

(b) $\rho_w g V_{sub} = \rho_{ice} g V_{ice}$ and

$$\frac{V_{sub}}{V_{ice}} = \frac{\rho_{ice}}{\rho_w} = \frac{0.92 \times 10^3 \text{ kg/m}^3}{1.00 \times 10^3 \text{ kg/m}^3} = 0.92$$

92.0% is submerged and 8.0% is above the surface.

*13.39. Set Up: $1 \text{ dyn/cm} = 10^{-3} \text{ N/m}$. The gauge pressure inside a bubble is

$$P_{gauge} = p - p_{atm} = \frac{4\gamma}{R}.$$

Solve: $p_{gauge} = \dfrac{4(25.0 \times 10^{-3} \text{ N/m})}{3.50 \times 10^{-2} \text{ m}} = 2.86 \text{ Pa}.$

*13.43. Set Up: $v_1 A_1 = v_2 A_2$

Solve: **(a)** $v_2 = v_1 \left(\dfrac{A_1}{A_2} \right) = (3.50 \text{ m/s}) \left(\dfrac{0.070 \text{ m}^2}{0.105 \text{ m}^2} \right) = 2.33 \text{ m/s}$

(b) $v_2 = v_1 \left(\dfrac{A_1}{A_2} \right) = (3.50 \text{ m/s}) \left(\dfrac{0.070 \text{ m}^2}{0.047 \text{ m}^2} \right) = 5.21 \text{ m/s}$

*13.45. Set Up: Apply the equation of continuity, $v_1 A_1 = v_2 A_2$, where $A = \pi r^2$

Solve: $v_2 = v_1(A_1/A_2)$, where $A_1 = \pi(0.80 \text{ cm})^2$, $A_2 = 20\pi(0.10 \text{ cm})^2$. $v_2 = (3.0 \text{ m/s}) \dfrac{\pi(0.80)^2}{20\pi(0.10)^2} = 9.6 \text{ m/s}.$

Reflect: The total area of the showerhead openings is less than the cross-section area of the pipe, and the speed of the water in the showerhead opening is greater than its speed in the pipe.

*13.49. Set Up: Let point 1 be in the mains and point 2 be in the emerging stream at the end of the fire hose. $v_1 \approx 0$. $y_1 - y_2 = 0$. $p_2 = p_{air}$. After the water emerges from the hose, with upward velocity, it moves in free fall.

Solve: $p_1 + \rho g y_1 + \frac{1}{2}\rho v_1^2 = p_2 + \rho g y_2 + \frac{1}{2}\rho v_2^2$ gives $p_1 - p_{air} = \frac{1}{2}\rho v_2^2$. The motion of a drop of water after it leaves the hose gives $\frac{1}{2}mv_2^2 = mgh$.

$$v_2^2 = 2gh = 2(9.80 \text{ m/s}^2)(15.0 \text{ m}) = 294 \text{ m}^2/\text{s}^2.$$

$$p_1 - p_{air} = \frac{1}{2}(1000 \text{ kg/m}^3)(294 \text{ m}^2/\text{s}^2) = 1.47 \times 10^5 \text{ Pa}$$

Reflect: Alternatively, the problem can be solved entirely by using Bernoulli's equation. In this case point 1 is in the mains and point 2 is at the maximum height of the stream. Thus, $v_1 = v_2 = 0$ and $p_1 + \rho g y_1 = p_2 + \rho g y_2$. This implies that $p_1 - p_{air} = \rho g \Delta y$, which gives the same answer as before. Note that this is the gauge pressure required to support a column of water 15.0 m high.

*13.51. Set Up: Let point 1 be at the top surface and point 2 be at the bottom surface. Neglect the $\rho g(y_1 - y_2)$ term in Bernoulli's equation. In calculating the net force, ΣF_y, take $+y$ to be upward.

Solve: $p_1 + \rho g y_1 + \frac{1}{2}\rho v_1^2 = p_2 + \rho g y_2 + \frac{1}{2}\rho v_2^2$.

$$p_2 - p_1 = \frac{1}{2}\rho(v_1^2 - v_2^2) = \frac{1}{2}(1.20 \text{ kg/m}^3)([70.0 \text{ m/s}]^2 - [60.0 \text{ m/s}]^2) = 780 \text{ Pa}$$

$$\Sigma F_y = p_2 A - p_1 A - mg = (780 \text{ Pa})(16.2 \text{ m}^2) - (1340 \text{ kg})(9.80 \text{ m/s}^2) = -500 \text{ N}.$$

The net force is 500 N, downward.
Reflect: The pressure is lower where the fluid speed is higher.

*13.53. Set Up: $y_1 = y_2$. $v_1 A_1 = v_2 A_2 = 465 \times 10^{-6} \text{ m}^3/\text{s}$.

Solve: $p_1 + \rho g y_1 + \frac{1}{2}\rho v_1^2 = p_2 + \rho g y_2 + \frac{1}{2}\rho v_2^2$.

$$v_1 = \frac{465 \times 10^{-6} \text{ m}^3/\text{s}}{\pi r_1^2} = \frac{465 \times 10^{-6} \text{ m}^3/\text{s}}{\pi(2.05 \times 10^{-2} \text{ m})^2} = 0.352 \text{ m/s}$$

Then $v_2 = \sqrt{\dfrac{2}{\rho}(p_1 - p_2) + v_1^2} = \sqrt{\dfrac{2}{1000 \text{ kg/m}^3}(0.40 \times 10^5 \text{ Pa}) + (0.352 \text{ m/s})^2} = 8.95 \text{ m/s}.$

$v_1 \pi r_1^2 = v_2 \pi r_2^2$ and

$$r_2 = r_1\sqrt{\frac{v_1}{v_2}} = (2.05 \text{ cm})\sqrt{\frac{0.352 \text{ m/s}}{8.95 \text{ m/s}}} = 4.07 \text{ mm}.$$

*13.57. Set Up: The viscous drag force is $F = 6\pi\eta r v_t$. The weight of the ball bearing is $mg = \frac{4}{3}\pi r^3 \rho g$.

Solve: $F = \frac{1}{4}mg$ gives $6\pi\eta r v_t = \frac{1}{3}\pi r^3 \rho g$.

$$v_t = \frac{r^2 g \rho}{18\eta} = \frac{(3.00 \times 10^{-3} \text{ m})^2(9.80 \text{ m/s}^2)(19.3 \times 10^3 \text{ kg/m}^3)}{18(0.986 \text{ N} \cdot \text{s/m}^2)} = 0.0959 \text{ m/s} = 9.59 \text{ cm/s}.$$

*13.59. Set Up: The flow rate, $\Delta V/\Delta t$, is related to the radius R or diameter D of the artery by Poiseuille's law:

$$\frac{\Delta V}{\Delta t} = \frac{\pi R^4}{8\eta}\left(\frac{p_1 - p_2}{L}\right) = \frac{\pi D^4}{128\eta}\left(\frac{p_1 - p_2}{L}\right).$$

Assume the pressure gradient $(p_1 - p_2)/L$ in the artery remains the same.

Solve: $(\Delta V/\Delta t)/D^4 = \dfrac{\pi}{128\eta}\left(\dfrac{p_1 - p_2}{L}\right) = \text{constant}$, so $(\Delta V/\Delta t)_{old}/D_{old}^4 = (\Delta V/\Delta t)_{new}/D_{new}^4$. $(\Delta V/\Delta t)_{new} = 2(\Delta V/\Delta t)_{old}$ and $D_{old} = D$. This gives

$$D_{new} = D_{old}\left[\frac{(\Delta V/\Delta t)_{new}}{(\Delta V/\Delta t)_{old}}\right]^{1/4} = 2^{1/4}D = 1.19D.$$

Reflect: Since the flow rate is proportional to D^4, a 19% increase in D doubles the flow rate.

*13.63. Set Up: The density of lead is $11.3 \times 10^3 \text{ kg/m}^3$. The buoyant force when the wood sinks is $F_B = \rho_{water} V_{tot} g$, where V_{tot} is the volume of the wood plus the volume of the lead. $\rho = m/V$.

Solve: $V_{wood} = (0.600 \text{ m})(0.250 \text{ m})(0.080 \text{ m}) = 0.0120 \text{ m}^3$.

$$m_{wood} = \rho_{wood} V_{wood} = (600 \text{ kg/m}^3)(0.0120 \text{ m}^3) = 7.20 \text{ kg}.$$

$F_B = (m_{wood} + m_{lead})g$. Using $F_B = \rho_{water} V_{tot} g$ and $V_{tot} = V_{wood} + V_{lead}$ gives

$$\rho_{water}(V_{wood} + V_{lead})g = (m_{wood} + m_{lead})g.$$

$m_{lead} = \rho_{lead} V_{lead}$ then gives $\rho_{water} V_{wood} + \rho_{water} V_{lead} = m_{wood} + \rho_{lead} V_{lead}$.

$$V_{lead} = \frac{\rho_{water} V_{wood} - m_{wood}}{\rho_{lead} - \rho_{water}} = \frac{(1000 \text{ kg/m}^3)(0.0120 \text{ m}^3) - 7.20 \text{ kg}}{11.3 \times 10^3 \text{ kg/m}^3 - 1000 \text{ kg/m}^3} = 4.66 \times 10^{-4} \text{ m}^3.$$

$m_{lead} = \rho_{lead} V_{lead} = 5.27 \text{ kg}$.

Reflect: The volume of the lead is only 3.9% of the volume of the wood. If the contribution of the volume of the lead to F_B is neglected, the calculation is simplified: $\rho_{water} V_{wood} g = (m_{wood} + m_{lead})g$ and $m_{lead} = 4.8 \text{ kg}$. The result of this calculation is in error by about 9%.

*13.65. Set Up: Seawater has density $\rho_{sw} = 1.03 \times 10^3 \text{ kg/m}^3$. $F_B = \rho_{sw} V_{sub} g$. V_{sub} is the submerged volume: $V_{sub} = V_{lp} + 0.80 V_p$, where $V_{lp} = 0.0400 \text{ m}^3$ is the volume of the life preserver and V_p is the volume of the person.

Solve: $F_B = m_{tot} g$. $m_{tot} = m_{lp} + m_p = \rho_{lp} V_{lp} + m_p$.

$$V_p = \frac{m_p}{\rho_p} = \frac{75.0 \text{ kg}}{980 \text{ kg/m}^3} = 0.0765 \text{ m}^3.$$

$F_B = m_{tot} g$ gives $\rho_{sw}(V_{lp} + 0.80 V_p)g = (\rho_{lp} V_{lp} + m_p)g$.

$$\rho_{lp} = \frac{\rho_{sw}(V_{lp} + 0.80 V_p) - m_p}{V_{lp}}$$

$$\rho_{lp} = \frac{(1.03 \times 10^3 \text{ kg/m}^3)(0.0400 \text{ m}^3 + [0.80][0.0765 \text{ m}^3]) - 75.0 \text{ kg}}{0.0400 \text{ m}^3} = 731 \text{ kg/m}^3$$

*13.69. Set Up: Apply $p_1 + \rho g y_1 + \frac{1}{2}\rho v_1^2 = p_2 + \rho g y_2 + \frac{1}{2}\rho v_2^2$, with point 1 at the surface of the acid in the tank and point 2 in the stream as it emerges from the hole. $p_1 = p_2 = p_{air}$. Since the hole is small the level in the tank drops slowly and $v_1 \approx 0$. After a drop of acid exits the hole the only force on it is gravity and it moves in projectile motion. For the projectile motion take $+y$ downward, so $a_x = 0$ and $a_y = +9.80 \text{ m/s}^2$.

Solve: Bernoulli's equation with $p_1 = p_2$ and $v_1 = 0$ gives

$$v_2 = \sqrt{2g(y_1 - y_2)} = \sqrt{2(9.80 \text{ m/s}^2)(0.75 \text{ m})} = 3.83 \text{ m/s}.$$

projectile motion: Use the vertical motion to find the time in the air. $v_{0y} = 0$, $a_y = +9.80 \text{ m/s}^2$, $y - y_0 = 1.4 \text{ m}$.

$y - y_0 = v_{0y} t + \frac{1}{2}a_y t^2$ gives

$$t = \sqrt{\frac{2(y - y_0)}{a_y}} = \sqrt{\frac{2(1.4 \text{ m})}{9.80 \text{ m/s}^2}} = 0.535 \text{ s}$$

The horizontal distance a drop travels in this time is $x - x_0 = v_{0x} t + \frac{1}{2}a_x t^2 = (3.83 \text{ m/s})(0.535 \text{ s}) = 2.05 \text{ m}$.

Reflect: If the depth of acid in the tank is increased, then the velocity of the stream as it emerges from the hole increases and the horizontal range of the stream increases.

*13.71. Set Up: For a spherical astronomical object, $g = G\dfrac{m}{R^2}$. $p = p_{air} + \rho g h$.

Solve: **(a)** Find g on Europa:

$$g = (6.673 \times 10^{-11}\,\text{N} \cdot \text{m}^2/\text{kg}^2)\frac{4.78 \times 10^{22}\,\text{kg}}{(1.565 \times 10^6\,\text{m})^2} = 1.30\,\text{m/s}^2$$

The gauge pressure at a depth of 100 m would be

$$p - p_{air} = \rho g h = (1.00 \times 10^3\,\text{kg/m}^3)(1.30\,\text{m/s}^2)(100\,\text{m}) = 1.30 \times 10^5\,\text{Pa}$$

(b) Solve for h on earth that gives $p - p_{air} = 1.30 \times 10^5\,\text{Pa}$:

$$h = \frac{p - p_{air}}{\rho g} = \frac{1.30 \times 10^5\,\text{Pa}}{(1.00 \times 10^3\,\text{kg/m}^3)(9.80\,\text{m/s}^2)} = 13.3\,\text{m}$$

Reflect: g on Europa is less than on earth so the pressure at a water depth of 100 m is much less on Europa than it would be on earth.

Solutions to Passage Problems

*13.75. Set Up: According to Bernoulli's equation $p_1 + \rho g y_1 + \frac{1}{2}\rho v_1^2 = p_2 + \rho g y_2 + \frac{1}{2}\rho v_2^2$.

Solve: Since the veins are flexible and do not collapse, we may assume that the pressure in the venous system is higher than atmospheric pressure. According to Bernoulli's equation, a large pressure differential would cause blood to exit a wound with a high velocity, which is contrary to experience. Thus, we may assume that the pressure inside an artery is only slightly greater than atmospheric pressure.

*13.77. Set Up: According to Bernoulli's equation $p_1 + \rho g y_1 + \frac{1}{2}\rho v_1^2 = p_2 + \rho g y_2 + \frac{1}{2}\rho v_2^2$. According to the continuity equation $A_1 v_1 = A_2 v_2$.

Solve: According to the continuity equation, we expect v to be constant if A does not change. In that case, Bernoulli's equation gives $\Delta p = -\rho g \Delta y$. Assuming that Δp does not change as the acceleration of gravity increases from g_1 to g_2, the maximum height that the heart can pump blood will decrease from Δy_1 to Δy_2. Using $\rho g_{earth}\Delta y_{earth} = \rho g_{max}\Delta y_{head}$ we find that $g_{max} = g_{earth}\dfrac{1.3\,\text{m}}{0.5\,\text{m}} = 2.6 g_{earth} \approx 3 g_{earth}$. Thus, the correct answer is B.

TEMPERATURE AND HEAT

Problems 1, 5, 7, 11, 13, 15, 19, 21, 25, 27, 31, 35, 39, 41, 43, 47, 49, 51, 55, 57, 61, 65, 67, 69, 73, 77, 81, 83, 87

Solutions to Problems

*14.1. Set Up and Solve: (a) $T_F = \frac{9}{5}(T_C + 32°) = \frac{9}{5}(40.2°) + 32° = 104.4°F$. Yes, you should be concerned.

(b) $T_F = \frac{9}{5}(T_C + 32°) = \frac{9}{5}(12°C) + 32° = 54°F$. Yes, bring a jacket. $T_C = T_K - 273.15 = 350 - 273.15 = 77°C$.
$T = \frac{9}{5}(77°) + 32° = 171°F$. This would be unpleasantly warm.

*14.5. Set Up and Solve: (a) $T_F = \frac{9}{5}T_C + 32$. Set $T_F = T_C = T$. $T = \frac{9}{5}T + 32$. $\frac{4}{5}T = -32$ and $T = -40°$.
$-40°C = -40°F$.

(b) $T_K = T_C + 273.15$. It is not possible to have $T_K = T_C$.

Reflect: Any Celsius temperature is less than the corresponding Kelvin temperature.

*14.7. Set Up: (a) $T_0 = -8.0°C$ and $T = 40.0°C$. $L_0 = 984$ ft and we are asked to solve for ΔL. From Table 14.1, the coefficient of linear expansion is $1.2 \times 10^{-5} (C°)^{-1}$. **(b)** Use $9 \text{ F°} = 5 \text{ C°}$.

Solve: (a) $\Delta L = \alpha L_0 \Delta T = [1.2 \times 10^{-5} (C°)^{-1}][984 \text{ ft}][40.0°C - (-8.0°C)] = 0.57 \text{ ft} = 6.8 \text{ in.}$

(b) $\alpha = [1.2 \times 10^{-5} (C°)^{-1}][5 \text{ C°}/9 \text{ F°}] = 6.7 \times 10^{-6} (F°)^{-1}$

Reflect: In (a) note that the fractional change in length, $\Delta L/L_0$, is very small. In (b), α is smaller in $(F°)^{-1}$ than in $(C°)^{-1}$. A Fahrenheit degree is smaller than a Celsius degree and a temperature difference is a larger number when expressed in F° than in C°. So, to give the same $\alpha \Delta T$ value, α is smaller in $(F°)^{-1}$.

*14.11. Set Up: For ethanol, $\beta = 75 \times 10^{-5} (C°)^{-1}$.

Solve: $\Delta V = V_0 \beta \Delta T = (1700 \text{ L})(75 \times 10^{-5} (C°)^{-1})(-9.0 \text{ C°}) = -11.5 \text{ L}$. The air space will have volume 11 L.

*14.13. Set Up: $\Delta V = \beta V_0 \Delta T$. Use the diameter at $-15°C$ to calculate the value of V_0 at that temperature. For a hemisphere of radius R, the volume is $V = \frac{2}{3}\pi R^3$. Table 17.2 gives $\beta = 7.2 \times 10^{-5} (°C)^{-1}$ for aluminum.

Solve: $V_0 = \frac{2}{3}\pi R^3 = \frac{2}{3}\pi(27.5 \text{ m})^3 = 4.356 \times 10^4 \text{ m}^3$.
$\Delta V = (7.2 \times 10^{-5} (°C)^{-1})(4.356 \times 10^4 \text{ m}^3)(35°C - [-15°C]) = 160 \text{ m}^3$

Reflect: We could also calculate $R = R_0(1 + \alpha\Delta T)$ and calculate the new V from R. The increase in volume is $V - V_0$, but we would have to be careful to avoid round-off errors when two large volumes of nearly the same size are subtracted.

***14.15. Set Up:** For mercury, $\beta_{Hg} = 18 \times 10^{-5}\,(C°)^{-1}$. When heated, both the volume of the flask and the volume of the mercury increase. 8.95 cm^3 of mercury overflows, so $\Delta V_{Hg} - \Delta V_{glass} = 8.95$ cm^3.

Solve: $\Delta V_{Hg} = V_0\beta_{Hg}\,\Delta T = (1000.00\text{ cm}^3)(18 \times 10^{-5}\,(C°)^{-1})(55.0\text{ C}°) = 9.9\text{ cm}^3$.

$$\Delta V_{glass} = \Delta V_{Hg} - 8.95\text{ cm}^3 = 0.95\text{ cm}^3.$$

$$\beta_{glass} = \frac{\Delta V_{glass}}{V_0\Delta T} = \frac{0.95\text{ cm}^3}{(1000.00\text{ cm}^3)(55.0\text{ C}°)} = 1.7 \times 10^{-5}\,(C°)^{-1}.$$

Reflect: The coefficient of volume expansion for the mercury is larger than for glass. When they are heated, both the volume of the mercury and the inside volume of the flask increase. But the increase for the mercury is greater and it no longer all fits inside the flask.

***14.19. Set Up:** For aluminum, $c_a = 0.91 \times 10^3$ J/kg $\cdot$ K. For iron, $c_i = 0.47 \times 10^3$ J/kg $\cdot$ K.
Solve: $Q = m_a c_a\,\Delta T + m_i c_i\,\Delta T$

$$Q = [(1.60\text{ kg})(0.91 \times 10^3\text{ J/kg} \cdot \text{K}) + (0.300\text{ kg})(0.47 \times 10^3\text{ J/kg} \cdot \text{K})][190\text{ C}°] = 3.03 \times 10^5\text{ J}$$

***14.21. Set Up:** 0.50 L of air has mass 0.65×10^{-3} kg.
Solve: (a) $Q = mc\Delta T = (0.65 \times 10^{-3}\text{ kg})(1020\text{ J/kg} \cdot \text{K})(57\text{ C}°) = 38\text{ J}$

(b) $Q = (38\text{ J/breath})(20\text{ breaths/min})(60\text{ min/h}) = 4.6 \times 10^4\text{ J/h}$.

Reflect: Air has a small density, so one liter of air has little mass. But the specific heat capacity of air is rather large, larger than for most metals and about one-fourth the value for water.

***14.25. Set Up:** $Q = mc\Delta T$. The mass of n moles is $m = nM$. For iron, $M = 55.845 \times 10^{-3}$ kg/mol and $c = 470$ J/kg $\cdot$ K.
Solve: (a) The mass of 3.00 mol is $m = nM = (3.00\text{ mol})(55.845 \times 10^{-3}\text{ kg/mol}) = 0.1675$ kg.

$$\Delta T = Q/mc = (8950\text{ J})/[(0.1675\text{ kg})(470\text{ J/kg} \cdot \text{K})] = 114\text{ K} = 114\text{ C}°.$$

(b) For $m = 3.00$ kg, $\Delta T = Q/mc = 6.35$ C°.
Reflect: (c) The result of part (a) is much larger; 3.00 kg is more material than 3.00 mol.

***14.27. Set Up:** Set the loss of kinetic energy of the bullet equal to the heat energy Q transferred to the water. From Table 14.3, the specific heat capacity of water is 4.19×10^3 J/kg $\cdot$ C°.
Solve: The kinetic energy lost by the bullet is

$$K_i - K_f = \tfrac{1}{2}m(v_i^2 - v_f^2) = \tfrac{1}{2}(15.0 \times 10^{-3}\text{ kg})([865\text{ m/s}]^2 - [534\text{ m/s}]^2) = 3.47 \times 10^3\text{ J},$$

so for the water $Q = 3.47 \times 10^3$ J. $Q = mc\,\Delta T$ gives

$$\Delta T = \frac{Q}{mc} = \frac{3.47 \times 10^3\text{ J}}{(13.5\text{ kg})(4.19 \times 10^3\text{ J/kg} \cdot \text{C}°)} = 0.0613\text{ C}°$$

Reflect: The heat energy required to change the temperature of ordinary size objects is very large compared to the typical kinetic energies of moving objects.

*14.31. Set Up: From Table 14.4, for lead $L_f = 24.5 \times 10^3$ J/kg and $L_v = 871 \times 10^3$ J/kg and for water $L_f = 334 \times 10^3$ J/kg. The temperatures given in the problem are the transition temperatures for each respective phase transition, so $Q = mL$.

Solve: (a) $Q = mL_f = (0.150 \text{ kg})(24.5 \times 10^3 \text{ J/kg}) = 3.68 \times 10^3$ J

(b) $Q = mL_v = (0.150 \text{ kg})(871 \times 10^3 \text{ J/kg}) = 1.31 \times 10^5$ J

(c) $Q = 3.68 \times 10^3 \text{ J} + 1.31 \times 10^5 \text{ J} = 1.35 \times 10^5$ J. $Q = mL_f$ so

$$m = \frac{Q}{L_f} = \frac{1.35 \times 10^5 \text{ J}}{334 \times 10^3 \text{ J}} = 0.404 \text{ kg}$$

*14.35. Set Up: The asteroid's kinetic energy is $K = \frac{1}{2}mv^2$. To boil the water, its temperature must be raised to $100.0°C$ and the heat needed for the phase change must be added to the water. For water, $c = 4190$ J/kg·K and $L_v = 2256 \times 10^3$ J/kg.

Solve: $K = \frac{1}{2}(2.60 \times 10^{15} \text{ kg})(32.0 \times 10^3 \text{ m/s})^2 = 1.33 \times 10^{24}$ J. $Q = mc\Delta T + mL_v$.

$$m = \frac{Q}{c\Delta T + L_v} = \frac{1.33 \times 10^{22} \text{ J}}{(4190 \text{ J/kg·K})(90.0 \text{ K}) + 2256 \times 10^3 \text{ J/kg}} = 5.05 \times 10^{15} \text{ kg.}$$

Reflect: The mass of water boiled is 2.5 times the mass of water in Lake Superior.

*14.39. Set Up: For water, $L_v = 2.256 \times 10^6$ J/kg and $c = 4.19 \times 10^3$ J/kg·K.

Solve: (a) $Q = + m(-L_v + c \, \Delta T)$

$$Q = + (25.0 \times 10^{-3} \text{ kg})(-2.256 \times 10^6 \text{ J/kg} + [4.19 \times 10^3 \text{ J/kg·K}][-66.0 \text{ C°}]) = -6.33 \times 10^4 \text{ J}$$

(b) $Q = mc \, \Delta T = (25.0 \times 10^{-3} \text{ kg})(4.19 \times 10^3 \text{ J/kg·K})(-66.0 \text{ C°}) = -6.91 \times 10^3$ J.

(c) The total heat released by the water that starts as steam is nearly a factor of ten larger than the heat released by water that starts at 100°C. Steam burns are much more severe than hot-water burns.

Reflect: For a given amount of material, the heat for a phase change is typically much more than the heat for a temperature change.

*14.41. Set Up: From Problem 14.40, $Q = 1.44 \times 10^6$ J and $m = 70$ kg. Convert the temperature change in C° to F° using that 9 F° = 5 C°.

Solve: (a) $Q = mc \, \Delta T$ so $\Delta T = \dfrac{Q}{mc} = \dfrac{1.44 \times 10^6 \text{ J}}{(70 \text{ kg})(3500 \text{ J/kg·C°})} = 5.9$ C°

(b) $\Delta T = (5.9°\text{C})\left(\dfrac{9 \text{ F°}}{5 \text{ C°}}\right) = 10.6°$F. $T = 98.6°\text{F} + 10.6 \text{ F°} = 109°$F. A fever this high can be lethal (heat stroke).

*14.43. Set Up: For water, $c = 4.19 \times 10^3$ J/kg·K.

Solve: (a) $Q_{\text{water}} = mc \, \Delta T = (1.00 \text{ kg})(4.19 \times 10^3 \text{ J/kg·K})(2.0 \text{ C°}) = +8.38 \times 10^3$ J.

$Q_{\text{metal}} = mc \, \Delta T = (0.500 \text{ kg})c(-78 \text{ C°})$. $\Sigma Q = 0$ says $(0.500 \text{ kg})c(-78 \text{ C°}) + 8.38 \times 10^3 \text{ J} = 0$.

$$c = \frac{8.38 \times 10^3 \text{ J}}{(0.500 \text{ kg})(78 \text{ C°})} = 215 \text{ J/kg·K.}$$

(b) Water has a much larger value of c so stores more heat for the same ΔT.

(c) If some of the heat went into the Styrofoam™ then more would have to come out of the metal and c would be found to be greater. The value calculated in (a) would be too small.

Reflect: The amount of heat that comes out of the metal when it cools equals the amount of heat that goes into the water to produce its temperature increase.

*14.47. Set Up: Since some ice remains, the ice and water from the melted ice remain at $0°C$. For silver, $c_s = 230 \text{ J/kg} \cdot \text{K}$. For water, $L_f = 3.34 \times 10^5 \text{ J/kg}$.

Solve: For the silver, $Q_s = m_s c_s \, \Delta T_s = (4.00 \text{ kg})(230 \text{ J/kg} \cdot \text{K})(-750 \text{ C°}) = -6.90 \times 10^5 \text{ J}$.

For the ice, $Q_i = +mL_f$, where m is the mass that melts.

$\Sigma Q = 0$ gives $m(3.34 \times 10^5 \text{ J/kg}) - 6.90 \times 10^5 \text{ J} = 0$ and $m = 2.07 \text{ kg}$.

Reflect: The heat that comes out of the ingot when it cools goes into the ice to produce a phase change.

*14.49. Set Up: The final amount of ice is less than the initial mass of water, so water remains and the final temperature is $0°C$. The ice added warms to $0°C$ and heat comes out of water to convert it to ice. Conservation of energy says $Q_i + Q_w = 0$, where Q_i and Q_w are the heat flows for the ice that is added and for the water that freezes. Let m_i be the mass of ice that is added and m_w is the mass of water that freezes. The mass of ice increases by 0.328 kg, so $m_i + m_w = 0.328 \text{ kg}$. For water, $L_f = 334 \times 10^3 \text{ J/kg}$ and for ice $c_i = 2100 \text{ J/kg} \cdot \text{K}$. Heat comes out of the water when it freezes, so $Q_w = -mL_f$

Solve: $Q_i + Q_w = 0$ gives $m_i c_i (15.0°C) + (-m_w L_f) = 0$, $m_w = 0.328 \text{ kg} - m_i$, so $m_i c_i (15.0°C) + (-0.328 \text{kg} + m_i)L_f = 0$.

$$m_i = \frac{(0.328 \text{ kg})L_f}{c_i(15.0°C) + L_f} = \frac{(0.328 \text{ kg})(334 \times 10^3 \text{ J/kg})}{(2100 \text{ J/kg} \cdot \text{K})(15.0 \text{ K}) + 334 \times 10^3 \text{ J/kg}} = 0.300 \text{ kg}.$$

0.300 kg of ice was added.

Reflect: The mass of water that froze when the ice at $-15.0°C$ was added was $0.778 \text{ kg} - 0.450 \text{ kg} - 0.300 \text{ kg} = 0.028 \text{ kg}$.

*14.51. Set Up: Apply Eq. (14.12) and solve for A. The area of each circular end of a cylinder is related to the diameter D by $A = \pi R^2 = \pi (D/2)^2$. For steel we have $k = 50.2 \text{ W/m} \cdot \text{K}$. The boiling water has $T = 100°C$, so $\Delta T = 300 \text{ K}$.

Solve: $\dfrac{Q}{t} = kA\dfrac{\Delta T}{L}$ and $150 \text{ J/s} = (50.2 \text{ W/m} \cdot \text{K})A\left(\dfrac{300 \text{ K}}{0.500 \text{ m}}\right)$. This gives $A = 4.98 \times 10^{-3} \text{ m}^2$, and $D = \sqrt{4A/\pi} = \sqrt{4(4.98 \times 10^{-3} \text{ m}^2)/\pi} = 8.0 \times 10^{-2} \text{ m} = 8.0 \text{ cm}$.

Reflect: H increases when A increases.

*14.55. Set Up: Assume the temperatures of the surfaces of the window are the outside and inside temperatures. Use the concept of thermal resistance. For part (b) use the fact that when insulating materials are in layers, the R values are additive. From Table 14.5 we obtain $k = 0.8 \text{ W/m} \cdot \text{K}$ for glass. $R = L/k$.

Solve: (a) For the glass, $R_{glass} = \dfrac{5.20 \times 10^{-3} \text{ m}}{0.8 \text{ W/m} \cdot \text{K}} = 6.50 \times 10^{-3} \text{ m}^2 \cdot \text{K/W}$.

$$H = \frac{A(T_H - T_C)}{R} = \frac{(1.40 \text{ m})(2.50 \text{ m})(39.5 \text{ K})}{6.50 \times 10^{-3} \text{ m}^2 \cdot \text{K/W}} = 2.1 \times 10^4 \text{ W}$$

(b) For the paper, $R_{paper} = \dfrac{0.750 \times 10^{-3} \text{ m}}{0.05 \text{ W/m} \cdot \text{K}} = 0.015 \text{ m}^2 \cdot \text{K/W}$. The total R is $R = R_{glass} + R_{paper} = 0.0215 \text{ m}^2 \cdot \text{K/W}$.

$$H = \frac{A(T_H - T_C)}{R} = \frac{(1.40 \text{ m})(2.50 \text{ m})(39.5 \text{ K})}{0.0215 \text{ m}^2 \cdot \text{K/W}} = 6.4 \times 10^3 \text{ W}.$$

Reflect: The layer of paper decreases the rate of heat loss by a factor of about 3.

*14.57. Set Up: Let the temperature of the fat-air boundary be T. A section of the two layers is sketched in the figure below. A Kelvin degree is the same size as a Celsius degree, so $W/m \cdot K$ and $W/m \cdot C°$ are equivalent units.

At steady state the heat current through each layer is the same, equal to 50 W. The area of each layer is $A = 4\pi r^2$, with $r = 0.75$ m.

```
                                  ——————————————— 2.7 °C
            L_air      air     k = 0.024 W/m · K
                                  ——————————————— T
            4.0 cm     fat     k = 0.20 W/m · K
                                  ——————————————— 31 °C
```

Solve: (a) Apply $H = kA\dfrac{T_H - T_C}{L}$ to the fat layer and solve for $T_C = T$. For the fat layer $T_H = 31°C$.

$$T = T_H - \frac{HL}{kA} = 31°C - \frac{(50 \text{ W})(4.0 \times 10^{-2} \text{ m})}{(0.20 \text{ W})(4\pi)(0.75 \text{ m})^2} = 31°C - 1.4°C = 29.6°C$$

(b) Apply $H = kA\dfrac{T_H - T_C}{L}$ to the air layer and solve for $L = L_{air}$. For the air layer $T_H = T = 29.6°C$ and $T_C = 2.7°C$.

$$L = \frac{kA(T_H - T_C)}{H} = \frac{(0.024 \text{ W})(4\pi)(0.75 \text{ m})^2(29.6°C - 2.7°C)}{50 \text{ W}} = 9.1 \text{ cm}$$

Reflect: The thermal conductivity of air is much lass than the thermal conductivity of fat, so the temperature gradient for the air must be much larger to achieve the same heat current. So, most of the temperature difference is across the air layer.

*14.61. Set Up: In the radiation equation the temperatures must be in kelvins; $T = 30°C = 303$ K and $T_s = 18°C = 291$ K. Call the basal metabolic rate BMR.

Solve: (a) $H_{net} = Ae\sigma(T^4 - T_s^{\ 4})$

$$H_{net} = (2.0 \text{ m}^2)(1.00)(5.67 \times 10^{-8} \text{ W/m}^2 \cdot \text{K}^4)([303 \text{ K}]^4 - [291 \text{ K}]^4) = 140 \text{ W}$$

(b) $(0.80)\text{BMR} = 140$ W, so BMR $= 180$ W

*14.65. Set Up: The heat energy generated by friction work equals the loss of kinetic energy. The heat Q_{ice} that goes into the ice is related to the mass m_{melts} of ice that melts by $Q_{ice} = m_{melts}L_f$. For ice, $L_f = 334 \times 10^3$ J/kg. The maximum amount of ice melts when the block loses all its kinetic energy and comes to rest.

Solve: (a) For $v_i = 15.0$ m/s and $v_f = 10.0$ m/s,

$$Q = K_i - K_f = \tfrac{1}{2}m(v_i^{\ 2} - v_f^{\ 2}) = \tfrac{1}{2}(8.5 \text{ kg})([15.0 \text{ m/s}]^2 - [10.0 \text{ m/s}]^2) = 531 \text{ J}.$$

$Q_{ice} = \tfrac{1}{2}Q = 265$ J. Then $Q_{ice} = mL_f$ gives

$$m = \frac{Q_{ice}}{L_f} = \frac{265 \text{ J}}{334 \times 10^3 \text{ J/kg}} = 7.93 \times 10^{-4} \text{ kg} = 0.793 \text{ g}$$

(b) For $v_f = 0$, $Q = K_i = 956$ J. $Q_{ice} = 478$ J and $m = \dfrac{Q_{ice}}{L_f} = 1.43$ g

Reflect: Since the ice is at $0°C$, any heat that goes into it causes a phase transition. 15.0 m/s is 34 mph. When brought to rest from this large initial speed only a very small fraction of the ice melts. Also note the irreversible nature of this; the water won't refreeze and release energy to get the block moving again.

*14.67. Set Up: Write the volume of the oceans as $V = Ad$, where d is the average depth and A is the average surface area of the oceans. Assume that only d changes when V changes, so $\Delta V = A \Delta d$.

Solve: $\Delta V = V_0 \beta \Delta T$, $\Delta V = A \Delta d$ and $V_0 = Ad_0$, with $d_0 = 4000$ m. This gives

$$\Delta d = d_0 \beta \Delta T = (4000 \text{ m})(0.207 \times 10^{-3} (\text{C}°)^{-1})(3.5 \text{ C}°) = 2.9 \text{ m.}$$

Reflect: This rise in sea level would have serious effects on coastal areas.

*14.69. Set Up: 1.00 L of water has a mass of 1.00 kg, so

$$9.46 \text{ L/min} = (9.46 \text{ L/min})(1.00 \text{ kg/L})(1 \text{ min}/60 \text{ s}) = 0.158 \text{ kg/s.}$$

For water, $c = 4190$ J/kg $\cdot$ C°.

Solve: $Q = mc \Delta T$ so

$$H = (Q/t) = (m/t)c \Delta T = (0.158 \text{ kg/s})(4190 \text{ J/kg} \cdot \text{C}°)(49°\text{C} - 10°\text{C}) = 2.6 \times 10^4 \text{ W} = 26 \text{ kW}$$

Reflect: The power requirement is large, the equivalent of 260 light bulbs that are 100-W, but this large power is needed only for short periods of time.

*14.73. Set Up: Let the cross sectional area of the cup be A. For the cup, $V_{0,\text{cup}} = A(10.0 \text{ cm})$. For the oil, $V_{0,\text{oil}} = A(9.9 \text{ cm})$. Oil starts to overflow when $\Delta V_{\text{oil}} = \Delta V_{\text{cup}} + (0.100 \text{ cm})A$.

Solve: $\Delta V_{\text{cup}} = V_{0,\text{cup}} \beta_{\text{glass}} \Delta T$. $\Delta V_{\text{oil}} = V_{0,\text{oil}} \beta_{\text{oil}} \Delta T$.

$(9.9 \text{ cm})A \beta_{\text{oil}} \Delta T = (10.0 \text{ cm})A \beta_{\text{glass}} \Delta T + (0.100 \text{ cm})A.$

$$\Delta T = \frac{0.100 \text{ cm}}{(9.9 \text{ cm})\beta_{\text{oil}} - (10.0 \text{ cm})\beta_{\text{glass}}}$$

$$\Delta T = \frac{0.100 \text{ cm}}{(9.9 \text{ cm})(6.8 \times 10^{-4} (\text{C}°)^{-1}) - (10.0 \text{ cm})(2.7 \times 10^{-5} (\text{C}°)^{-1})} = 15.5 \text{ C}°.$$

$$T_f = T_i + \Delta T = 37.5 °\text{C.}$$

Reflect: The olive oil expands more than the capacity of the cup does because it has a much larger coefficient of volume expansion.

*14.77. Set Up: Use $Q = Mc\Delta T$ to find Q for a temperature rise from $34.0°$C to $40.0°$C. Set this equal to $Q = mL_v$ and solve for m, where m is the mass of water the camel would have to drink.

$c = 3480$ J/kg $\cdot$ K and $L_v = 2.42 \times 10^6$ J/kg. For water, 1.00 kg has a volume 1.00 L.

$M = 400$ kg is the mass of the camel.

Solve: The mass of water that the camel saves is

$$m = \frac{Mc\Delta T}{L_v} = \frac{(400 \text{ kg})(3480 \text{ J/kg} \cdot \text{K})(6.0 \text{ K})}{(2.42 \times 10^6 \text{ J/kg})} = 3.45 \text{ kg} \text{ which is a volume of } 3.45 \text{ L.}$$

Reflect: This is nearly a gallon of water, so it is an appreciable savings.

*14.81. Set Up: Problem 14.83 calculates that the net rate of heat input to the person is 1130 W. 9 F° = 5 C°.

Solve: (a) $Q = Pt = (1130 \text{ W})(1800 \text{ s}) = 2.03 \times 10^6 \text{ J}$

$Q = mc \Delta T$ so

$$\Delta T = \frac{Q}{mc} = \frac{2.03 \times 10^6 \text{ J}}{(68 \text{ kg})(3480 \text{ J/kg} \cdot \text{C}°)} = 8.6 \text{ C}°$$

(b) $\Delta T = (8.6 \text{ C}°)(9 \text{ F}°/5 \text{ C}°) = 15.5 \text{ F}°$. $T = 98.6°\text{F} + 15.5 \text{ F}° = 114°\text{F}$

This body temperature is lethal.

*14.83. Set Up: Apply $H = kA\dfrac{\Delta T}{L}$ and solve for k. H equals the power input required to maintain a constant interior temperature

Solve: $k = H\dfrac{L}{A\Delta T} = (180 \text{ W})\dfrac{(3.9 \times 10^{-2} \text{ m})}{(2.18 \text{ m}^2)(65.0 \text{ K})} = 5.0 \times 10^{-2} \text{ W/m} \cdot \text{K}.$

Reflect: Our result is consistent with the values for insulating solids in Table 14.5.

*14.87. Set Up: Use Eq. (14.14) to find the net heat current into the can due to radiation. Use $Q = Ht$ to find the heat that goes into the liquid helium, set this equal to mL and solve for the mass m of helium that changes phase.

Solve: Calculate the net rate of radiation of heat from the can. $H_{\text{net}} = Ae\sigma(T^4 - T_s^4)$. The surface area of the cylindrical can is $A = 2\pi rh + 2\pi r^2$ (see the figure below).

$$A = 2\pi r(h + r) = 2\pi(0.045 \text{ m})(0.250 \text{ m} + 0.045 \text{ m}) = 0.08341 \text{ m}^2.$$

$$H_{\text{net}} = (0.08341 \text{ m}^2)(0.200)(5.67 \times 10^{-8} \text{ W/m}^2 \cdot \text{K}^4)((4.22 \text{ K})^4 - (77.3 \text{ K})^4)$$

$H_{\text{net}} = -0.0338 \text{ W}$ (the minus sign says that the net heat current is into the can). The heat that is put into the can by radiation in one hour is $Q = -(H_{\text{net}})t = (0.0338 \text{ W})(3600 \text{ s}) = 121.7 \text{ J}$. This heat boils a mass m of helium according to the equation $Q = mL_f$, so

$$m = \frac{Q}{L_f} = \frac{121.7 \text{ J}}{2.09 \times 10^4 \text{ J/kg}} = 5.82 \times 10^{-3} \text{ kg} = 5.82 \text{ g}.$$

Reflect: In the expression for the net heat current into the can the temperature of the surroundings is raised to the fourth power. The rate at which the helium boils away increases by about a factor of $(293/77)^4 = 210$ if the walls surrounding the can are at room temperature rather than at the temperature of the liquid nitrogen.

THERMAL PROPERTIES OF MATTER

Problems 1, 3, 7, 11, 15, 17, 19, 21, 25, 27, 31, 33, 37, 41, 45, 47, 49, 53, 55, 59, 61, 65, 67, 73, 75, 79, 81, 85, 87, 89

Solutions to Problems

*15.1. Set Up: $pV = nRT$.

Solve: nRT is constant so $p_1 V_1 = p_2 V_2$. $p_2 = p_1 \left(\dfrac{V_1}{V_2} \right) = (3.40 \text{ atm}) \left(\dfrac{0.110 \text{ m}^3}{0.390 \text{ m}^3} \right) = 0.959 \text{ atm}$.

*15.3. Set Up: $pV = nRT$. $T_1 = 20.0°\text{C} = 293 \text{ K}$.

Solve: **(a)** n, R, and V are constant. $\dfrac{p}{T} = \dfrac{nR}{V} = \text{constant}$. $\dfrac{p_1}{T_1} = \dfrac{p_2}{T_2}$.

$$T_2 = T_1 \left(\frac{p_2}{p_1} \right) = (293 \text{ K}) \left(\frac{1.00 \text{ atm}}{3.00 \text{ atm}} \right) = 97.7 \text{ K} = -175°\text{C}.$$

(b) $p_2 = 1.00 \text{ atm}$, $V_2 = 3.00 \text{ L}$. $p_3 = 3.00 \text{ atm}$. n, R, and T are constant so $pV = nRT = \text{constant}$. $p_2 V_2 = p_3 V_3$.

$$V_3 = V_2 \left(\frac{p_2}{p_3} \right) = (3.00 \text{ L}) \left(\frac{1.00 \text{ atm}}{3.00 \text{ atm}} \right) = 1.00 \text{ L}.$$

Reflect: The final volume is one-third the initial volume. The initial and final pressures are the same, but the final temperature is one-third the initial temperature.

*15.7. Set Up: $pV = nRT$. $T_1 = 300 \text{ K}$, $T_2 = 430 \text{ K}$.

Solve: **(a)** n, R are constant so $\dfrac{pV}{T} = nR = \text{constant}$. $\dfrac{p_1 V_1}{T_1} = \dfrac{p_2 V_2}{T_2}$.

$$p_2 = p_1 \left(\frac{V_1}{V_2} \right) \left(\frac{T_2}{T_1} \right) = (1.50 \times 10^5 \text{ Pa}) \left(\frac{0.750 \text{ m}^3}{0.480 \text{ m}^3} \right) \left(\frac{430 \text{ K}}{300 \text{ K}} \right) = 3.36 \times 10^5 \text{ Pa}.$$

*15.11. Set Up: $pV = nRT$. $T_1 = 300 \text{ K}$. $p_1 = 1.01 \times 10^5 \text{ Pa}$.

$$p_2 = 1.01 \times 10^5 \text{ Pa} + 2.72 \times 10^6 \text{ Pa} = 2.82 \times 10^6 \text{ Pa}.$$

Solve: n, R are constant so $\dfrac{pV}{T} = nR = \text{constant}$. $\dfrac{p_1 V_1}{T_1} = \dfrac{p_2 V_2}{T_2}$ and

$$T_2 = T_1 \left(\frac{p_2}{p_1} \right) \left(\frac{V_2}{V_1} \right) = (300 \text{ K}) \left(\frac{2.82 \times 10^6 \text{ Pa}}{1.01 \times 10^5 \text{ Pa}} \right) \left(\frac{46.2 \text{ cm}^3}{499 \text{ cm}^3} \right) = 775.5 \text{ K} = 502°\text{C}.$$

Reflect: Even though the pressures enter in a ratio, we must use absolute pressures. The ratio of the absolute pressures is different from the ratio of gauge pressures.

*15.15. Set Up: The volume per mole for water at its critical point is $(V/\text{mole})_c = 56.0 \times 10^{-6} \text{ m}^3/\text{mole}$. 1 mole of water has mass $18.0 \times 10^{-3} \text{ kg}$.

Solve: $V = \frac{m}{\rho} = \frac{18.0 \times 10^{-3} \text{ kg}}{998 \text{ kg/m}^3} = 1.80 \times 10^{-5} \text{ m}^3 = 18.0 \text{ cm}^3$. This is about one-third of the volume of 1 mole at the critical point.

*15.17. Set Up: Figure 15.7 in the textbook shows that there is no liquid phase below the triple point pressure. Table 15.1 gives the triple point pressure to be 610 Pa for water and 5.17×10^5 Pa for CO_2.

Solve: The atmospheric pressure is below the triple point pressure of water, and there can be no liquid water on Mars. The same holds true for CO_2.

Reflect: On earth $p_{\text{atm}} = 1 \times 10^5$ Pa, so on the surface of the earth there can be liquid water but not liquid CO_2.

*15.19. Set Up: The density of water is $1.00 \times 10^3 \text{ kg/m}^3$. $1.00 \text{ L} = 1.00 \times 10^{-3} \text{ m}^3$. $N_A = 6.022 \times 10^{23}$ molecules/mol. For water, $M = 18 \times 10^{-3}$ kg/mol.

Solve: $m = \rho V = (1.00 \times 10^3 \text{ kg/m}^3)(1.00 \times 10^{-3} \text{ m}^3) = 1.00 \text{ kg}$.

$$n = \frac{m}{M} = \frac{1.00 \text{ kg}}{18 \times 10^{-3} \text{ kg/mol}} = 55.6 \text{ mol}.$$

$$N = nN_A = (55.6 \text{ mol})(6.022 \times 10^{23} \text{ molecules/mol}) = 3.35 \times 10^{25} \text{ molecules}.$$

Reflect: Note that we converted M to kg/mol.

*15.21. Set Up: $pV = nRT = \frac{N}{N_A} RT = \frac{m_{\text{tot}}}{M} RT$. We known that $V_A = V_B$ and that $T_A > T_B$.

Solve: **(a)** $p = nRT/V$; we don't know n for each box, so either pressure could be higher.

(b) $pV = \left(\frac{N}{N_A} \right) RT$ so $N = \frac{pVN_A}{RT}$, where N_A is Avogadro's number. We don't know how the pressures compare, so either N could be larger.

(c) $pV = (m_{\text{tot}}/M)RT$. We don't know the mass of the gas in each box, so they could contain the same gas or different gases.

(d) $\frac{1}{2} m(v^2)_{\text{av}} = \frac{3}{2} kT$. $T_A > T_B$ and the average kinetic energy per molecule depends only on T, so the statement **must** be true.

(e) $v_{\text{rms}} = \sqrt{3kT/m}$. We don't know anything about the masses of the atoms of the gas in each box, so either set of molecules could have a larger v_{rms}.

Reflect: Only statement (d) must be true. We need more information in order to determine whether the other statements are true or false.

*15.25. Set Up: At STP, $T = 273 \text{ K}$, $p = 1.01 \times 10^5 \text{ Pa}$. $N = 6 \times 10^9$ molecules.

Solve: $pV = nRT$.

$$V = \frac{NkT}{p} = \frac{(6 \times 10^9 \text{ molecules})(1.381 \times 10^{-23} \text{ J/molecule} \cdot \text{K})(273 \text{ K})}{1.01 \times 10^5 \text{ Pa}} = 2.24 \times 10^{-16} \text{ m}^3.$$

$L^3 = V$ so $L = V^{1/3} = 6.1 \times 10^{-6} \text{ m}$.

*15.27. Set Up: $v_{rms} = \sqrt{3RT/M}$. $T = 20°C = 293$ K. $M_{H_2} = 2.02 \times 10^{-3}$ kg/mol. $M_{O_2} = 32.0 \times 10^{-3}$ kg/mol.

$M_{N_2} = 28.0 \times 10^{-3}$ kg/mol

Solve: (a) $v_{rms} = \sqrt{\dfrac{3(8.315 \text{ J/mol} \cdot \text{K})(293 \text{ K})}{2.02 \times 10^{-3} \text{ kg/mol}}} = 1.90 \times 10^3$ m/s $= 1.90$ km/s

(b) $v_{rms} < 11$ km/s so no, the average H_2 molecule is not moving fast enough to escape.

(c) O_2: $v_{rms, O_2} = v_{rms, H_2}\sqrt{\dfrac{M_{H_2}}{M_{O_2}}} = (1.90 \text{ km/s})\sqrt{\dfrac{2.02 \times 10^{-3} \text{ kg/mol}}{32.0 \times 10^{-3} \text{ kg/mol}}} = 0.477$ km/s

N_2: $v_{rms, N_2} = v_{rms, H_2}\sqrt{\dfrac{M_{H_2}}{M_{N_2}}} = (1.90 \text{ km/s})\sqrt{\dfrac{2.02 \times 10^{-3} \text{ kg/mol}}{28.0 \times 10^{-3} \text{ kg/mol}}} = 0.510$ km/s

v_{rms} for H_2 is about four times larger.

(d) Figure 15.12 in Chapter 15 illustrates that some molecules move faster than v_{rms}, so some have a speed greater than the escape speed and can escape. Since the rms speed for H_2 is greater than that of O_2 and N_2, a higher percent of H_2 molecules are moving fast enough to escape.

Reflect: At a given temperature all three species of molecules have the same average kinetic energy. To achieve this, the lighter H_2 molecules have a larger v_{rms}.

*15.31. Set Up: $20.9\% - 16.3\% = 4.6\%$ of the volume of air breathed in is O_2 that is absorbed by the body.

$T = 20°C = 293$ K. $p = 1$ atm $= 1.013 \times 10^5$ Pa. 1.0 L $= 1.0 \times 10^{-3}$ m^3.

Solve: (a) In one minute the person must absorb

$$\dfrac{14.5 \text{ L/hr}}{60 \text{ min/hr}} = 0.242 \text{ L/min of } O_2.$$

Since the volume of O_2 absorbed is 4.6% of the air breathed in, the person must have breathed in

$$\dfrac{0.242 \text{ L/min}}{0.046} = 5.26 \text{ L/m of air.}$$

The number of breaths per minute is

$$\dfrac{5.26 \text{ L/min}}{0.50 \text{ L/breath}} = 10.5 \text{ breaths/min.}$$

(b) With each breath the person inhales $(0.209)(0.50 \text{ L}) = 0.1045$ L of O_2. $pV = NkT$ gives the number of O_2 molecules in this volume to be

$$N = \dfrac{pV}{kT} = \dfrac{(1.013 \times 10^5 \text{ Pa})(0.1045 \times 10^{-3} \text{ m}^3)}{(1.381 \times 10^{-23} \text{ J/molecule} \cdot \text{K})(293 \text{ K})} = 2.62 \times 10^{21} \text{ molecules}$$

Reflect: $V = \dfrac{NkT}{p}$ says that the volume of O_2 is directly proportional to the number of O_2 molecules. About 10 breaths per minute is what we observe for a healthy, resting person; this result makes sense.

*15.33. Set Up: The rigid container means the process occurs at constant volume. For a monatomic ideal gas $C_V = \frac{3}{2}R$ and for a diatomic ideal gas $C_V = \frac{5}{2}R$.

Solve: (a) $Q = nC_V \Delta T$ so

$$\Delta T = \dfrac{Q}{nC_V} = \dfrac{1850 \text{ J}}{(2.25 \text{ mol})\left(\frac{5}{2}\right)(8.315 \text{ J/mol} \cdot \text{K})} = 39.6 \text{ C}°.$$

$T_f = T_i + \Delta T = 49.6°C$

(b) $Q = nC_V \, \Delta T = (2.25 \text{ mol})\left(\frac{3}{2}\right)(8.315 \text{ J/mol} \cdot \text{K})(39.6 \text{ C}°) = 1110 \text{ J}$

(c) V is constant. Then $pV = nRT$ says p increases when T increases. The process for either gas is sketched in the figure below.

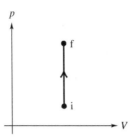

Reflect: More heat is required for the diatomic gas because not all of the energy added by Q goes into the translational kinetic energy that determines the temperature. Some of the energy that flows into the gas goes into the rotational kinetic energy of the diatomic molecule. The answer in (b) is smaller than 1850 J by the ratio

$$\frac{\frac{3}{2}R}{\frac{5}{2}R} = \frac{3}{5}.$$

*15.37. Set Up: The volume is constant.
Solve: (a) The pV diagram is given in the figure below.

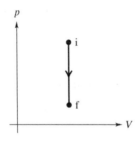

(b) Since $\Delta V = 0$, $W = 0$.
Reflect: For any constant volume process the work done is zero.

*15.41. Set Up: For a constant volume process, $W = 0$. For a constant pressure process, $W = p\,\Delta V$. 1 atm $= 1.013 \times 10^5$ Pa and 1 L $= 10^{-3}$ m^3. The work done by the system in a process is positive when the volume increases and negative when the volume decreases.

Solve: (a) Find W for each process in the cycle.

$1 \to 2$: $W = p\,\Delta V = (2.5 \text{ atm})(1.013 \times 10^5 \text{ Pa/atm})(8 \text{ L} - 2 \text{ L})(10^{-3} \text{ m}^3/1 \text{ L}) = 1.5 \times 10^3 \text{ J}$

$2 \to 3$: $W = 0$ since $\Delta V = 0$

$3 \to 4$: $W = p\,\Delta V = (0.5 \text{ atm})(1.013 \times 10^5 \text{ Pa/atm})(2 \text{ L} - 8 \text{ L})(10^{-3} \text{ m}^3/1 \text{ L}) = -3.0 \times 10^2 \text{ J}$

$4 \to 1$: $W = 0$ since $\Delta V = 0$

$W_{\text{cycle}} = 1.5 \times 10^3 \text{ J} + 0 + (-3.0 \times 10^2 \text{ J}) + 0 = 1.2 \times 10^3 \text{ J}$

The area enclosed by the cycle is

$$(2.0 \text{ atm})(6.0 \text{ L}) = (12.0 \text{ L} \cdot \text{atm})(10^{-3} \text{ m}^3/\text{L})(1.013 \times 10^5 \text{ Pa/atm}) = 1.2 \times 10^3 \text{ J}.$$

The total work done does equal the area enclosed by the cycle.

(b) For $1 \rightarrow 4 \rightarrow 3 \rightarrow 2 \rightarrow 1$, $W_{cycle} = -1.5 \times 10^3$ J $+ 3.0 \times 10^2$ J $= -1.2 \times 10^3$ J. The negative work done for $2 \rightarrow 1$ is greater in magnitude than the positive work done in $4 \rightarrow 3$ and the total work for the cycle has the opposite sign from what it had in (a).

(c) In the pV-diagram the cycle $3 \rightarrow 4 \rightarrow 2 \rightarrow 3$ is a triangle with area $\frac{1}{2}(2.0 \text{ atm})(6.0 \text{ L}) = 6.0 \text{ L} \cdot \text{atm} = 6.0 \times 10^2$ J. The work done is $\frac{1}{2}$ that done in the cycle of part (a). The positive work done in process $4 \rightarrow 2$ is larger in magnitude than the negative work done in process $3 \rightarrow 4$ so the net work done in the cycle is positive.

Reflect: For the cycles in parts (a) and (b), the initial and final states are the same but the work done is different. This illustrates the general result that the work done in a process depends not only on the initial and final states but also on the path.

*15.45. Set Up: $\Delta U = Q - W$. $Q > 0$ if heat flows into the gas. For a constant pressure process, $W = p \, \Delta V$.

Solve: (a) $W = p \, \Delta V = (2.30 \times 10^5 \text{ Pa})(1.20 \text{ m}^3 - 1.70 \text{ m}^3) = -1.15 \times 10^5$ J

(b) $Q = \Delta U + W = -1.40 \times 10^5$ J $+ (-1.15 \times 10^5 \text{ J}) = -2.55 \times 10^5$ J. $Q < 0$ so this amount of heat flows out of the gas.

Reflect: $\Delta V < 0$ and $W < 0$. The work done on the gas adds energy to it, but more energy flows out as heat than is added by the work and the internal energy decreases.

*15.47. Set Up: $\Delta U = Q - W$. ΔU is path independent. $Q > 0$ when the system absorbs heat.

Solve: (a) Use path acb to find $\Delta U = U_b - U_a$. $U_b - U_a = Q - W = 90.0$ J $- 60.0$ J $= 30.0$ J. For path adb, $\Delta U = 30.0$ J and $Q = \Delta U + W = 30.0$ J $+ 15.0$ J $= 45.0$ J.

(b) $\Delta U = U_a - U_b = -30.0$ J. For process $b \rightarrow a$, $\Delta V < 0$ and $W = -35.0$ J. $Q = \Delta U + W = -30.0$ J $+ (-35.0 \text{ J}) = -65.0$ J. The system liberated 65.0 J of heat.

(c) For process $a \rightarrow d$: $\Delta U = U_d - U_a = 8.0$ J. $W_{d \rightarrow b} = 0$ so $W_{a \rightarrow d} = W_{adb} = 15.0$ J.

$Q = \Delta U + W = 8.0$ J $+ 15.0$ J $= 23.0$ J.

For process $d \rightarrow b$: $\Delta U = U_b - U_d = 30.0$ J $- 8.0$ J $= 22.0$ J and $W = 0$. $Q = \Delta U + W = 22.0$ J.

Reflect: Both Q and W depend on the path.

*15.49. Set Up: For an ideal gas, $\Delta U = nC_V \, \Delta T$ so for an isothermal process $\Delta U = 0$. Use $W = nRT \ln (V_2/V_1)$ to calculate W for the isothermal process and then apply $\Delta U = Q - W$.

Solve: (a) $W = nRT \ln (V_2/V_1) = (1.75 \text{ mol})(8.315 \text{ J/mol} \cdot \text{K})(273 \text{ K}) \ln (1.35 \text{ L}/4.20 \text{ L}) = -4510$ J

(b) $\Delta U = Q - W$ and $\Delta U = 0$ says $Q = W = -4510$ J. 4510 J of heat comes out of the gas.

Reflect: When the gas is compressed, heat must be removed to keep the temperature constant.

*15.53. Set Up: The pV diagram shows that in the process the volume decreases while the pressure is constant. $1 \text{ L} = 10^{-3} \text{ m}^3$ and $1 \text{ atm} = 1.013 \times 10^5$ Pa

Solve: (a) $pV = nRT$. n, R, and p are constant so $\dfrac{V}{T} = \dfrac{nR}{p} = $ constant. $\dfrac{V_a}{T_a} = \dfrac{V_b}{T_b}$.

$$V_b = V_a \left(\frac{T_b}{T_a} \right) = (0.500 \text{ L}) \left(\frac{T_a/4}{T_a} \right) = 0.125 \text{ L}$$

(b) For a constant pressure process, $W = p \, \Delta V = (1.50 \text{ atm})(0.125 \text{ L} - 0.500 \text{ L})$ and

$$W = (-0.5625 \text{ L} \cdot \text{atm}) \left(\frac{10^{-3} \text{ m}^3}{1 \text{ L}} \right) \left(\frac{1.013 \times 10^5 \text{ Pa}}{1 \text{ atm}} \right) = -57.0 \text{ J}$$

W is negative since the volume decreases. Since W is negative, work is done on the gas.

(c) For an ideal gas, $U = nCT$ so U decreases when T decreases. The internal energy of the gas decreases because the temperature decreases.

(d) For a constant pressure process, $Q = nC_p \, \Delta T$. T decreases so ΔT is negative and Q is therefore negative. Negative Q means heat leaves the gas.

Reflect: $W = nR \, \Delta T$ and $Q = nC_p \, \Delta T$. $C_p > R$, so more energy leaves as heat than is added by work done on the gas, and the internal energy of the gas decreases.

*15.55. Set Up: Use $pV = nRT$ to calculate T_c/T_a. Calculate ΔU and W and use $\Delta U = Q - W$ to obtain Q. For path ac, the work done is the area under the line representing the process in the pV-diagram.

Solve: (a) $\dfrac{T_c}{T_a} = \dfrac{p_c V_c}{p_a V_a} = \dfrac{(1.0 \times 10^5 \text{ J})(0.060 \text{ m}^3)}{(3.0 \times 10^5 \text{ J})(0.020 \text{ m}^3)} = 1.00$. $T_c = T_a$.

(b) Since $T_c = T_a$, $\Delta U = 0$ for process abc. For ab, $\Delta V = 0$ and $W_{ab} = 0$. For bc, p is constant and $W_{bc} = p \Delta V = (1.0 \times 10^5 \text{ Pa})(0.040 \text{ m}^3) = 4.0 \times 10^3$ J. Therefore, $W_{abc} = +4.0 \times 10^3$ J. Since $\Delta U = 0$, $Q = W = +4.0 \times 10^3$ J. 4.0×10^3 J of heat flows into the gas during process abc.

(c) $W = \frac{1}{2}(3.0 \times 10^5 \text{ Pa} + 1.0 \times 10^5 \text{ Pa})(0.040 \text{ m}^3) = +8.0 \times 10^3$ J. $Q_{ac} = W_{ac} = +8.0 \times 10^3$ J.

Reflect: The work done is path dependent and is greater for process ac than for process abc, even though the initial and final states are the same.

*15.59. Set Up: O_2 is diatomic and if treated as ideal it has $C_V = \frac{5}{2}R$ and $C_p = \frac{7}{2}R$. $pV = nRT$. 1 atm $= 1.013 \times 10^5$ Pa. $Q = nC_V \, \Delta T$ for $\Delta V = 0$ and $Q = nC_p \, \Delta T$ for $\Delta p = 0$.

Solve: (a) *point a:* $T = \dfrac{pV}{RT} = \dfrac{(0.60 \text{ atm})(1.013 \times 10^5 \text{ Pa/atm})(0.10 \text{ m}^3)}{(1.10 \text{ mol})(8.315 \text{ J/mol} \cdot \text{K})} = 665$ K

point b: $T = \dfrac{pV}{RT} = 4(665 \text{ K}) = 2660$ K

point c: $T = \dfrac{pV}{RT} = \dfrac{2660 \text{ K}}{3} = 887$ K;

point d: $T = \dfrac{pV}{RT} = \dfrac{887 \text{ K}}{4} = 222$ K

(b) and **(c)** (i) *ab*

$Q = nC_p \, \Delta T = n\left(\frac{7}{2}R\right) \Delta T = (1.10 \text{ mol})\left(\frac{7}{2}\right)(8.315 \text{ J/mol} \cdot \text{K})(2660 \text{ K} - 665 \text{ K}) = 6.39 \times 10^4$ J

$Q > 0$ so heat enters the gas.

(ii) *bc*

$Q = nC_V \, \Delta T = n\left(\frac{5}{2}R\right) \Delta T = (1.10 \text{ mol})\left(\frac{5}{2}\right)(8.315 \text{ J/mol} \cdot \text{K})(887 \text{ K} - 2660 \text{ K}) = -4.05 \times 10^4$ J

$Q < 0$ so heat leaves the gas.

(iii) *cd*

$Q = nC_p \, \Delta T = n\left(\frac{7}{2}R\right) \Delta T = (1.10 \text{ mol})\left(\frac{7}{2}\right)(8.315 \text{ J/mol} \cdot \text{K})(222 \text{ K} - 887 \text{ K}) = -2.13 \times 10^4$ J

$Q < 0$ so heat leaves the gas.

(iv) *da*

$Q = nC_V \, \Delta T = n\left(\frac{5}{2}R\right) \Delta T = (1.10 \text{ mol})\left(\frac{5}{2}\right)(8.315 \text{ J/mol} \cdot \text{K})(665 \text{ K} - 222 \text{ K}) = 1.01 \times 10^4$ J

$Q > 0$ so heat enters the gas.

Reflect: The net heat flow for the complete cycle is $Q_{ab} + Q_{bc} + Q_{cd} + Q_{da} = 1.22 \times 10^4$ J. The work done in the cycle is positive and equal to the area enclosed by the cycle, so $W = (0.40 \text{ atm})(1.013 \times 10^5 \text{ Pa/atm})(0.30 \text{ m}^3) = 1.22 \times 10^4$ J. For a cycle, where the system returns to the initial state, $\Delta U = 0$ so $Q = W$.

*15.61. Set Up: $\Delta U = Q - W$. Apply $Q = nC_p\Delta T$ to calculate C_p. Apply $\Delta U = nC_V\Delta T$ to calculate C_V. $\gamma = C_p/C_V$. $\Delta T = 15.0°C = 15.0$ K. Since heat is added, $Q = +970$ J.

Solve: (a) $\Delta U = Q - W = +970$ J $- 223$ J $= 747$ J

(b) $C_p = \dfrac{Q}{n\Delta T} = \dfrac{970 \text{ J}}{(1.75 \text{ mol})(15.0 \text{ K})} = 37.0$ J/mol·K. $C_V = \dfrac{\Delta U}{n\Delta T} = \dfrac{747 \text{ J}}{(1.75 \text{ mol})(15.0 \text{ K})} = 28.5$ J/mol·K.

$\gamma = \dfrac{C_p}{C_V} = \dfrac{37.0 \text{ J/mol·K}}{28.5 \text{ J/mol·K}} = 1.30$

Reflect: The value of γ we calculated is similar to the values given in Tables 15.4 for polyatomic gases.

*15.65. Set Up: $p = 9.00 \times 10^{-14}$ atm $= 9.12 \times 10^{-9}$ Pa. $V = 1.00$ cm$^3 = 1.00 \times 10^{-6}$ m^3.

Solve: $pV = NkT$.

$$N = \frac{pV}{kT} = \frac{(9.12 \times 10^{-9} \text{ Pa})(1.00 \times 10^{-6} \text{ m}^3)}{(1.38 \times 10^{-23} \text{ J/molecule·K})(300 \text{ K})} = 2.20 \times 10^6 \text{ molecules.}$$

*15.67. Set Up: The pressure difference between two points in a fluid is $\Delta p = \rho g h$, where h is the difference in height of two points.

Solve: (a) $\Delta p = \rho g h = (1.2 \text{ kg/m}^3)(9.80 \text{ m/s}^2)(1000 \text{ m}) = 1.18 \times 10^4$ Pa

(b) At the bottom of the mountain, $p = 1.013 \times 10^5$ Pa. At the top, $p = 8.95 \times 10^4$ Pa. $pV = nRT =$ constant so $p_b V_b = p_t V_t$ and

$$V_t = V_b \left(\frac{p_b}{p_t} \right) = (0.50 \text{ L}) \left(\frac{1.013 \times 10^5 \text{ Pa}}{8.95 \times 10^4 \text{ Pa}} \right) = 0.566 \text{ L.}$$

Reflect: The pressure variation with altitude is affected by changes in air density and temperature and we have neglected those effects. The pressure decreases with altitude and the volume increases. You may have noticed this effect: bags of potato chips "puff-up" when taken to the top of a mountain.

*15.73. Set Up: For helium, $C_V = 12.47$ J/mol·K and $C_p = 20.78$ J/mol·K.

Solve: (a) $\Delta T = 0$. $\Delta U = 0$ and $Q = W = 300$ J.

(b) $Q = 0$. $\Delta U = -W = -300$ J.

(c) $\Delta p = 0$. $W = p\Delta V = nR\Delta T$.

$$Q = nC_p\Delta T = \frac{C_p}{R}W = \left(\frac{20.78 \text{ J/mol·K}}{8.315 \text{ J/mol·K}} \right)(300 \text{ J}) = 750 \text{ J.}$$

$\Delta U = Q - W = 750$ J $- 300$ J $= 450$ J. This is also $\left(\dfrac{C_V}{R} \right)W$.

Reflect: In any process of an ideal gas, $\Delta U = nC_V\Delta T$. ΔU is different for each of these processes because they have different values of ΔT.

*15.75. Set Up: The mass of one molecule is the molar mass, M, divided by the number of molecules in a mole, N_A. The average translational kinetic energy of a single molecule is $\frac{1}{2}m(v^2)_{av} = \frac{3}{2}kT$. Use $pV = NkT$ to calculate N, the number of molecules. $k = 1.381 \times 10^{-23}$ J/molecule·K. $M = 28.0 \times 10^{-3}$ kg/mol. $T = 295.15$ K. The volume of the balloon is $V = \frac{4}{3}\pi(0.250 \text{ m})^3 = 0.0654$ m^3. $p = 1.25$ atm $= 1.27 \times 10^5$ Pa.

Solve: (a) $m = \dfrac{M}{N_A} = \dfrac{28.0 \times 10^{-3} \text{ kg/mol}}{6.022 \times 10^{23} \text{ molecules/mol}} = 4.65 \times 10^{-26}$ kg

(b) $\frac{1}{2}m(v^2)_{av} = \frac{3}{2}kT = \frac{3}{2}(1.381 \times 10^{-23} \text{ J/molecule·K})(295.15 \text{ K}) = 6.11 \times 10^{-21}$ J

(c) $N = \dfrac{pV}{kT} = \dfrac{(1.27 \times 10^5 \text{ Pa})(0.0654 \text{ m}^3)}{(1.381 \times 10^{-23} \text{ J/molecule} \cdot \text{K})(295.15 \text{ K})} = 2.04 \times 10^{24}$ molecules

(d) The total average translational kinetic energy is

$N\left(\tfrac{1}{2}m(v^2)_{av}\right) = (2.04 \times 10^{24} \text{ molecules}) \ (6.11 \times 10^{-21} \text{ J/molecule}) = 1.25 \times 10^4$ J.

Reflect: The number of moles is $n = \dfrac{N}{N_A} = \dfrac{2.04 \times 10^{24} \text{ molecules}}{6.022 \times 10^{23} \text{ molecules/mol}} = 3.39$ mol. We have $K_{tr} = \tfrac{3}{2}nRT = $

$\tfrac{3}{2}(3.39 \text{ mol})(8.314 \text{ J/mol} \cdot \text{K})(295.15 \text{ K}) = 1.25 \times 10^4$ J, which agrees with our results in part (d).

*15.79. Set Up: ab is at constant volume, ca is at constant pressure and bc is at constant temperature. For $\Delta T = 0$, $\Delta U = 0$ and $Q = W = nRT \ln (V_c/V_b)$. For ideal H_2 (diatomic), $C_V = \tfrac{5}{2}R$ and $C_p = \tfrac{7}{2}R$. $\Delta U = nC_V \Delta T$ for any process of an ideal gas.

Solve: **(a)** $T_b = T_c$. For states b and c, $pV = nRT = $ constant so $p_b V_b = p_c V_c$ and

$$V_c = V_b \left(\frac{p_b}{p_c}\right) = (0.20 \text{ L})\left(\frac{2.0 \text{ atm}}{0.50 \text{ atm}}\right) = 0.80 \text{ L}$$

(b) $T_a = \dfrac{p_a V_a}{nR} = \dfrac{(0.50 \text{ atm})(1.013 \times 10^5 \text{ Pa/atm})(0.20 \times 10^{-3} \text{ m}^3)}{(0.0040 \text{ mol})(8.315 \text{ J/mol} \cdot \text{K})} = 305$ K

$V_a = V_b$ so for states a and b, $\dfrac{T}{p} = \dfrac{V}{nR} = $ constant so $\dfrac{T_a}{p_a} = \dfrac{T_b}{p_b}$.

$$T_b = T_c = T_a \left(\frac{p_b}{p_a}\right) = (305 \text{ K})\left(\frac{2.0 \text{ atm}}{0.50 \text{ atm}}\right) = 1220 \text{ K}; \ T_c = 1220 \text{ K}$$

(c) ab: $Q = nC_V \Delta T = n\left(\tfrac{5}{2}R\right)\Delta T = (0.0040 \text{ mol})\left(\tfrac{5}{2}\right)(8.315 \text{ J/mol} \cdot \text{K})(1220 \text{ K} - 305 \text{ K}) = +76$ J

Q is positive and heat goes into the gas.

ca: $Q = nC_p \Delta T = n\left(\tfrac{7}{2}R\right)\Delta T = (0.0040 \text{ mol})\left(\tfrac{7}{2}\right)(8.315 \text{ J/mol} \cdot \text{K})(305 \text{ K} - 1220 \text{ K}) = -107$ J

Q is negative and heat comes out of the gas.

bc: $Q = W = nRT \ln (V_c/V_b) = (0.0040 \text{ mol})(8.315 \text{ J/mol} \cdot \text{K})(1220 \text{ K})\ln(0.80 \text{ L} / 0.20 \text{ L}) = 56$ J

Q is positive and heat goes into the gas.

(d) ab: $\Delta U = nC_V \Delta T = n\left(\tfrac{5}{2}R\right)\Delta T = (0.0040 \text{ mol})\left(\tfrac{5}{2}\right)(8.315 \text{ J/mol} \cdot \text{K})(1220 \text{ K} - 305 \text{ K}) = +76$ J

The internal energy increased.

bc: $\Delta T = 0$ so $\Delta U = 0$. The internal energy does not change.

ca: $\Delta U = nC_V \Delta T = n\left(\tfrac{5}{2}R\right)\Delta T = (0.0040 \text{ mol})\left(\tfrac{5}{2}\right)(8.315 \text{ J/mol} \cdot \text{K})(305 \text{ K} - 1220 \text{ K}) = -76$ J

The internal energy decreased.

Reflect: The net internal energy change for the complete cycle $a \to b \to c \to a$ is $\Delta U_{tot} = +76 \text{ J} + 0 + (-76 \text{ J}) = 0$. For any complete cycle the final state is the same as the initial state and the net internal energy change is zero. For the cycle the net heat flow is $Q_{tot} = +76 \text{ J} + (-107 \text{ J}) + 56 \text{ J} = +25$ J. $\Delta U_{tot} = 0$ so $Q_{tot} = W_{tot}$. The net work done in the cycle is positive and this agrees with our result that the net heat flow is positive.

*15.81. Set Up: In part (a), apply $pV = nRT$ to the ethane in the flask. The volume is constant once the stopcock is in place. In part (b) apply $pV = \dfrac{m_{tot}}{M} RT$ to the ethane at its final temperature and pressure. $1.50 \text{ L} = 1.50 \times 10^{-3} \text{ m}^3$.

$M = 30.1 \times 10^{-3}$ kg/mol. Neglect the thermal expansion of the flask.

Solve: **(a)** $p_2 = p_1(T_2/T_1) = (1.013 \times 10^5 \text{ Pa})(300 \text{ K}/380 \text{ K}) = 8.00 \times 10^4$ Pa.

(b) $m_{tot} = \left(\dfrac{p_2 V}{R T_2}\right) M = \left(\dfrac{(8.00 \times 10^4 \text{ Pa})(1.50 \times 10^{-3} \text{ m}^3)}{(8.3145 \text{ J/mol} \cdot \text{K})(300 \text{ K})}\right)(30.1 \times 10^{-3} \text{ kg/mol}) = 1.45 \text{ g}.$

Reflect: We could also calculate m_{tot} with $p = 1.013 \times 10^5$ Pa and $T = 380$ K, and we would obtain the same result. Originally, before the system was warmed, the mass of ethane in the flask was

$m = (1.45 \text{ g}) \left(\dfrac{1.013 \times 10^5 \text{ Pa}}{8.00 \times 10^4 \text{ Pa}}\right) = 1.84 \text{ g}.$

*15.85. Set Up: Use $Q = n C_V \Delta T$ to calculate the temperature change in the constant volume process and use $pV = nRT$ to calculate the temperature change in the constant pressure process. The work done in the constant volume process is zero and the work done in the constant pressure process is $W = p \Delta V$. Use $Q = n C_p \Delta T$ to calculate the heat flow in the constant pressure process.
$\Delta U = n C_V \Delta T$, or $\Delta U = Q - W$. For N_2, $C_V = 20.76$ J/mol$\cdot$K and $C_p = 29.07$ J/mol$\cdot$K.

Solve: (a) For process ab, $\Delta T = \dfrac{Q}{n C_V} = \dfrac{1.52 \times 10^4 \text{ J}}{(2.50 \text{ mol})(20.76 \text{ J/mol} \cdot \text{K})} = 293$ K. $T_a = 293$ K, so $T_b = 586$ K. $pV = nRT$

says T doubles when V doubles and p is constant, so $T_c = 2(586 \text{ K}) = 1172 \text{ K} = 899°\text{C}$.

(b) For process ab, $W_{ab} = 0$. For process bc, $W_{bc} = p \Delta V = nR \Delta T = (2.50 \text{ mol})(8.314 \text{ J/mol} \cdot \text{K})(1172 \text{ K} - 586 \text{ K}) = 1.22 \times 10^4$ J. $W = W_{ab} + W_{bc} = 1.22 \times 10^4$ J.

(c) For process bc, $Q = n C_p \Delta T = (2.50 \text{ mol})(29.07 \text{ J/mol} \cdot \text{K})(1172 \text{ K} - 586 \text{ K}) = 4.26 \times 10^4$ J.

(d) $\Delta U = n C_V \Delta T = (2.50 \text{ mol})(20.76 \text{ J/mol} \cdot \text{K})(1172 \text{ K} - 293 \text{ K}) = 4.56 \times 10^4$ J.

Reflect: The total Q is 1.52×10^4 J $+ 4.26 \times 10^4$ J $= 5.78 \times 10^4$ J.
$\Delta U = Q - W = 5.78 \times 10^4$ J $- 1.22 \times 10^4$ J $= 4.56 \times 10^4$ J, which agrees with our results in part (d).

Solutions to Passage Problems

*15.87. Set Up: The internal energy of an ideal gas is proportional to nRT. The constant of proportionality depends on the number of degrees of freedom that the gas has at the given temperature.

Solve: Near room temperature a diatomic gas has 5 degrees of freedom and a monatomic gas has 3 degrees of freedom. For n moles of each gas the ratio of their internal energies will be $\left(\dfrac{5}{2} nRT\right) / \left(\dfrac{3}{2} nRT\right) = 5/3$. The correct answer is E.

*15.89. Set Up: As seen in the previous problem, the specific heat capacity of an ideal gas is proportional to $\dfrac{1}{2} R$. The constant of proportionality depends on the number of degrees of freedom that the gas has at the given temperature.

Solve: Since a monatomic ideal gas has 3 degrees of freedom at all temperatures, we expect its molar heat capacity at constant volume to be $\dfrac{3}{2} R$ at all temperatures. The correct answer is A.

THE SECOND LAW OF THERMODYNAMICS

Problems 1, 5, 9, 11, 15, 17, 19, 23, 25, 29, 31, 35, 37, 39, 43, 45, 49, 51

Solutions to Problems

*16.1. Set Up: We have $e = W/Q_H = 0.38$. Also, $W = Q_H + Q_C$ where W and Q_H are positive, and Q_C is negative. In one second the work done by the power plant is $W = 750$ MJ.

Solve: The power output of the plant is $\left(750 \times 10^6 \dfrac{J}{s} \right) \left(\dfrac{24\ h}{1\ day} \right) \left(\dfrac{3600\ s}{1\ h} \right) = 6.48 \times 10^{13}$ J/day. In one day the required

input heat is

$Q_H = \dfrac{W}{e} = \dfrac{6.48 \times 10^{13}\ J}{0.38} = \underline{1.71} \times 10^{14}\ J$. So we have

$Q_C = W - Q_H = 6.48 \times 10^{13}\ J - \underline{1.71} \times 10^{14}\ J = -1.1 \times 10^{14}\ J$. Thus, in one day $1.1 \times 10^{14}\ J$ of heat is discarded into the outside air.

Reflect: Q is positive when heat enters a system and negative when heat leaves a system. Thus, Q_C is counted as negative as it leaves the power plant (system 1) and positive as it enters the environment (system 2).

*16.5. Set Up: For a heat engine, $W = |Q_H| - |Q_C|$. $e = \dfrac{W}{Q_H}$. $Q_H > 0$, $Q_C < 0$. 1 hp = 746 W.

Solve: (a) $e = \dfrac{W}{Q_H} = \dfrac{3700\ J}{1.61 \times 10^4\ J} = 0.230 = 23.0\%$.

(b) $|Q_C| = |Q_H| - W = 1.61 \times 10^4\ J - 3700\ J = 1.24 \times 10^4\ J$.

(c) $m = \dfrac{1.61 \times 10^4\ J}{4.60 \times 10^4\ J/g} = 0.350$ g.

(d) $P = \dfrac{(60.0)(3700\ J)}{1.00\ s} = 2.22 \times 10^5\ W = 222\ kW = 298$ hp.

*16.9. Set Up: ca is at constant volume, ab has $Q = 0$, and bc is at constant pressure. For a constant pressure

process $W = p\,\Delta V$ and $Q = nC_p\,\Delta T$. $pV = nRT$ gives $n\,\Delta T = \dfrac{p\,\Delta V}{R}$ so $Q = \left(\dfrac{C_p}{R} \right) p\,\Delta V$. If $\gamma = 1.40$ the gas is

diatomic and $C_p = \tfrac{7}{2}R$. For a constant volume process $W = 0$ and $Q = nC_V\,\Delta T$. $pV = nRT$ gives $n\,\Delta T = \dfrac{V\,\Delta p}{R}$ so

$Q = \left(\dfrac{C_V}{R} \right) V\,\Delta p$. For a diatomic ideal gas $C_V = \tfrac{5}{2}R$. 1 atm = 1.013×10^5 Pa

Solve: (a) $V_b = 9.0 \times 10^{-3} \, \text{m}^3$, $p_b = 1.5 \, \text{atm}$ and $V_a = 2.0 \times 10^{-3} \, \text{m}^3$. For an adiabatic process $p_a V_a^\gamma = p_b V_b^\gamma$.

$$p_a = p_b \left(\frac{V_b}{V_a}\right)^\gamma = (1.5 \, \text{atm}) \left(\frac{9.0 \times 10^{-3} \, \text{m}^3}{2.0 \times 10^{-3} \, \text{m}^3}\right)^{1.4} = 12.3 \, \text{atm}$$

(b) Heat enters the gas in process ca, since T increases.

$$Q = \left(\frac{C_V}{R}\right) V \, \Delta p = \left(\tfrac{5}{2}\right)(2.0 \times 10^{-3} \, \text{m}^3)(12.3 \, \text{atm} - 1.5 \, \text{atm})(1.013 \times 10^5 \, \text{Pa/atm}) = 5470 \, \text{J}$$

$$Q_H = 5470 \, \text{J}$$

(c) Heat leaves the gas in process bc, since T increases.

$$Q = \left(\frac{C_p}{R}\right) p \, \Delta V = \left(\tfrac{7}{2}\right)(1.5 \, \text{atm})(1.013 \times 10^5 \, \text{Pa/atm})(-7.0 \times 10^{-3} \, \text{m}^3) = -3723 \, \text{J}$$

$$Q_C = -3723 \, \text{J}$$

(d) $W = Q_H + Q_C = +5470 \, \text{J} + (-3723 \, \text{J}) = 1747 \, \text{J}$

(e) $e = \dfrac{W}{Q_H} = \dfrac{1747 \, \text{J}}{5470 \, \text{J}} = 0.319 = 31.9\%$

Reflect: We did not use the number of moles of the gas.

*16.11. Set Up and Solve: $e = 1 - \dfrac{1}{r^{\gamma-1}}$. $\dfrac{1}{r^{\gamma-1}} = 1 - e = 0.350$. $r^{0.40} = \dfrac{1}{0.350}$ and $r = 13.8$.

*16.15. Set Up: For a refrigerator, $|Q_H| = |Q_C| + |W|$. $Q_C > 0$, $Q_H < 0$. $K = \dfrac{|Q_C|}{|W|}$.

Solve: (a) $W = \dfrac{|Q_C|}{K} = \dfrac{3.40 \times 10^4 \, \text{J}}{2.10} = 1.62 \times 10^4 \, \text{J}$.

(b) $|Q_H| = 3.40 \times 10^4 \, \text{J} + 1.62 \times 10^4 \, \text{J} = 5.02 \times 10^4 \, \text{J}$.

Reflect: More heat is discarded to the high-temperature reservoir than is absorbed from the cold reservoir.

*16.17. Set Up: $|Q_H| = |Q_C| + |W|$. $K = \dfrac{|Q_C|}{W}$. For water, $c_w = 4190 \, \text{J/kg} \cdot \text{K}$ and $L_f = 3.34 \times 10^5 \, \text{J/kg}$. For ice, $c_{ice} = 2010 \, \text{J/kg} \cdot \text{K}$.

Solve: (a)

$Q = m c_{ice} \, \Delta T_{ice} - m L_f + m c_w \, \Delta T_w$

$Q = (1.80 \, \text{kg})([2010 \, \text{J/kg} \cdot \text{K}][-5.0 \, \text{C}^\circ] - 3.34 \times 10^5 \, \text{J/kg} + [4190 \, \text{J/kg} \cdot \text{K}][-25.0 \, \text{C}^\circ]) = -8.08 \times 10^5 \, \text{J}$

$Q = -8.08 \times 10^5 \, \text{J}$. Q is negative for the water since heat is removed from it.

(b) $|Q_C| = 8.08 \times 10^5 \, \text{J}$.

$$W = \frac{|Q_C|}{K} = \frac{8.08 \times 10^5 \, \text{J}}{2.40} = 3.37 \times 10^5 \, \text{J}.$$

(c) $|Q_H| = 8.08 \times 10^5 \, \text{J} + 3.37 \times 10^5 \, \text{J} = 1.14 \times 10^6 \, \text{J}$.

Reflect: For this device, $Q_C > 0$ and $Q_H < 0$. More heat is rejected to the room than is removed from the water.

*16.19. Set Up: $|W| = |Q_H| - |Q_C|$. $\dfrac{Q_C}{Q_H} = -\dfrac{T_C}{T_H}$. $Q_C < 0$, $Q_H > 0$. $e = \dfrac{W}{Q_H}$.

Solve: (a) $|W| = |Q_H| - |Q_C| = 550 \, \text{J} - 335 \, \text{J} = 215 \, \text{J}$.

(b) $\dfrac{T_C}{T_H} = -\dfrac{Q_C}{Q_H}$. $T_C = -T_H \dfrac{Q_C}{Q_H} = -(620 \text{ K})\left(\dfrac{-335 \text{ J}}{550 \text{ J}}\right) = 378 \text{ J}.$

(c) $e = \dfrac{W}{Q_H} = \dfrac{215 \text{ J}}{550 \text{ J}} = 0.391 = 39.1\%.$

*16.23. Set Up: $T_C = 273$ K, $T_H = 297$ K. For water, $L_f = 3.34 \times 10^5$ J/kg. $Q_C > 0$, $Q_H < 0$. $\dfrac{Q_C}{Q_H} = -\dfrac{T_C}{T_H}$.

$|Q_H| = |W| + |Q_C|.$

Solve: (a) $|Q_C|$ is the heat removed from the water at $0°C$ to make ice at $0°C$.

$$|Q_C| = mL_f = (85.0 \text{ kg})(3.34 \times 10^5 \text{ J/kg}) = 2.84 \times 10^7 \text{ J}.$$

$$Q_H = -\left(\dfrac{T_H}{T_C}\right)Q_C = -\left(\dfrac{297 \text{ K}}{273 \text{ K}}\right)(2.84 \times 10^7 \text{ J}) = -3.09 \times 10^7 \text{ J}$$

(b) $|W| = |Q_H| - |Q_C| = 3.09 \times 10^7 \text{ J} - 2.84 \times 10^7 \text{ J} = 2.5 \times 10^6 \text{ J}.$

*16.25. Set Up: For an engine, W and Q_H are positive and Q_C is negative. For a refrigerator, Q_C is positive and W and Q_H are negative. A heat pump air conditioner takes in heat at a cool place and expels heat into a warm place. A heat pump house heater takes in heat at a cool place and expels heat into a warm place. For both types of heat pumps, W is negative and energy must be supplied to operate the device. bc and da are adiabatic, so the heat flow for these processes is zero.

Solve: (a) The cycle is clockwise. More positive work is done during ab and bc than the magnitude of the negative work done in cd and da, so the net work done in the cycle is positive. Heat enters the gas in ab and leaves the gas in cd.

(b) The cycle is counterclockwise and the net work done in the cycle is negative. Heat enters the gas in dc and leaves the gas in ba. dc occurs inside the food compartment and ba occurs in the air of the room.

(c) The cycle is counterclockwise and the net work done in the cycle is negative. Heat enters the gas in dc and leaves the gas in ba. dc occurs inside the house and ba occurs outside.

(d) The cycle is counterclockwise and the net work done in the cycle is negative. Heat enters the gas in dc and leaves the gas in ba. dc occurs outside and ba occurs inside the house.

Reflect: For a refrigerator the signs of W, Q_H, and Q_C are all opposite to what they are for an engine.

*16.29. Set Up: Apply $Q_{\text{system}} = 0$ to calculate the final temperature. The equation $Q = mc\Delta T$ gives the heat required to warm or cool the water (for water we have $c = 4190$ J/kg$\cdot$K). The temperature of boiling water is $T = 100.0°C = 373$ K. For an isothermal process we have $\Delta S = \dfrac{Q}{T}$.

Solve: (a) The heat transfer between $100°C$ water and $30°C$ water occurs over a finite temperature difference and the process is irreversible.

(b) $(270 \text{ kg})c(T_2 - 30.0°C) + (5.00 \text{ kg})c(T_2 - 100°C) = 0$. $T_2 = 31.27°C = 304.42$ K.

(c) As explained in Example 16.8, when the temperature of an object changes, an exact calculation of its entropy change requires calculus; however, we can make a rough estimate by assuming a sequence of nearly isothermal steps. For the water in the tub, we have added $(270 \text{ kg})(4190 \text{ J/kg}\cdot\text{K})(31.27°C - 30.0°C) = 1.44 \times 10^6$ J. Since the temperature of the bathwater only increases by $1.27°C$ we can treat this process as isothermal. Thus, the entropy of the bathwater increases by approximately $\Delta S_1 = \dfrac{Q}{T} = \dfrac{1.44 \times 10^6 \text{ J}}{303 \text{ K}} = +4.75 \times 10^3$ J/K. The change in entropy of the boiling water is more difficult to estimate due to its large temperature change. Following Example 16.8, we must *remove* $(5.00 \text{ kg})(4190 \text{ J/kg}\cdot\text{K})(1°C) = 20,950$ J of heat for each degree the temperature of the boiling water is reduced. reduced. Thus, we can form a chain of nearly isothermal steps starting from 373 K $\rightarrow 372$ K and ending with 305 K $\rightarrow 304$ K. The approximate change in entropy is

$\Delta S_2 \approx (-20{,}950 \text{ J})\left(\dfrac{1}{373} + \dfrac{1}{372} + \dfrac{1}{371} \cdots \dfrac{1}{305}\right) = (-20{,}950 \text{ J})(0.2042) = -4.28 \times 10^3 \text{ J/K}$. The total change in entropy

is approximately $\Delta S = \Delta S_1 + \Delta S_2 \approx +4.75 \times 10^3 \text{ J/K} + -4.28 \times 10^3 \text{ J/K} = 470 \text{ J/K}$.

Reflect: $\Delta S_{\text{system}} > 0$, as it should for an irreversible process.

*16.31. Set Up: $\Delta S = \dfrac{Q}{T}$. $\Delta U = Q - W$. For an isothermal process of an ideal gas, $\Delta U = 0$ and $Q = W$. For a

compression, $\Delta V < 0$ and $W < 0$.

Solve: $Q = W = -1850 \text{ J}$. $\Delta S = \dfrac{-1850 \text{ J}}{293 \text{ K}} = -6.31 \text{ J/K}$.

*16.35. Set Up and **Solve:** $(0.600)(300 \text{ W/m}^2)A = 20.0 \times 10^3 \text{ W}$. $A = 111 \text{ m}^2$.

*16.37. Set Up: $e_{\text{max}} = e_{\text{Carnot}} = 1 - T_C/T_H$. $e = \dfrac{W}{Q_H} = \dfrac{W/t}{Q_H/t}$. $W = Q_H + Q_C$ so $\dfrac{W}{t} = \dfrac{Q_C}{t} + \dfrac{Q_H}{t}$. For a temperature

change $Q = mc\Delta T$. $T_H = 300.15 \text{ K}$, $T_C = 279.15 \text{ K}$. For water, $\rho = 1000 \text{ kg/m}^3$, so a mass of 1 kg has a volume of
1 L. For water, $c = 4190 \text{ J/kg} \cdot \text{K}$.

Solve: **(a)** $e = 1 - \dfrac{279.15 \text{ K}}{300.15 \text{ K}} = 7.0\%$.

(b) $\dfrac{Q_H}{t} = \dfrac{P_{\text{out}}}{e} = \dfrac{210 \text{ kW}}{0.070} = 3.0 \text{ MW}$. $\dfrac{Q_C}{t} = \dfrac{Q_H}{t} - \dfrac{W}{t} = 3.0 \text{ MW} - 210 \text{ kW} = 2.8 \text{ MW}$.

(c) $\dfrac{m}{t} = \dfrac{|Q_C|/t}{c\Delta T} = \dfrac{(2.8 \times 10^6 \text{ W})(3600 \text{ s/h})}{(4190 \text{ J/kg} \cdot \text{K})(4 \text{ K})} = 6 \times 10^5 \text{ kg/h} = 6 \times 10^5 \text{ L/h}$.

Reflect: The efficiency is small since T_C and T_H don't differ greatly.

*16.39. Set Up: $W = Q_C + Q_H$. Since it is a Carnot cycle, $\dfrac{Q_C}{Q_H} = -\dfrac{T_C}{T_H}$. The heat required to melt the ice is

$Q = mL_f$. For water, $L_f = 334 \times 10^3 \text{ J/kg}$. $Q_H > 0$, $Q_C < 0$. $Q_C = -mL_f$. $T_H = 527°C = 800.15 \text{ K}$.

Solve: **(a)** $Q_H = +400 \text{ J}$, $W = +300 \text{ J}$. $Q_C = W - Q_H = -100 \text{ J}$.

$\qquad T_C = -T_H(Q_C/Q_H) = -(800.15 \text{ K})[(-100 \text{ J})/(400 \text{ J})] = +200 \text{ K} = -73°C$

(b) The total Q_C required is $-mL_f = -(10.0 \text{ kg})(334 \times 10^3 \text{ J/kg}) = -3.34 \times 10^6 \text{ J}$. Q_C for one cycle is -100 J, so the

number of cycles required is $\dfrac{-3.34 \times 10^6 \text{ J}}{-100 \text{ J/cycle}} = 3.34 \times 10^4$ cycles.

Reflect: The results depend only on the maximum temperature of the gas, not on the number of moles or the
maximum pressure.

*16.43. Set Up: The efficiency of the composite engine is $e_{12} = \dfrac{W_1 + W_2}{Q_{H1}}$, where Q_{H1} is the heat input to the first

engine and W_1 and W_2 are the work outputs of the two engines. For any heat engine, $W = Q_C + Q_H$, and for a Carnot

engine, $\dfrac{Q_{\text{low}}}{Q_{\text{high}}} = -\dfrac{T_{\text{low}}}{T_{\text{high}}}$, where Q_{low} and Q_{high} are the heat flows at the two reservoirs that have temperatures T_{low} and

T_{high}. We know that $Q_{\text{high},2} = -Q_{\text{low},1}$, $T_{\text{low},1} = T'$, $T_{\text{high},1} = T_H$, $T_{\text{low},2} = T_C$ and $T_{\text{high},2} = T'$.

Solve: $e_{12} = \dfrac{W_1 + W_2}{Q_{H1}} = \dfrac{Q_{high,1} + Q_{low,1} + Q_{high,2} + Q_{low,2}}{Q_{high,1}}$. Since $Q_{high,2} = -Q_{low,1}$, this reduces to $e_{12} = 1 + \dfrac{Q_{low,2}}{Q_{high,1}}$.

$Q_{low,2} = -Q_{high,2} \dfrac{T_{low,2}}{T_{high,2}} = Q_{low,1} \dfrac{T_C}{T'} = -Q_{high,1}\left(\dfrac{T_{low,1}}{T_{high,1}}\right)\dfrac{T_C}{T'} = -Q_{high,1}\left(\dfrac{T'}{T_H}\right)\dfrac{T_C}{T'}$. This gives $e_{12} = 1 - \dfrac{T_C}{T_H}$. The efficiency

of the composite system is the same as that of the original engine.

Reflect: The overall efficiency is independent of the value of the intermediate temperature T'.

***16.45. Set Up:** A person with surface area A and surface temperature $T = 303$ K radiates at a rate $H = e\sigma T^4$. The person absorbs heat from the room at a rate $H_s = Ae\sigma T_s^4$, where $T_s = 293$ K is the temperature of the room. In $t = 1.0$ s, heat $Ae\sigma t T^4$ flows into the room and heat $Ae\sigma t T_s^4$ flows out of the room. The heat flows into and out of the room occur at a temperature of T_s.

Solve: For the room,

$$\Delta S = \frac{Ae\sigma t T^4}{T_s} - \frac{Ae\sigma t T_s^4}{T_s} = \frac{Ae\sigma t(T^4 - T_s^4)}{T_s}$$

$$\Delta S = \frac{(1.85 \text{ m}^2)(1.00)(5.67 \times 10^{-8} \text{ W/m}^2 \cdot \text{K}^4)(1.0)([303 \text{ K}]^4 - [293 \text{ K}]^4)}{293 \text{ K}} = 0.379 \text{ J/K}$$

***16.49. Set Up:** For water, $L_f = 3.34 \times 10^5$ J/kg. $\Delta S = \dfrac{Q}{T}$.

Solve: (a) The heat that goes into the ice-water mixture is $Q = L_f = (0.160 \text{ kg})(3.34 \times 10^5 \text{ J/kg}) = 5.34 \times 10^4$ J. This is same amount of heat leaves the boiling water, so

$$\Delta S = \frac{Q}{T} = \frac{-5.34 \times 10^4 \text{ J}}{373 \text{ K}} = -143 \text{ J/K}.$$

(b) $\Delta S = \dfrac{Q}{T} = \dfrac{5.34 \times 10^4 \text{ J}}{273 \text{ K}} = +196$ J/K

(c) For any segment of the rod, the net heat flow is zero, so $\Delta S = 0$.

(d) $\Delta S_{tot} = -143$ J/K $+ 196$ J/K $= +53$ J/K.

***16.51. Set Up:** For a refrigerator, $K = \dfrac{|Q_C|}{|W|}$. If the cycle is run in reverse, W has the same magnitude and opposite sign. Heat enters the refrigerator in processes ac and cb and leaves the refrigerator in process ba. From Problem 16.57, Q for ab is $+1.72 \times 10^4$ J, so for ba, $Q = -1.72 \times 10^4$ J. This is Q_H for the refrigerator. In Problem 16.57, $W = 2.40 \times 10^3$ J, so here $W = -2.40 \times 10^3$ J. For the refrigerator, $W < 0$, $Q_H < 0$ and $Q_C > 0$. $W = Q_C + Q_H$

Solve: $Q_C = W - Q_H = -2.40 \times 10^3 \text{ J} - (-1.72 \times 10^4 \text{ J}) = 1.48 \times 10^4$ J

$$K = \frac{1.48 \times 10^4 \text{ J}}{2.40 \times 10^3 \text{ J}} = 6.17$$

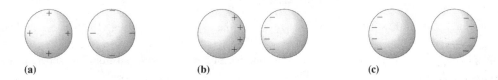

17

ELECTRIC CHARGE AND ELECTRIC FIELD

Problems 3, 5, 7, 11, 15, 19, 23, 25, 29, 31, 33, 37, 39, 43, 45, 47, 51, 53, 57, 61, 63, 67, 69, 71, 75, 77, 79

Solutions to Problems

*17.3. Set Up: For an isolated sphere, the excess charge is uniformly distributed over the surface of the conductor. Unlike charges attract and like charges repel, and in a conductor the excess charge is free to move.
Solve: **(a)** The uniform distribution of charge over the surface of each sphere is sketched in Figure (a) below.
(b) When the spheres are close to each other, the negative and positive excess charges are drawn toward each other, as shown in Figure (b) below.
(c) When the spheres are close to each other, the excess negative charges on each sphere repel, as shown in Figure (c) below.

(a)   (b)   (c)

Reflect: We will learn later in the chapter that the excess charge on a conductor is on the surface of the conductor.

*17.5. Set Up: The charge of one electron is $-e = -1.60 \times 10^{-19}$ C. $1\,\mu C = 10^{-6}$ C; 1 nC $= 10^{-9}$ C.

Solve: **(a)** $N = \dfrac{|Q|}{e} = \dfrac{2.50 \times 10^{-6}\ \text{C}}{1.60 \times 10^{-19}\ \text{C}} = 1.56 \times 10^{13}$ electrons

(b) $N = \dfrac{|Q|}{e} = \dfrac{2.50 \times 10^{-9}\ \text{C}}{1.60 \times 10^{-19}\ \text{C}} = 1.56 \times 10^{10}$ electrons

*17.7. Set Up: Use the mass m of the ring and the atomic mass M of gold to calculate the number of gold atoms. Each atom has 79 protons and an equal number of electrons. $N_A = 6.02 \times 10^{23}$ atoms/mol. A proton has charge $+e$.

Solve: The mass of gold is 17.7 g and the atomic weight of gold is 197 g/mol. So the number of atoms is

$$N_A n = (6.02 \times 10^{23} \text{ atoms/mol}) \left(\frac{17.7 \text{ g}}{197 \text{ g/mol}} \right) = 5.41 \times 10^{22} \text{ atoms. The number of protons is } n_p = (79 \text{ protons/atom})$$

$(5.41 \times 10^{22} \text{ atoms}) = 4.27 \times 10^{24} \text{ protons.}$

$Q = (n_p)(1.60 \times 10^{-19} \text{ C/proton}) = 6.83 \times 10^5 \text{ C.}$

(b) The number of electrons is $n_e = n_p = 4.27 \times 10^{24}$.

Reflect: The total amount of positive charge in the ring is very large, but there is an equal amount of negative charge.

*17.11. Set Up: The electrical force is given by Coulomb's law, with $k = 8.99 \times 10^9 \text{ N} \cdot \text{m}^2/\text{C}^2$. A proton has charge $+e$ and an electron has charge $-e$.

Solve: (a) $F = k \frac{|q_1 q_2|}{r^2} = \frac{(8.99 \times 10^9 \text{ N} \cdot \text{m}^2/\text{C}^2)(1.60 \times 10^{-19} \text{ C})^2}{(1.00 \times 10^{-15} \text{ m})^2} = 230 \text{ N.}$ Yes, this force is about 52 lbs.

(b) The force is smaller than in part (a) by a factor of $\left(\frac{1.00 \times 10^{-15} \text{ m}}{1.00 \times 10^{-10} \text{ m}} \right)^2$ so it is $2.30 \times 10^{-8} \text{ N.}$ No, this force is very small.

*17.15. Set Up: Use the mass of a sphere and the atomic mass of aluminum to find the number of aluminum atoms in one sphere. Each atom has 13 electrons. Apply Coulomb's law and calculate the magnitude of charge $|q|$ on each sphere. We have $N_A = 6.02 \times 10^{23}$ atoms/mol and $|q| = n_e' e$, where n_e' is the number of electrons removed from one sphere and added to the other.

Solve: (a) Assuming that the two spheres are initially neutral, the total number of electrons on each sphere equals the number of protons on each sphere:

$$n_e = n_p = (13)(N_A) \left(\frac{0.0250 \text{ kg}}{0.026982 \text{ kg/mol}} \right) = 7.25 \times 10^{24} \text{ electrons.}$$

(b) For a force of 1.00×10^4 N to act between the spheres, $F = 1.00 \times 10^4 \text{ N} = k \frac{q^2}{r^2}$. This gives

$|q| = \sqrt{(1.00 \times 10^4 \text{ N})(0.800 \text{ m})^2/k} = 8.44 \times 10^{-4} \text{ C.}$ The number of electrons removed from one sphere and added to the other is $n_e' = |q|/e = 5.27 \times 10^{15}$ electrons.

(c) $n_e'/n_e = 7.27 \times 10^{-10}$.

Reflect: When ordinary objects receive a net charge the fractional change in the total number of electrons in the object is very small.

*17.19. Set Up: Apply Coulomb's law. The two forces on q_3 must have equal magnitudes and opposite directions. Like charges repel and unlike charges attract.

Solve: The force $\vec{F}_2$ that q_2 exerts on q_3 has magnitude $F_2 = k \frac{|q_2 q_3|}{r_2^2}$ and is in the $+x$ direction. The force $\vec{F}_1$ must

be in the $-x$ direction, so q_1 must be positive. The condition $F_1 = F_2$ gives $k \frac{|q_1||q_3|}{r_1^2} = k \frac{|q_2||q_3|}{r_2^2}$.

$|q_1| = |q_2| \left(\frac{r_1}{r_2} \right)^2 = (3.00 \text{ nC}) \left(\frac{2.00 \text{ cm}}{4.00 \text{ cm}} \right)^2 = 0.750 \text{ nC.}$

Reflect: The result for the magnitude of q_1 doesn't depend on the magnitude of q_2.

*17.23. Set Up: $F = k\dfrac{|qq_1|}{r^2}$. Like charges repel and unlike charges attract. The charges and the forces on the charges q_1 and q_2 in the dipole are shown in the figure below. Use the coordinates shown.

$\sin\theta = \dfrac{1.50 \text{ cm}}{2.00 \text{ cm}}$ and $\theta = 48.6°$.

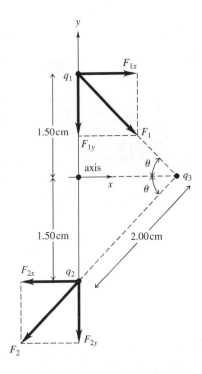

Solve: $F_1 = F_2 = (8.99 \times 10^9 \text{ N}\cdot\text{m}^2/\text{C}^2)\dfrac{(5.00 \times 10^{-6} \text{ C})(10.0 \times 10^{-6} \text{ C})}{(0.0200 \text{ m})^2} = 1.124 \times 10^3 \text{ N}.$

$F_x = F_{1x} + F_{2x} = 0.$ $F_y = F_{1y} + F_{2y} = -2F_1 \sin\theta = -2(1.124 \times 10^3 \text{ N})\left(\dfrac{1.50 \text{ cm}}{2.00 \text{ cm}}\right) = -1.69 \times 10^3 \text{ N}.$

The net force has magnitude 1.69×10^3 N and is in the direction from the $+5.00\ \mu\text{C}$ to the $-5.00\ \mu\text{C}$ charge.

(b) F_{1x} and F_{2x} each produce clockwise torques and each have a moment arm of 1.50 cm. F_{1y} and F_{2y} each have zero moment arm and produce no torque.

$\tau = 2F_{1x}(0.0150 \text{ m}) = 2F_1 \cos\theta(0.0150 \text{ m}) = 2(1.124 \times 10^3 \text{ N})(\cos 48.6°)(0.0150 \text{ m}) = 22.3 \text{ N}\cdot\text{m}$

The torque is clockwise.

Reflect: The x components of the two forces are in opposite directions so they cancel and the net force has no x component. But the torques of F_{1x} and F_{2x} are in the same direction and therefore produce a net torque.

*17.25. Set Up: In the O-H-O combination the O^- is 0.180 nm from the H^+ and 0.290 nm from the other O^-. In the N-H-N combination the N^- is 0.190 nm from the H^+ and 0.300 nm from the other N^-. In the O-H-N combination the O^- is 0.180 nm from the H^+ and 0.290 nm from the other N^-. Like charges repel and unlike charges attract. The net force is the vector sum of the individual forces.

Solve: $F = k\dfrac{|q_1 q_2|}{r^2} = k\dfrac{e^2}{r^2}$. The attractive forces are: $O^- - H^+$, 7.10×10^{-9} N; $N^- - H^+$, 6.37×10^{-9} N; $O^- - H^+$,

7.10×10^{-9} N. The total attractive force is 2.06×10^{-8} N. The repulsive forces are: $O^- - O^-$, 2.74×10^{-9} N; $N^- - N^-$, 2.56×10^{-9} N; $O^- - N^-$, 2.74×10^{-9} N. The total repulsive force is 8.04×10^{-9} N. The net force is attractive and has magnitude 1.26×10^{-8} N.

*17.29. Set Up: In space we assume that the only force accelerating the free proton is the electrical repulsion of the other proton. Coulomb's law gives the force, and Newton's second law gives the acceleration: $a = \dfrac{F}{m} = k\dfrac{e^2}{mr^2}$.

Solve: (a) $a = (8.99 \times 10^9 \text{ N} \cdot \text{m}^2/\text{C}^2)\dfrac{(1.60 \times 10^{-19} \text{ C})^2}{(1.67 \times 10^{-27} \text{ kg})(0.00250 \text{ m})^2} = 2.20 \times 10^4 \text{ m/s}^2$.

(b) The graphs are sketched in the figure below.
Reflect: The electrical force of a single stationary proton gives the moving proton an initial acceleration about 20,000 times as great as the acceleration caused by the gravity of the entire earth. As the protons move farther apart, the electrical force gets weaker, so the acceleration decreases. Since the protons continue to repel, the velocity keeps increasing, but at a decreasing rate.

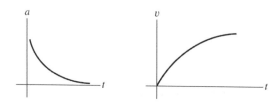

*17.31. Set Up: $\vec{F} = q\vec{E}$. A proton has charge $q = +e = +1.60 \times 10^{-19}$ C.

Solve: (a) $F = |q|E$. $E = \dfrac{F}{|q|} = \dfrac{20.0 \times 10^{-9} \text{ N}}{8.00 \times 10^{-9} \text{ C}} = 2.50$ N/C. Since the charge is negative, the force and electric field are in opposite directions and the electric field is upward.

(b) $F = |q|E = eE = (1.60 \times 10^{-19} \text{ C})(2.50 \text{ N}) = 4.00 \times 10^{-19}$ N. The charge is positive so the force is in the same direction as the electric field; the force is upward.

*17.33. Set Up: Use the coordinates shown in the figure below. Since the electric field is uniform, the force is constant and the acceleration is constant. $\vec{F} = q\vec{E}$ and $\vec{F} = m\vec{a}$. $v_{0y} = 0$. For an electron, $q = -e = -1.60 \times 10^{-19}$ C and $m = 9.11 \times 10^{-31}$ kg.

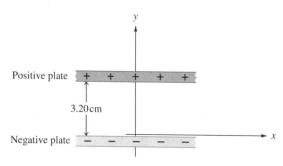

Solve: (a) $y = v_{0y}t + \frac{1}{2}a_y t^2$. $a_y = \dfrac{2y}{t^2} = \dfrac{2(3.20 \times 10^{-2} \text{ m})}{(1.5 \times 10^{-8} \text{ s})^2} = 2.84 \times 10^{14} \text{ m/s}^2$.

$$F_y = ma_y = (9.11 \times 10^{-31} \text{ kg})(2.84 \times 10^{14} \text{ m/s}^2) = 2.59 \times 10^{-16} \text{ N}.$$

$$E_y = \frac{F_y}{|q|} = \frac{2.59 \times 10^{-16} \text{ N}}{1.60 \times 10^{-19} \text{ C}} = 1.62 \times 10^3 \text{ N/C} \text{ and } E = 1.62 \times 10^3 \text{ N/C}.$$

(b) $v_y = v + a_y t = 0 + (2.84 \times 10^{14} \text{ m/s}^2)(1.5 \times 10^{-8} \text{s}) = 4.26 \times 10^6 \text{ m/s}$

Reflect: We could also use the work-energy theorem and set the work done by the force equal to the gain in kinetic energy of the electron.

*17.37. Set Up: The electric field of a point charge has magnitude $E = k \dfrac{|q|}{r^2}$. A proton has charge

$q = e = 1.60 \times 10^{-19}$ C.

Solve: (a) $E = (8.99 \times 10^9 \text{ N} \cdot \text{m}^2/\text{C}^2) \dfrac{1.60 \times 10^{-19} \text{ C}}{(5.0 \times 10^{-15} \text{ m})^2} = 5.8 \times 10^{19}$ N/C

(b) $E = (8.99 \times 10^9 \text{ N} \cdot \text{m}^2/\text{C}^2) \dfrac{1.60 \times 10^{-19} \text{ C}}{(5.0 \times 10^{-10} \text{ m})^2} = 5.8 \times 10^9$ N/C

Reflect: The electric fields inside atoms are very large.

*17.39. Set Up: If the axon is modeled as a point charge, its electric field is $E = k \dfrac{q}{r^2}$. The electric field of a point

charge is directed away from the charge if it is positive.

Solve: (a) 5.6×10^{11} Na$^+$ ions enter per meter so in a $0.10 \text{ mm} = 1.0 \times 10^{-4}$ m section, 5.6×10^7 Na$^+$ ions enter.

This number of ions has charge $q = (5.6 \times 10^7)(1.60 \times 10^{-19} \text{ C}) = 9.0 \times 10^{-12}$ C.

(b) $E = k \dfrac{|q|}{r^2} = (8.99 \times 10^9 \text{ N} \cdot \text{m}^2/\text{C}^2) \dfrac{9.0 \times 10^{-12} \text{ C}}{(5.00 \times 10^{-2} \text{ m})^2} = 32$ N/C, directed away from the axon.

(c) $r = \sqrt{\dfrac{k|q|}{E}} = \sqrt{\dfrac{(8.99 \times 10^9 \text{ N} \cdot \text{m}^2/\text{C}^2)(9.0 \times 10^{-12} \text{ C})}{1.0 \times 10^{-6} \text{ N/C}}} = 280$ m

*17.43. Set Up: For a point charge, $E = k \dfrac{|q|}{r^2}$. $\vec{E}$ is toward a negative charge and away from a positive charge. Let

$q_1 = +0.500$ nC and $q_2 = +8.00$ nC. For the net electric field to be zero, $\vec{E}_1$ and $\vec{E}_2$ must have equal magnitudes and opposite directions.

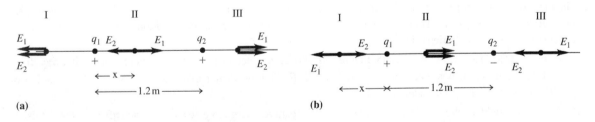

(a)   (b)

Solve: The two charges and the directions of their electric fields in three regions are shown in Figure (a) above. Only in region II are the two electric fields in opposite directions. Consider a point a distance x from q_1 so a distance

$1.20 \text{ m} - x$ from q_2. $E_1 = E_2$ gives $k \dfrac{0.500 \text{ nC}}{x^2} = k \dfrac{8.00 \text{ nC}}{(1.20 - x)^2}$. $16x^2 = (1.20 - x)^2$. $4x = \pm(1.20 \text{ m} - x)$ and $x = 0.24$ m

is the positive solution. The electric field is zero at a point between the two charges, 0.24 m from the 0.500 nC charge.

(b) Let $q_2 = -8.00$ nC be the negative charge. The two charges and the directions of their electric fields in three regions are shown in Figure (b) above. $\vec{E}_1$ and $\vec{E}_2$ are in opposite directions in regions I and III. But for the magnitudes of the fields to be equal the point must be closer to the charge q_1 that has smaller magnitude, and that occurs only for region I. Consider a point a distance x to the left of q_1 so 1.20 m $+ x$ from q_2. $E_1 = E_2$ gives

$k\dfrac{0.500 \text{ nC}}{x^2} = k\dfrac{8.00 \text{ nC}}{(1.20 + x)^2}$. $16x^2 = (1.20 \text{ m} + x)^2$. $4x = \pm(1.20 \text{ m} + x)$ and $x = 0.40$ m is the positive solution. The electric field is zero at a point 0.40 m from q_1 and 1.60 m from q_2.

Reflect: In each case there is only one point along the line connecting the two charges where the net electric field is zero.

*17.45. Set Up: The force on a charge q that is in an electric field $\vec{E}$ is $\vec{F} = q\vec{E}$.

Solve: (a) The force on $+q$ is $F_+ = qE$, to the right. The force on $-q$ is $F_- = qE$, to the left. The net force on the dipole is zero.

(b) The axis is at the midpoint of the line connecting the charges, and perpendicular to the plane of the figure. The torque is zero when each force, F_+ and F_-, has zero moment arm. This is the case for $\theta = 0°$ and $\theta = 180°$, as shown in Figure (a) and (b) below.

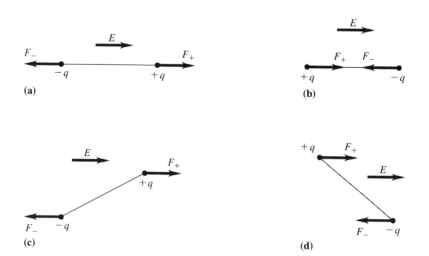

(c) For θ slightly greater than zero, as shown in Figure (c) above, the torque on the dipole is clockwise and is directed so as to return the dipole to the equilibrium position. For θ slightly less than zero the torque is counterclockwise and again tends to rotate the dipole back to its equilibrium position. The $\theta = 0°$ position of Figure (a) above is a stable orientation.

For θ slightly less than 180°, as in Figure (d) above, the net torque on the dipole rotates it away from the $\theta = 180°$ equilibrium position. The same is true for θ slightly greater than 180°. The $\theta = 180°$ position of Figure (b) above is an unstable orientation.

(d) The electric field of the dipole is directed from the positive charge and toward the negative charge. Thus, in Figure (a) above the electric field of the dipole is to the left, opposite to the direction of the external field.

Reflect: For any orientation of the dipole the net force on the dipole is zero. But only for $\theta = 0°$ and $\theta = 180°$ is the net torque zero.

*17.47. Set Up: $E = k\dfrac{|q|}{r^2}$

Solve: **(a)** For $r_1 = 1.0$ cm, $E_1 = E = k\dfrac{|q|}{r_1^2}$. $r_2 = 2r_1$. $E_2 = k\dfrac{|q|}{(2r_1)^2} = \dfrac{1}{4}k\dfrac{|q|}{r_1^2} = E/4$

(b) Now $r_3 = 3r_1$. $E_3 = k\dfrac{|q|}{(3r_1)^2} = \dfrac{1}{9}k\dfrac{|q|}{r_1^2} = E/9$

*17.51. Set Up: The electric field lines point away from the positive charge and into the negative charge. To indicate the relative size of each charge, we should draw roughly four times as many field lines terminating on the $-4Q$ charge as originate from the $+Q$ charge.

Solve: The sketch is shown in the figure below.

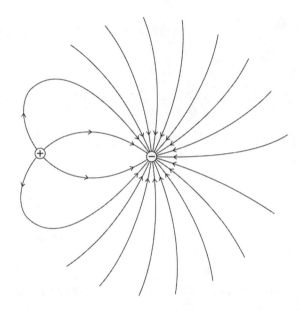

Reflect: The sketch is similar to that of Figure 17.22b in the textbook—except that in that figure the two charges are equal in magnitude so that the number of field lines terminating on the negative charge is equal to the number of field lines originating from the positive charge.

*17.53. Set Up: Electric field is directed away from positive charge and toward negative charge. By symmetry, far from the edges of the sheets the field lines are perpendicular to the sheets; there is no reason to prefer to the left or to the right for a component of electric field.

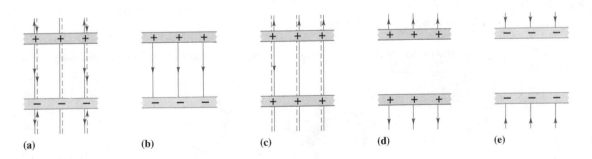

(a) (b) (c) (d) (e)

Solve: **(a)** The fields of each sheet are sketched in Figure (a) above. The solid lines are the field due to the upper positive sheet and the dashed lines are the field due to the lower negative sheet. Between the two sheets the fields are in the same direction and add. Outside the two sheets the fields are in opposite directions and cancel. The net electric field is sketched in Figure (b) above. This result is consistent with the field in the similar case described in Example 17.5.

(b) The fields of each sheet are sketched in Figure (c) above. The solid lines are the field due to the upper sheet and the dashed lines are the field due to the lower sheet. Since both sheets are positive, the field of each sheet is directed away from that sheet. Between the two sheets the fields are in opposite directions and cancel. Outside the two sheets the fields are in the same direction and add. The net field is sketched in Figure (d) above.

(c) The fields are the same as in part (b), except they are toward each sheet. The net field of the two sheets is sketched in Figure (e) above.

Reflect: More advanced treatments show that the electric field of an infinite sheet is uniform and does not depend on the distance from the sheet.

*17.57. Set Up: Example 17.10 shows that $E = 0$ inside a uniform spherical shell and that $E = k\dfrac{|q|}{r^2}$ outside the shell.
Solve: (a) $E = 0$

(b) $r = 0.060$ m and $E = (8.99 \times 10^9 \text{ N} \cdot \text{m}^2/\text{C}^2)\dfrac{15.0 \times 10^{-6} \text{ C}}{(0.060 \text{ m})^2} = 3.75 \times 10^7$ N/C

(c) $r = 0.110$ m and $E = (8.99 \times 10^9 \text{ N} \cdot \text{m}^2/\text{C}^2)\dfrac{15.0 \times 10^{-6} \text{ C}}{(0.110 \text{ m})^2} = 1.11 \times 10^7$ N/C

Reflect: Outside the shell the electric field is the same as if all the charge were concentrated at the center of the shell. But inside the shell the field is not the same as for a point charge at the center of the shell; inside the shell the electric field is zero.

*17.61. Set Up: The charge distribution has spherical symmetry, so the electric field, if nonzero, is radial and depends only on the distance from the center of the shell.
Solve: (a) Apply Gauss's law to a sphere of radius $r < a$ and concentric with the shell. The electric field, if nonzero, is constant over the Gaussian surface and perpendicular to it, so $\Phi_E = E(4\pi r^2)$. But no charge is enclosed by the Gaussian surface, so $Q_{encl} = 0$. Gauss's law gives $E(4\pi r^2) = 0$ and $E = 0$.

(b) Apply Gauss's law to a sphere of radius $r > b$. $\Phi_E = E(4\pi r^2)$. $Q_{encl} = Q$. Gauss's law gives $E(4\pi r^2) = \dfrac{Q}{\epsilon_0}$ and

$E = \dfrac{Q}{4\pi\epsilon_0 r^2}.$

(c) The thick shell can be constructed from a series of concentric thin shells. $E = 0$ inside each of these thin shells, so $E = 0$ inside the thick shell.

Reflect: In using Gauss's law it is very helpful to select a Gaussian surface of appropriate symmetry, so it is simple to express the flux in terms of the electric field at the surface.

*17.63. Set Up: $E = 0$ everywhere within the conductor. Any net charge must be on the inner and outer surfaces of the conductor.

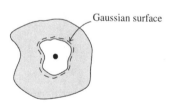

Gaussian surface

Solve: **(a)** and **(b)** Apply Gauss's law to a surface that is within the conductor, just outside the cavity, as shown in the figure above. $E = 0$ everywhere on the Gaussian surface so $\Phi_E = 0$ for that surface. Gauss's law then says that Q_{encl} for this surface is zero. The conductor is neutral, so if the outer surface has charge $-12 \, \mu C$ the inner surface must have charge $+12 \, \mu C$. To make $Q_{encl} = 0$ there must be $-12 \, \mu C$ within the hole.

Reflect: The charge in the hole creates the charge separation in the conductor. It pulls $+12 \, \mu C$ to the inner surface and that leaves $-12 \, \mu C$ on the outer surface.

*17.67. Set Up: $F = k \dfrac{|qq'|}{r^2}$. Like charges repel and unlike charges attract. The three charges and the forces on q_3 are shown in the figure below.

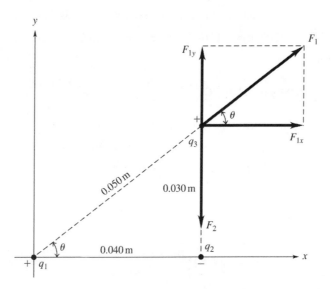

Solve: **(a)** $F_1 = k \dfrac{|q_1 q_3|}{r_1^2} = (8.99 \times 10^9 \ \text{N} \cdot \text{m}^2/\text{C}^2) \dfrac{(5.00 \times 10^{-9} \ \text{C})(6.00 \times 10^{-9} \ \text{C})}{(0.0500 \ \text{m})^2} = 1.079 \times 10^{-4} \ \text{C}.$

$\theta = 36.9°$. $F_{1x} = +F_1 \cos \theta = 8.63 \times 10^{-5} \ \text{N}$. $F_{1y} = +F_1 \sin \theta = 6.48 \times 10^{-5} \ \text{N}$.

$F_2 = k \dfrac{|q_2 q_3|}{r_2^2} = (8.99 \times 10^9 \ \text{N} \cdot \text{m}^2/\text{C}^2) \dfrac{(2.00 \times 10^{-9} \ \text{C})(6.00 \times 10^{-9} \ \text{C})}{(0.0300 \ \text{m})^2} = 1.20 \times 10^{-4} \ \text{C}.$

$F_{2x} = 0$, $F_{2y} = -F_2 = -1.20 \times 10^{-4} \ \text{N}$. $F_x = F_{1x} + F_{2x} = 8.63 \times 10^{-5} \ \text{N}$.

$F_y = F_{1y} + F_{2y} = 6.48 \times 10^{-5} \ \text{N} + (-1.20 \times 10^{-4} \ \text{N}) = -5.52 \times 10^{-5} \ \text{N}$.

(b) $F = \sqrt{F_x^2 + F_y^2} = 1.02 \times 10^{-4} \ \text{N}$. $\tan \phi = \left| \dfrac{F_y}{F_x} \right| = 0.640$. $\phi = 32.6°$, below the $+x$ axis.

Reflect: The individual forces on q_3 are computed from Coulomb's law and then added as vectors, using components.

*17.69. Set Up: $F = k \dfrac{|qq'|}{r^2}$. Like charges repel and unlike charges attract.

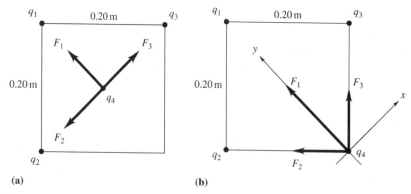

q_1 0.20 m q_3 q_1 0.20 m q_3

(a)   (b)

Solve: (a) The charges and the forces on the $-1.00\,\mu\text{C}$ charge are shown in Figure (a) above. The distance r_{14} between q_1 and q_4 is $r = \left(\frac{1}{2}\right)\sqrt{2}(0.200\text{ m}) = 0.1414$ m. $\vec{F}_2$ and $\vec{F}_3$ are equal in magnitude and opposite in direction, so $\vec{F}_2 + \vec{F}_3 = 0$ and $\vec{F}_{\text{net}} = \vec{F}_1$.

$$F_1 = k\frac{|q_1 q_4|}{r_{14}^{\,2}} = (8.99 \times 10^9 \text{ N} \cdot \text{m}^2/\text{C}^2)\frac{(3.00 \times 10^{-9}\text{ C})(1.00 \times 10^{-6}\text{ C})}{(0.1414\text{ m})^2} = 1.35 \times 10^{-3}\text{ N}.$$

The resultant force has magnitude 1.35×10^{-3} N and is directed away from the vacant corner.

(b) The charges and the forces on the $-1.00\,\mu\text{C}$ charge are shown in Figure (b) above. The distance r_{14} between q_1 and q_4 is $\sqrt{2}(0.200\text{ m}) = 0.2828$ m. Use coordinates as shown in the figure. $F_{2x} = -F_{3x}$ and $F_{1x} = 0$, so $F_x = 0$. $F_y = F_{1y} + F_{2y} + F_{3y} = F_1 + 2F_2\cos 45°$.

$$F_1 = k\frac{|q_1 q_4|}{r_{14}^{\,2}} = (8.99 \times 10^9 \text{ N} \cdot \text{m}^2/\text{C}^2)\frac{(3.00 \times 10^{-9}\text{ C})(1.00 \times 10^{-6}\text{ C})}{(0.2828\text{ m})^2} = 3.372 \times 10^{-4}\text{ N}.$$

$$F_2 = k\frac{|q_2 q_4|}{r_{24}^{\,2}} = (8.99 \times 10^9 \text{ N} \cdot \text{m}^2/\text{C}^2)\frac{(3.00 \times 10^{-9}\text{ C})(1.00 \times 10^{-6}\text{ C})}{(0.200\text{ m})^2} = 6.742 \times 10^{-4}\text{ N}$$

$$F_y = 3.372 \times 10^{-4}\text{ N} + 2(6.742 \times 10^{-4}\text{ N})\cos 45° = 1.29 \times 10^{-3}\text{ N}.$$

The resultant force has magnitude 1.29×10^{-3} N and is directed toward the center of the square.

*17.71. Set Up: The ball is in equilibrium, so for it $\Sigma F_x = 0$ and $\Sigma F_y = 0$. The force diagram for the ball is given in the figure below. F_E is the force exerted by the electric field. $\vec{F} = q\vec{E}$. Since the electric field is horizontal, $\vec{F}_E$ is horizontal. Use the coordinates shown in the figure. The tension in the string has been replaced by its x and y components.

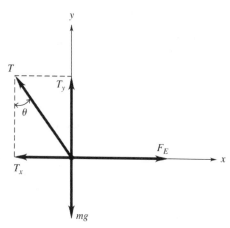

Solve: $\sum F_y = 0$ gives $T_y - mg = 0$. $T\cos\theta - mg = 0$ and $T = \dfrac{mg}{\cos\theta}$.

$\sum F_x = 0$ gives $F_E - T_x = 0$. $F_E - T\sin\theta = 0$.

$$F_E = \left(\frac{mg}{\cos\theta}\right)\sin\theta = mg\tan\theta = (12.3\times10^{-3}\text{ kg})(9.80\text{ m/s}^2)\tan17.4° = 3.78\times10^{-2}\text{ N}$$

$F_E = |q|E$ so $E = \dfrac{F_E}{|q|} = \dfrac{3.78\times10^{-2}\text{ N}}{1.11\times10^{-6}\text{ C}} = 3.41\times10^4$ N/C

q is negative and $\vec{F}_E$ is to the right, so $\vec{E}$ is to the left in the figure.

Reflect: The larger the electric field E the greater the angle the string makes with the wall.

*17.75. Set Up: The force on the electron is upward in the figure so the electric field must be downward. To produce a net electric field that is downward, it must be that q_1 is positive, q_2 is negative, and $|q_1| = |q_2|$. The field due to q_1 and q_2 at the location of the electron are sketched in the figure below. The electron is $r = 3.61$ cm from each charge. $\tan\theta = \dfrac{3.00\text{ cm}}{2.00\text{ cm}}$ and $\theta = 56.3°$.

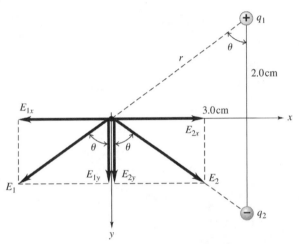

Solve: The net electric field is $E = 2E_{1y} = 2k\dfrac{q_1}{r^2}\cos\theta$. $F = eE = \dfrac{2keq_1}{r^2}\cos\theta$. $F = ma$ so $\dfrac{2keq_1}{r^2}\cos\theta = ma$ and

$$q_1 = \frac{r^2 ma}{2ke\cos\theta} = \frac{(3.61\times10^{-2}\text{ m})^2(9.11\times10^{-31}\text{ kg})(8.25\times10^{18}\text{ m/s}^2)}{2(8.99\times10^9\text{ N}\cdot\text{m}^2/\text{C}^2)(1.60\times10^{-19}\text{ C})\cos56.3°} = 6.14\times10^{-6}\text{ C}.$$

$q_2 = -6.14\times10^{-6}$ C.

Reflect: The force that an electric field exerts on a negative charge is opposite to the direction of the electric field. We could also do this problem by considering the force that each charge exerts on the electron.

*17.77. Set Up: The force between the proton and electron is given by $F_E = k\dfrac{q_1 q_2}{r^2} = k\dfrac{e^2}{r^2}$. For a circular orbit we have $a = \dfrac{v^2}{r}$.

Solve: Applying $F = ma$ gives $k\dfrac{e^2}{r^2} = m_e\dfrac{v^2}{r}$. Solving for v we obtain

$$v = \sqrt{\frac{ke^2}{m_e r}} = \sqrt{\frac{(8.99\times10^9\text{ N}\cdot\text{m}^2/\text{C}^2)(1.602\times10^{-19}\text{ C})^2}{(9.109\times10^{-31}\text{ kg})(5.29\times10^{-11}\text{ m})}} = 2.2\times10^6\text{ m/s}.$$

Reflect: Although this speed is fast, it is only about 1% the speed of light.

Solutions to Passage Problems

*17.79. Set Up: Molecules are in a constant state of agitation and tend to diffuse randomly in the absence of a perturbing force. Although an external electric field can alter the distribution of the proteins, the total number of protein molecules will remain fixed.

Solve: Once the external electric field is removed, the random motion of the proteins will cause them to distribute themselves randomly within the membrane. Since the total number of proteins must remain fixed, the final density of proteins must be intermediate between the minimum value shown at $0°$ and the maximum value shown at $180°$. The correct answer is C.

ELECTRIC POTENTIAL AND CAPACITANCE

Problems 1, 5, 9, 11, 13, 17, 21, 23, 27, 29, 31, 35, 39, 41, 43, 47, 49, 53, 57, 61, 63, 65, 69, 73, 75, 77, 81, 83, 85, 89, 91

Solutions to Problems

*18.1. Set Up: Since the charge is positive the force on it is in the same direction as the electric field. Since the field is uniform the force is constant and $W = Fs\cos\phi$.

Solve: (a) $\vec{F}$ is upward and $\vec{s}$ is to the right, so $\phi = 90°$ and $W = 0$.

(b) $\vec{F}$ is upward and $\vec{s}$ is upward, so $\phi = 0°$

$$W = Fs = qEs = (28.0 \times 10^{-9} \text{ C})(4.00 \times 10^{4} \text{ N/C})(0.670 \text{ m}) = 7.50 \times 10^{-4} \text{ J}.$$

(c) $\vec{F}$ is upward and $\vec{s}$ is at $45.0°$ below the horizontal, so $\phi = 135.0°$.

$$W = Fs\cos\phi = qEs\cos\phi = (28.0 \times 10^{-9} \text{ C})(4.00 \times 10^{4} \text{ N/C})(2.60 \text{ m})\cos135.0° = -2.06 \times 10^{-3} \text{ J}.$$

Reflect: The work is positive when the displacement has a component in the direction of the force and it is negative when the displacement has a component opposite to the direction of the force. When the displacement is perpendicular to the force the work done is zero.

*18.5. Set Up: An electron has charge $q = -e = -1.60 \times 10^{-19}$ C. For a pair of point charges, $U = k\dfrac{qq'}{r}$. Let $q_1 = +3.00$ nC and $q_2 = +2.00$ nC. Let q_3 be the electron. The potential energy of the electron is the sum of U_{13} and U_{23}.

Solve: (a) $r_1 = r_2 = 0.250$ m.

$$U_{13} = k\frac{q_1 q_3}{r_1} = (8.99 \times 10^{9} \text{ N} \cdot \text{m}^2/\text{C}^2)\left(\frac{[3.00 \times 10^{-9} \text{ C}][-1.60 \times 10^{-19} \text{ C}]}{0.250 \text{ m}}\right) = -1.73 \times 10^{-17} \text{ J}.$$

$$U_{23} = k\frac{q_2 q_3}{r_2} = (8.99 \times 10^{9} \text{ N} \cdot \text{m}^2/\text{C}^2)\left(\frac{[2.00 \times 10^{-9} \text{ C}][-1.60 \times 10^{-19} \text{ C}]}{0.250 \text{ m}}\right) = -1.15 \times 10^{-17} \text{ J}.$$

$$U = U_{13} + U_{23} = -2.88 \times 10^{-17} \text{ J}$$

(b) $r_1 = 0.100$ m and $r_2 = 0.400$ m.

$$U_{13} = k\frac{q_1 q_3}{r_1} = (8.99 \times 10^{9} \text{ N} \cdot \text{m}^2/\text{C}^2)\left(\frac{[3.00 \times 10^{-9} \text{ C}][-1.60 \times 10^{-19} \text{ C}]}{0.100 \text{ m}}\right) = -4.32 \times 10^{-17} \text{ J}.$$

$$U_{23} = k\frac{q_2 q_3}{r_2} = (8.99 \times 10^9 \text{ N} \cdot \text{m}^2/\text{C}^2) \left(\frac{[2.00 \times 10^{-9} \text{ C}][-1.60 \times 10^{-19} \text{ C}]}{0.400 \text{ m}} \right) = -7.19 \times 10^{-18} \text{ J}.$$

$$U = U_{13} + U_{23} = -5.04 \times 10^{-17} \text{ J}.$$

Reflect: The potential energy is negative since the charge of the electron has opposite sign from the charge of each of the other two particles. The magnitude of the potential energy increases when the electron moves toward the larger charge.

*18.9. Set Up: Call the three charges 1, 2, and 3. $U = U_{12} + U_{13} + U_{23}$. $U_{12} = U_{23} = U_{13}$ because the charges are equal and each pair of charges has the same separation, 0.500 m.

Solve: $U = \dfrac{3kq^2}{0.500 \text{ m}} = \dfrac{3k(1.2 \times 10^{-6} \text{ C})^2}{0.500 \text{ m}} = 0.078 \text{ J}.$

Reflect: When the three charges are brought in from infinity to the corners of the triangle, the repulsive electrical forces between each pair of charges do negative work and electrical potential energy is stored.

*18.11. Set Up: For a pair of oppositely charged parallel metal plates, $V_{ab} = Ed$. $F = |q|E$.

Solve: (a) $E = \dfrac{V_{ab}}{d} = \dfrac{360 \text{ V}}{45.0 \times 10^{-3} \text{ m}} = 8000 \text{ V/m} = 8000 \text{ N/C}$

(b) $F = |q|E = (2.40 \times 10^{-9} \text{ C})(8000 \text{ N/C}) = 1.92 \times 10^{-5} \text{ N}$

*18.13. Set Up: For two oppositely charged parallel plates, $V_{ab} = Ed$, where V_{ab} is the potential difference between the two plates, E is the uniform electric field between the plates, and d is the separation of the plates. An electric field $\vec{E}$ exerts a force $\vec{F} = |q|\vec{E}$ on a charge placed in the field. $a = F/m$. An electron has charge $-e$ and mass 9.11×10^{-31} kg.

Solve: (a) $E = \dfrac{V_{ab}}{d} = \dfrac{25 \text{ V}}{2.0 \times 10^{-3} \text{ m}} = 1.25 \times 10^4 \text{ V/m}$

(b) $F = |q|E = eE.$ $a = \dfrac{F}{m} = \dfrac{eE}{m} = \dfrac{(1.60 \times 10^{-19} \text{ C})(1.25 \times 10^4 \text{ V/m})}{9.11 \times 10^{-31} \text{ kg}} = 2.20 \times 10^{15} \text{ m/s}^2$

Reflect: The electric field is the same at all points between the plates (away from the edges) so the acceleration would be the same at all points between the plates as it is for a point midway between the plates.

*18.17. Set Up: From Example 18.4, with $V_b = 0$ and x rather than y as the distance from the negative plate, $V_x = Ex$.

Solve: $E = \dfrac{V_x}{x} = \dfrac{5.0 \text{ V}}{0.200 \text{ m}} = 25 \text{ V/m}$

Reflect: $V_x = Ex$ says the potential increases linearly with x and the graph of V_x versus x should be a straight line, in agreement with the figure in the problem.

*18.21. Set Up: Apply conservation of energy to points A and B.

Solve: We know that $K_A + U_A = K_B + U_B$ and $U = qV$, so $K_A + qV_A = K_B + qV_B$. Thus,

$$K_B = K_A + q(V_A - V_B) = 0.00250 \text{ J} + (-5.00 \times 10^{-6} \text{ C})(200 \text{ V} - 800 \text{ V}) = 0.00550 \text{ J}.$$

Which gives

$$v_B = \sqrt{2K_B/m} = 7.42 \text{ m/s}.$$

Reflect: It is faster at B; a negative charge gains speed when it moves to higher potential.

*18.23. Set Up: $U = k\dfrac{qQ}{r}$. Conservation of energy says $U_a + K_a = U_b + K_b$. $K = \frac{1}{2}mv^2$.

Solve: (a) $U = (8.99 \times 10^9 \text{ N} \cdot \text{m}^2/\text{C}^2)\dfrac{(1.20 \times 10^{-6} \text{ C})(4.60 \times 10^{-6} \text{ C})}{0.250 \text{ m}} = 0.198 \text{ J}$

(b) (i) $U_a = +0.198 \text{ J}$, $K_a = 0$. $r_b = 0.500 \text{ m}$ so $U_b = \frac{1}{2}U_a = 0.099 \text{ J}$.

$$K_b = U_a + K_a - U_b = 0.198 \text{ J} - 0.099 \text{ J} = 0.099 \text{ J}. \quad v_b = \sqrt{\dfrac{2K_b}{m}} = \sqrt{\dfrac{2(0.099 \text{ J})}{2.80 \times 10^{-4} \text{ kg}}} = 26.6 \text{ m/s}.$$

(ii) $r_b = 5.00 \text{ m}$ so $U_b = 0.0099 \text{ J}$. $K_b = 0.198 \text{ J} - 0.0099 \text{ J} = 0.188 \text{ J}$. $v_b = 36.7 \text{ m/s}$.

(iii) $r_b = 50.0 \text{ m}$ so $U_b = 0.00099 \text{ J}$. $K_b = 0.198 \text{ J} - 0.00099 \text{ J} = 0.197 \text{ J}$. $v_b = 37.5 \text{ m/s}$.

Reflect: As the charge q moves away from Q the repulsive force does positive work on q and its kinetic energy increases.

*18.27. Set Up: Treat the gold nucleus as a point charge so that $V = k\dfrac{q}{r}$. According to conservation of energy we have $K_a + U_a = K_b + U_b$, where $U = qV$.

Solve: Assume that the alpha particle is at rest before it is accelerated and that it momentarily stops when it arrives at its closest approach to the surface of the gold nucleus. Thus we have $K_a = K_b = 0$, which implies that $U_a = U_b$. Since $U = qV$ we conclude that the accelerating voltage must be equal to the voltage at its point of closest approach to the surface of the gold nucleus:

$$V_a = V_b = k\dfrac{q}{r} = (8.99 \times 10^9 \text{ N} \cdot \text{m}^2/\text{C}^2)\dfrac{79(1.60 \times 10^{-19} \text{ C})}{(7.3 \times 10^{-15} \text{ m} + 2.0 \times 10^{-14} \text{ m})} = 4.2 \times 10^6 \text{ V}.$$

Reflect: Although the alpha particle has kinetic energy as it approaches the gold nucleus this is irrelevant to our solution since energy is conserved for the whole process.

*18.29. Set Up: Example 18.4 shows that $V = Ey$, where y is the distance from the negative plate. $V = 0$ at the negative plate. For $y = 25 \text{ cm}$, $V = 50.0 \text{ V}$.

Solve: (a) $E = \dfrac{50.0 \text{ V}}{25 \text{ cm}} = 2.0 \text{ V/cm}$. So $V = (2.0 \text{ V/cm})y$.

$V = +10.0 \text{ V}$ for $y = 5 \text{ cm}$, $V = +20.0 \text{ V}$ for $y = 10 \text{ cm}$, $V = +30.0 \text{ V}$ for $y = 15 \text{ cm}$, $V = +40.0 \text{ V}$ for $y = 20 \text{ cm}$ and $V = +50.0 \text{ V}$ for $y = 25 \text{ cm}$. The equipotential surfaces are drawn in the figure below.

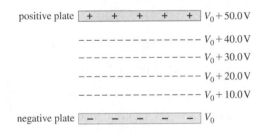

(b) Yes, they are separated by 5 cm.
(c) The equipotential surfaces are flat sheets parallel to the plates.
Reflect: The electric field lines are straight lines perpendicular to the plates, so are perpendicular to the equipotential surfaces, as they must be. The electric field is uniform so the equipotential lines of constant potential difference are equally spaced.

*18.31. Set Up: For a uniformly charged sphere $V = k\dfrac{q}{r}$.

Solve: (a) At the surface of the sphere we have

$$V = k\frac{q}{r} = \frac{(8.99 \times 10^9 \ \text{N} \cdot \text{m}^2/\text{C}^2)(1.50 \times 10^{-6} \ \text{C})}{0.500 \ \text{m}} = \underline{26.970} \ \text{kV} = \ 27.0 \ \text{kV}.$$

(b) Solving $V = k\dfrac{q}{r}$ for r we obtain $r = \dfrac{kq}{V}$. This allows us to find the distance between any two equipotential

surfaces V_1 and V_2: $r_2 - r_1 = \dfrac{kq}{V_2} - \dfrac{kq}{V_1} = \dfrac{kq(V_1 - V_2)}{V_1 V_2}$.

We desire to space the equipotentials in 500 V increments outside the sphere. Thus, the first equipotential is at $\underline{26.970} \ \text{kV} - 500 \ \text{V} = \underline{26.470} \ \text{kV}$ and the second equipotential is at $\underline{26.970} \ \text{kV} - 2(500 \ \text{V}) = \underline{25.970} \ \text{kV}$. The required spacing of the first two equipotentials is

$$r_2 - r_1 = \frac{kq(V_1 - V_2)}{V_1 V_2} = \frac{(8.99 \times 10^9 \ \text{N} \cdot \text{m}^2/\text{C}^2)(1.50 \times 10^{-6} \ \text{C})(500 \ \text{V})}{(\underline{26.470} \times 10^3 \ \text{V})(\underline{25.970} \times 10^3 \ \text{V})} = 0.0098 \ \text{m} = 0.98 \ \text{cm}.$$

The 20th equipotential is at $\underline{26.970} \ \text{kV} - 20(500 \ \text{V}) = \underline{16.970} \ \text{kV}$ and the 21st equipotential is at $\underline{26.970} \ \text{kV} - 21(500 \ \text{V}) = \underline{16.470} \ \text{kV}$. Thus, the required spacing is

$$r_2 - r_1 = \frac{kq(V_1 - V_2)}{V_1 V_2} = \frac{(8.99 \times 10^9 \ \text{N} \cdot \text{m}^2/\text{C}^2)(1.50 \times 10^{-6} \ \text{C})(500 \ \text{V})}{(\underline{16.970} \times 10^3 \ \text{V})(\underline{16.470} \times 10^3 \ \text{V})} = 0.024 \ \text{m} = 2.4 \ \text{cm}.$$

(c) The increased spacing of the equipotentials is due to the fact that the electric field is weaker at greater distances from the charged sphere.

*18.35. Set Up: $1 \ \text{eV} = 1.60 \times 10^{-19} \ \text{J}$. $K = \tfrac{1}{2} mv^2$. $c = 3.00 \times 10^8 \ \text{m/s}$. An electron has mass $9.11 \times 10^{-31} \ \text{kg}$ and a proton has mass $1.67 \times 10^{-27} \ \text{kg}$.

Solve: (a) $K = 1.00 \ \text{eV} = 1.60 \times 10^{-19} \ \text{J}$. $v = \sqrt{\dfrac{2K}{m}}$

electron: $v_e = \sqrt{\dfrac{2(1.60 \times 10^{-19} \ \text{J})}{9.11 \times 10^{-31} \ \text{kg}}} = 5.93 \times 10^5 \ \text{m/s}$

proton: $v_p = \sqrt{\dfrac{2(1.60 \times 10^{-19} \ \text{J})}{1.67 \times 10^{-27} \ \text{kg}}} = 1.38 \times 10^4 \ \text{m/s}$. $\dfrac{v_e}{v_p} = \sqrt{\dfrac{m_p}{m_e}} = \sqrt{1836}$

(b) $K = 1.00 \ \text{keV} = 1.60 \times 10^{-16} \ \text{J}$

electron: $v_e = 1.87 \times 10^7 \ \text{m/s}$; proton: $v_p = 4.38 \times 10^5 \ \text{m/s}$

(c) $v_e = v_p = (0.0100)c = 3.00 \times 10^6 \ \text{m/s}$

$$K_e = \tfrac{1}{2} m_e v_e^2 = \tfrac{1}{2}(9.11 \times 10^{-31} \ \text{kg})(3.00 \times 10^6 \ \text{m/s})^2 = 4.10 \times 10^{-18} \ \text{J} = 0.0256 \ \text{keV}$$

$$K_p = \tfrac{1}{2} m_p v_p^2 = \tfrac{1}{2}(1.67 \times 10^{-27} \ \text{kg})(3.00 \times 10^6 \ \text{m/s})^2 = 7.52 \times 10^{-15} \ \text{J} = 47.0 \ \text{keV}$$

The proton energy is larger by a factor of m_p/m_e.

Reflect: When we use $K = \tfrac{1}{2} mv^2$ we should express all quantities in SI units.

*18.39. Set Up: For a parallel-plate capacitor $C = \dfrac{Q}{V_{ab}}$. $C = \dfrac{\epsilon_0 A}{d}$. $V_{ab} = Ed$. The surface charge density is $\sigma = \dfrac{Q}{A}$.

Solve: (a) $V_{ab} = \dfrac{Q}{C} = \dfrac{0.200 \times 10^{-6} \ \text{C}}{500.0 \times 10^{-12} \ \text{F}} = 400 \ \text{V}$

(b) $A = \dfrac{Cd}{\epsilon_0} = \dfrac{(500.0 \times 10^{-12} \text{ F})(0.600 \times 10^{-3} \text{ m})}{8.854 \times 10^{-12} \text{ C}^2/(\text{N} \cdot \text{m}^2)} = 0.0339 \text{ m}^2$

(c) $E = \dfrac{V_{ab}}{d} = \dfrac{400 \text{ V}}{0.600 \times 10^{-3} \text{ m}} = 6.67 \times 10^5 \text{ V/m}$

(d) $\sigma = \dfrac{Q}{A} = \dfrac{0.200 \times 10^{-6} \text{ C}}{0.0339 \text{ m}^2} = 5.90 \times 10^{-6} \text{ C/m}^2$

Reflect: If the plates are square, each side is about 18 cm in length. We could also calculate σ from $E = \sigma/\epsilon_0$.

*18.41. Set Up: $C = \dfrac{Q}{V_{ab}}$. $C = \dfrac{\epsilon_0 A}{d}$.

Solve: (a) $Q = CV_{ab} = (10.0 \times 10^{-6} \text{ F})(12.0 \text{ V}) = 1.20 \times 10^{-4} \text{ C} = 120 \ \mu\text{C}$

(b) When d is doubled C is halved, so Q is halved. $Q = 60 \ \mu\text{C}$.

(c) If r is doubled, A increases by a factor of 4. C increases by a factor of 4 and Q increases by a factor of 4. $Q = 480 \ \mu\text{C}$.

Reflect: When the plates are moved apart, less charge on the plates is required to produce the same potential difference. With the separation of the plates constant, the electric field must remain constant to produce the same potential difference. The electric field depends on the surface charge density, σ. To produce the same σ, more charge is required when the area increases.

*18.43. Set Up: For parallel disks we have $C = \epsilon_0 \dfrac{A}{d}$, where $A = \pi r^2$. The voltage between the charged plates is given by $V = \dfrac{Q}{C} = \dfrac{Qd}{\epsilon_0 A}$.

Solve: (a) $C = (8.854 \times 10^{-12} \text{ F/m}) \dfrac{\pi (0.0750 \text{ m})^2}{(0.00350 \text{ m})} = 44.7 \text{ pF}$

(b) Once the battery is disconnected, the charge remains fixed. Use the equation $V = \dfrac{Qd}{\epsilon_0 A}$ and look at the ratio of the voltage for the two different plate separations: $\dfrac{V}{6.00 \text{ volts}} = \dfrac{3.50 \text{ cm}}{3.50 \text{ mm}} = 10.0$. Thus the voltage is increased by a factor of 10 to a final value of 600 volts.

Reflect: The work you do in pulling the oppositely charged plates apart provides the energy needed to increase the voltage between the plates.

*18.47. Set Up: The capacitors between b and c are in parallel. This combination is in series with the 15 pF capacitor. $C_1 = 15$ pF, $C_2 = 9.0$ pF and $C_3 = 11$ pF.

Solve: (a) For capacitors in parallel, $C_{eq} = C_1 + C_2 + \cdots$ so $C_{23} = C_2 + C_3 = 20$ pF

(b) $C_1 = 15$ pF is in series with $C_{23} = 20$ pF. For capacitors in series, $\dfrac{1}{C_{eq}} = \dfrac{1}{C_1} + \dfrac{1}{C_2} + \cdots$ so

$$\dfrac{1}{C_{123}} = \dfrac{1}{C_1} + \dfrac{1}{C_{23}} \text{ and } C_{123} = \dfrac{C_1 C_{23}}{C_1 + C_{23}} = \dfrac{(15 \text{ pF})(20 \text{ pF})}{15 \text{ pF} + 20 \text{ pF}} = 8.6 \text{ pF}.$$

Reflect: For capacitors in parallel the equivalent capacitance is larger than any of the individual capacitors. For capacitors in series the equivalent capacitance is smaller than any of the individual capacitors.

*18.49. Set Up: For capacitors in parallel the voltages are the same and the charges add. For capacitors in series, the charges are the same and the voltages add. $C = Q/V$. C_1 and C_2 are in parallel and C_3 is in series with the parallel combination of C_1 and C_2.

Solve: (a) C_1 and C_2 are in parallel and so have the same potential across them: $V_1 = V_2 = \dfrac{Q_2}{C_2} = \dfrac{40.0 \times 10^{-6} \text{ C}}{3.00 \times 10^{-6} \text{ F}} =$

13.33 V. Therefore we have $Q_1 = V_1 C_1 = (13.33 \text{ V})(6.00 \times 10^{-6} \text{ F}) = 80.0 \times 10^{-6}$ C. Since C_3 is in series with the parallel combination of C_1 and C_2, its charge must be equal to their combined charge:

$Q_3 = 40.0 \times 10^{-6}$ C $+ 80.0 \times 10^{-6}$ C $= 120.0 \times 10^{-6}$ C.

(b) The total capacitance is found from $\dfrac{1}{C_{tot}} = \dfrac{1}{C_{12}} + \dfrac{1}{C_3} = \dfrac{1}{9.00 \times 10^{-6} \text{ F}} + \dfrac{1}{5.00 \times 10^{-6} \text{ F}}$ and $C_{tot} = 3.21 \, \mu\text{F}$.

$$V_{ab} = \frac{Q_{tot}}{C_{tot}} = \frac{120.0 \times 10^{-6} \text{ C}}{3.21 \times 10^{-6} \text{ F}} = 37.4 \text{ V}.$$

Reflect: $V_3 = \dfrac{Q_3}{C_3} = \dfrac{120.0 \times 10^{-6} \text{ C}}{5.00 \times 10^{-6} \text{ F}} = 24.0$ V. $V_{ab} = V_1 + V_3$.

*18.53. Set Up: $C = \dfrac{Q}{V}$. For two capacitors in parallel, $C_{eq} = C_1 + C_2$. For two capacitors in series,

$$\frac{1}{C_{eq}} = \frac{1}{C_1} + \frac{1}{C_2} \quad \text{and} \quad C_{eq} = \frac{C_1 C_2}{C_1 + C_2}.$$

For capacitors in parallel, the voltages are the same and the charges add. For capacitors in series, the charges are the same and the voltages add. Let $C_1 = 3.00 \, \mu\text{F}$, $C_2 = 5.00 \, \mu\text{F}$ and $C_3 = 6.00 \, \mu\text{F}$.

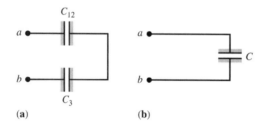

(a)   (b)

Solve: (a) The equivalent capacitance for C_1 and C_2 in parallel is $C_{12} = C_1 + C_2 = 8.00 \, \mu\text{F}$. This gives the circuit shown in Figure (a) above. In that circuit the equivalent capacitance is

$$C = \frac{C_{12} C_3}{C_{12} + C_3} = \frac{(8.00 \, \mu\text{F})(6.00 \, \mu\text{F})}{8.00 \, \mu\text{F} + 6.00 \, \mu\text{F}} = 3.43 \, \mu\text{F}.$$

This gives the circuit shown in Figure (b) above. In Figure (b), $Q = CV = (3.43 \times 10^{-6} \text{ F})(24.0 \text{ V}) = 8.23 \times 10^{-5}$ C. In Figure (a) each capacitor therefore has charge 8.23×10^{-5} C. The potential differences are

$$V_3 = \frac{Q_3}{C_3} = \frac{8.23 \times 10^{-5} \text{ C}}{6.00 \times 10^{-6} \text{ F}} = 13.7 \text{ V} \quad \text{and} \quad V_{12} = \frac{Q_{12}}{C_{12}} = \frac{8.23 \times 10^{-5} \text{ C}}{8.00 \times 10^{-6} \text{ F}} = 10.3 \text{ V}.$$

Note that $V_3 + V_{12} = 24.0$ V. Then in the original circuit, $V_1 = V_2 = V_{12} = 10.3$ V.

$$Q_1 = V_1 C_1 = (10.3 \text{ V})(3.00 \times 10^{-6} \text{ F}) = 3.09 \times 10^{-5} \text{ C}.$$

$$Q_2 = V_2 C_2 = (10.3 \text{ V})(5.00 \times 10^{-6} \text{ F}) = 5.15 \times 10^{-5} \text{ C}.$$

$Q_1 = 30.9 \, \mu\text{C}$, $Q_2 = 51.5 \, \mu\text{C}$ and $Q_3 = 82.3 \, \mu\text{C}$. Note that $Q_1 + Q_2 = Q_3$.

(b) $V_1 = 10.3$ V, $V_2 = 10.3$ V and $V_3 = 13.7$ V

Reflect: Note that $Q_1 + Q_2 = Q_3$, $V_1 = V_2$ and $V_1 + V_3 = 24.0$ V

*18.57. **Set Up:** The charge across a capacitor charged to a voltage V is $Q = CV$. The energy stored in a charged capacitor is $U = \frac{1}{2}CV^2 = \frac{1}{2}QV = \frac{1}{2}\frac{Q^2}{C}$. For capacitors in parallel we have $C_{eq} = C_1 + C_2$ and for capacitors in series we have $C_{eq} = \frac{C_1 C_2}{C_1 + C_2}$.

Solve: **(a)** For the capacitors in parallel we have $C_{eq} = C_1 + C_2 = 2.5\ \mu\text{F} + 5.0\ \mu\text{F} = 7.5\ \mu\text{F}$. The charge provided by the battery is $Q = CV = (7.5\ \mu\text{F})(12\ \text{V}) = 90\ \mu\text{C}$.

(b) For the capacitors in series we have $C_{eq} = \frac{C_1 C_2}{C_1 + C_2} = \frac{(2.5\ \mu\text{F})(5.0\ \mu\text{F})}{2.5\ \mu\text{F} + 5.0\ \mu\text{F}} = 1.67\ \mu\text{F}$. The charge provided by the battery is $Q = CV = (1.67\ \mu\text{F})(12\ \text{V}) = 20\ \mu\text{C}$.

(c) For the capacitors in parallel we have $U = \frac{1}{2}QV = \frac{1}{2}(90\ \mu\text{C})(12\ \text{V}) = 5.4 \times 10^{-4}$ J. For the capacitors in series we have $U = \frac{1}{2}QV = \frac{1}{2}(20\ \mu\text{C})(12\ \text{V}) = 1.2 \times 10^{-4}$ J.

Reflect: Since we also know the charge on each equivalent capacitor we could also use either $U = \frac{1}{2}\frac{Q^2}{C_{eq}}$ or $U = \frac{1}{2}C_{eq}V^2$ to calculate the energy supplied by the battery.

*18.61. **Set Up:** For a parallel plate capacitor we have $C = \frac{\epsilon_0 A}{d}$. The stored energy can be expressed either as $\frac{Q^2}{2C}$ or as $\frac{CV^2}{2}$, whichever is more convenient for the calculation. Since d is halved, C doubles.

Solve: **(a)** If the separation distance is halved while the charge is kept fixed, then the capacitance increases and the stored energy, which was 8.38 J, decreases since $U = Q^2/2C$. Therefore the new energy is 4.19 J.

(b) If the voltage is kept fixed while the separation is decreased by one half, then the doubling of the capacitance leads to a doubling of the stored energy to 16.8 J, using $U = CV^2/2$, when V is held constant throughout.

Reflect: When the capacitor is disconnected, the stored energy decreases because of the positive work done by the attractive force between the plates. When the capacitor remains connected to the battery, $Q = CV$ tells us that the charge on the plates increases. The increased stored energy comes from the battery when it puts more charge onto the plates.

*18.63. **Set Up:** The two capacitors are in series. For capacitors in series the voltages add and the charges are the same. $\frac{1}{C_{eq}} = \frac{1}{C_1} + \frac{1}{C_2} + \cdots$ $C = \frac{Q}{V}$. $U = \frac{1}{2}CV^2$.

Solve: **(a)** $\frac{1}{C_{eq}} = \frac{1}{C_1} + \frac{1}{C_2}$ so $C_{eq} = \frac{C_1 C_2}{C_1 + C_2} = \frac{(150\ \text{nF})(120\ \text{nF})}{150\ \text{nF} + 120\ \text{nF}} = 66.7\ \text{nF}$.

$$Q = CV = (66.7\ \text{nF})(36\ \text{V}) = 2.4 \times 10^{-6}\ \text{C} = 2.4\ \mu\text{C}$$

(b) $Q = 2.4\ \mu\text{C}$ for each capacitor.

(c) $U = \frac{1}{2}C_{eq}V^2 = \frac{1}{2}(66.7 \times 10^{-9}\ \text{F})(36\ \text{V})^2 = 43.2\ \mu\text{J}$

(d) We know C and Q for each capacitor so rewrite U in terms of these quantities.

$$U = \frac{1}{2}CV^2 = \frac{1}{2}C(Q/C)^2 = Q^2/2C$$

$$150\ \text{nF}:\ U = \frac{(2.4 \times 10^{-6}\ \text{C})^2}{2(150 \times 10^{-9}\ \text{F})} = 19.2\ \mu\text{J};\ \ 120\ \text{nF}:\ U = \frac{(2.4 \times 10^{-6}\ \text{C})^2}{2(120 \times 10^{-9}\ \text{F})} = 24.0\ \mu\text{J}$$

Note that $19.2 \, \mu J + 24.0 \, \mu J = 43.2 \, \mu J$, the total stored energy calculated in part (c).

$$150 \text{ nF: } V = \frac{Q}{C} = \frac{2.4 \times 10^{-6} \text{ C}}{150 \times 10^{-9} \text{ F}} = 16 \text{ V}; \quad 120 \text{ nF: } V = \frac{Q}{C} = \frac{2.4 \times 10^{-6} \text{ C}}{120 \times 10^{-9} \text{ F}} = 20 \text{ V}$$

Note that these two voltages sum to 36 V, the voltage applied across the network.

Reflect: Since Q is the same the capacitor with smaller C stores more energy $(U = Q^2/2C)$ and has a larger voltage $(V = Q/C)$.

*18.65. Set Up: $C = \dfrac{Q}{V_{ab}}$. Let $C_1 = 20.0 \, \mu F$ and $C_2 = 10.0 \, \mu F$. The energy stored in a capacitor is

$$\tfrac{1}{2}QV_{ab} = \tfrac{1}{2}CV_{ab}^2 = \tfrac{1}{2}\frac{Q^2}{V_{ab}}.$$

Solve: (a) The initial charge on the $20.0 \, \mu F$ capacitor is

$$Q = C_1(800 \text{ V}) = (20.0 \times 10^{-6} \text{ F})(800 \text{ V}) = 0.0160 \text{ C}.$$

(b) In the final circuit, charge Q is distributed between the two capacitors and $Q_1 + Q_2 = Q$. The final circuit contains only the two capacitors, so the voltage across each is the same, $V_1 = V_2$.

$$V = \frac{Q}{C} \text{ so } V_1 = V_2 \text{ gives } \frac{Q_1}{C_1} = \frac{Q_2}{C_2}. \quad Q_1 = \frac{C_1}{C_2}Q_2 = 2Q_2.$$

Using this in $Q_1 + Q_2 = 0.0160$ C gives $3Q_2 = 0.0160$ C and $Q_2 = 5.33 \times 10^{-3}$ C. $Q = 2Q_2 = 1.066 \times 10^{-2}$ C.

$$V_1 = \frac{Q_1}{C_1} = \frac{1.066 \times 10^{-2} \text{ C}}{20.0 \times 10^{-6} \text{ F}} = 533 \text{ V}. \quad V_2 = \frac{Q_2}{C_2} = \frac{5.33 \times 10^{-3} \text{ C}}{10.0 \times 10^{-6} \text{ F}} = 533 \text{ V}.$$

The potential differences across the capacitors are the same, as they should be.

(c) Energy $= \tfrac{1}{2}C_1V^2 + \tfrac{1}{2}C_2V^2 = \tfrac{1}{2}(C_1 + C_2)V^2 = \tfrac{1}{2}(20.0 \times 10^{-6} \text{ F} + 10.0 \times 10^{-6} \text{ F})(533 \text{ V})^2 = 4.26 \text{ J}.$

(d) The $20.0 \, \mu F$ capacitor initially has energy $= \tfrac{1}{2}C_1V^2 = \tfrac{1}{2}(20.0 \times 10^{-6} \text{ F})(800 \text{ V})^2 = 6.40 \text{ J}$. The decrease in stored energy that occurs when the capacitors are connected is $6.40 \text{ J} - 4.26 \text{ J} = 2.14 \text{ J}$.

Reflect: The decrease in stored energy is because of conversion of electrical energy to other forms during the motion of the charge when it becomes distributed between the two capacitors. Thermal energy is generated by the current in the wires and energy is emitted in electromagnetic waves.

*18.69. Set Up: $C = KC_0 = K\epsilon_0 \dfrac{A}{d}$. $A = 1.0 \text{ cm}^2 = 1.0 \times 10^{-4} \text{ m}^2$. $V = Ed$ for a parallel plate capacitor; this equation applies whether or not a dielectric is present.

Solve: (a) $C = (10)\dfrac{(8.85 \times 10^{-12} \text{ F/m})(1.0 \times 10^{-4} \text{ m}^2)}{7.5 \times 10^{-9} \text{ m}} = 1.18 \, \mu F$ per cm^2.

(b) $E = \dfrac{V}{Kd} = \dfrac{85 \text{ mV}}{(10)(7.5 \times 10^{-9} \text{ m})} = 1.13 \times 10^6 \text{ V/m}.$

Reflect: The dielectric material increases the capacitance and decreases the electric field that corresponds to a given potential difference.

*18.73. Set Up: We know that $C = Q/V$, $C = KC_0$, and $V = Ed$. Table 18.1 gives the dielectric constant $K = 3.1$ for Mylar.

Solve: (a) $\Delta Q = Q - Q_0 = (K-1)Q_0 = (K-1)C_0V_0 = (2.1)(2.5 \times 10^{-7} \text{ F})(12 \text{ V}) = 6.3 \times 10^{-6} \text{ C}.$

(b) $\sigma_i = \sigma(1 - 1/K)$ so $Q_i = Q(1 - 1/K) = (9.3 \times 10^{-6} \text{ C})(1 - 1/3.1) = 6.3 \times 10^{-6} \text{ C}.$

(c) The addition of the Mylar doesn't affect the electric field since the induced charge cancels the additional charge drawn to the plates.

Reflect: $E = V/d$ and V is constant so E doesn't change when the dielectric is inserted.

*18.75. Set Up: For a point charge, $E = \dfrac{k|q|}{r^2}$ and $V = \dfrac{kq}{r}$. The electric field is directed toward a negative charge and away from a positive charge.

Solve: (a) $V > 0$ so $q > 0$. $\dfrac{V}{E} = \dfrac{kq/r}{k|q|/r^2} = \left(\dfrac{kq}{r}\right)\left(\dfrac{r^2}{kq}\right) = r$. $\quad r = \dfrac{4.98 \text{ V}}{12.0 \text{ V/m}} = 0.415 \text{ m}$.

(b) $q = \dfrac{rV}{k} = \dfrac{(0.415 \text{ m})(4.98 \text{ V})}{8.99 \times 10^9 \text{ N} \cdot \text{m}^2/\text{C}^2} = 2.30 \times 10^{-10} \text{ C}$

(c) $q > 0$, so the electric field is directed away from the charge.

Reflect: The ratio of V to E due to a point charge increases as the distance r from the charge increases, because E falls off as $1/r^2$ and V falls off as $1/r$.

*18.77. Set Up: With air between the layers, $E_0 = \dfrac{Q}{\epsilon_0 A} = \dfrac{\sigma}{\epsilon_0}$ and $V_0 = E_0 d$. The energy density in the electric field is $u = \frac{1}{2}\epsilon_0 E^2$. The volume of a shell of thickness t and average radius R is $4\pi R^2 t$. The volume of a solid sphere of radius R is $\frac{4}{3}\pi R^3$. With the dielectric present, $E = \dfrac{E_0}{K}$ and $V = \dfrac{V_0}{K}$.

Solve: (a) $E_0 = \dfrac{\sigma}{\epsilon_0} = \dfrac{0.50 \times 10^{-3} \text{ C/m}^2}{8.854 \times 10^{-12} \text{ C}^2/(\text{N} \cdot \text{m}^2)} = 5.6 \times 10^7 \text{ V/m}$

(b) $V_0 = E_0 d = (5.6 \times 10^7 \text{ V/m})(5.0 \times 10^{-9} \text{ m}) = 0.28 \text{ V}$. The outer wall of the cell is at higher potential, since it has positive charge.

(c) For the cell, $V_{\text{cell}} = \frac{4}{3}\pi R^3$ and

$$R = \left(\dfrac{3V_{\text{cell}}}{4\pi}\right)^{1/3} = \left(\dfrac{3(10^{-16} \text{ m}^3)}{4\pi}\right)^{1/3} = 2.9 \times 10^{-6} \text{ m}.$$

The volume of the cell wall is $V_{\text{wall}} = 4\pi R^2 t = 4\pi(2.9 \times 10^{-6} \text{ m})^2(5.0 \times 10^{-9} \text{ m}) = 5.3 \times 10^{-19} \text{ m}^3$.

$$u_0 = \frac{1}{2}\epsilon_0 E_0^2 = \frac{1}{2}(8.854 \times 10^{-12} \text{ C}^2/(\text{N} \cdot \text{m}^2))(5.6 \times 10^7 \text{ V/m})^2 = 1.39 \times 10^4 \text{ V/m}^3.$$

The total electric field in the cell wall is $(1.39 \times 10^4 \text{ V/m}^3)(5.3 \times 10^{-19} \text{ m}^3) = 7 \times 10^{-15} \text{ V}$.

(d) $E = \dfrac{E_0}{K} = \dfrac{5.6 \times 10^7 \text{ V/m}}{5.4} = 1.0 \times 10^7 \text{ V/m}$ and $V = \dfrac{V_0}{K} = \dfrac{0.28 \text{ V}}{5.4} = 0.052 \text{ V}$.

*18.81. Set Up: $C = \dfrac{\epsilon_0 A}{d}$. $C = \dfrac{Q}{V_{ab}}$. $V_{ab} = Ed$ and $E = \dfrac{Q}{\epsilon_0 A}$. The stored energy is $\frac{1}{2}QV_{ab} = \frac{1}{2}CV_{ab}^2 = \frac{1}{2}\dfrac{Q^2}{C}$.

Solve: (a) $C = \dfrac{\epsilon_0 A}{d} = \dfrac{(8.854 \times 10^{-12} \text{ C}^2/(\text{N} \cdot \text{m}^2))(0.200 \text{ m})^2}{0.800 \times 10^{-2} \text{ m}} = 4.43 \times 10^{-11} \text{ F}$

(b) $Q = CV_{ab} = (4.43 \times 10^{-11} \text{ F})(120 \text{ V}) = 5.32 \times 10^{-9} \text{ C}$

(c) $E = \dfrac{V_{ab}}{d} = \dfrac{120 \text{ V}}{0.800 \times 10^{-2} \text{ m}} = 1.50 \times 10^4 \text{ V/m}$

(d) Energy $= \frac{1}{2}QV = \frac{1}{2}(5.32 \times 10^{-9} \text{ C})(120 \text{ V}) = 3.19 \times 10^{-7} \text{ J}$

(e) Since the battery is disconnected, the charge Q on the capacitor stays constant.

(a) $C = \dfrac{\epsilon_0 A}{d}$ so $C = \frac{1}{2}(4.43 \times 10^{-11} \text{ F}) = 2.22 \times 10^{-11} \text{ F}$.

(b) The charge can't change, so $Q = 5.32 \times 10^{-9}$ C.

(c) $E = \dfrac{Q}{\epsilon_0 A}$. Since Q doesn't change, E doesn't change and $E = 1.50 \times 10^4$ V/m.

(d) Energy $= \dfrac{Q^2}{2C}$.

Q doesn't change and C changes by a factor of $\frac{1}{2}$, so the stored energy doubles and becomes 6.38×10^{-7} J.

Reflect: Since the stored energy increases, work must be done by the force that pulls the plates apart.

*18.83. Set Up: $C = \dfrac{K\epsilon_0 A}{d}$. $C = \dfrac{Q}{V_{ab}}$. Energy $= \frac{1}{2}QV_{ab} = \frac{1}{2}CV_{ab}^2 = \frac{1}{2}\dfrac{Q^2}{C}$.

Solve: (a) $C = (5.00)\dfrac{(8.854 \times 10^{-12} \ C^2/(N \cdot m^2))(16.0 \times 10^{-4} \ m^2)}{0.200 \times 10^{-2} \ m} = 3.54 \times 10^{-11}$ F

(b) $Q = CV_{ab} = (3.54 \times 10^{-11} \ F)(300 \ V) = 1.06 \times 10^{-8}$ C

(c) Energy $= \frac{1}{2}CV^2 = \frac{1}{2}(3.54 \times 10^{-11} \ F)(300 \ V)^2 = 1.59 \times 10^{-6}$ J

*18.85. Set Up: For capacitors in series, the equivalent resistance C_{eq} is given by $\dfrac{1}{C_{eq}} = \dfrac{1}{C_1} + \dfrac{1}{C_2} + \cdots$. For capacitors in parallel, the equivalent capacitance C_{eq} is given by $C_{eq} = C_1 + C_2 + \cdots$

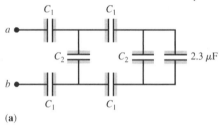

(a)

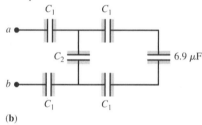

(b)

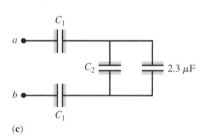

(c)

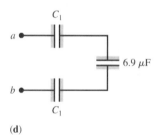

(d)

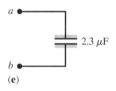

(e)

Solve: (a) Using the rules for combining capacitors in series and in parallel gives the sequence of equivalent networks shown in the figure above. The equivalent capacitance of the network is $2.3 \ \mu F$.

(b) In Figure (d) above the three capacitors in series have the same capacitance, so the voltage across each is $(420 \text{ V})/3 = 140 \text{ V}$. $Q_1 = C_1 V_1 = (6.9 \text{ } \mu\text{F})(140 \text{ V}) = 966 \text{ } \mu\text{C}$. The voltage across C_2 in Figure (c) above is 140 V, so $Q_2 = C_2 V_2 = (4.6 \text{ } \mu\text{F})(140 \text{ V}) = 644 \text{ } \mu\text{C}$.

Reflect: We could continue to analyze the networks in the figure above and find V and Q for each capacitor in the network.

Solutions to Passage Problems

*18.89. Set Up: Assume that the egg cell is spherical with a radius of $100 \text{ } \mu\text{m} = 10^{-2}$ cm. The volume of the cell is then $V = \frac{4}{3}\pi r^3$ and its surface area is $A = 4\pi r^2$. We are told that the initial concentration of Na^+ ions is 30 mmoles/liter within the cell.

Solve: From the previous problem we know that 10^{-12} mole/cm^2 of Na^+ ions move into the egg during fertilization. Thus, the change in concentration is

$$\Delta c = \frac{4\pi r^2 (10^{-12} \text{ mole/cm}^2)}{\frac{4}{3}\pi r^3} = \frac{3(10^{-12} \text{ mole/cm}^2)}{10^{-2} \text{ cm}} = 3 \times 10^{-10} \text{ mole/cm}^3.$$

The fractional change in concentration is $\dfrac{\Delta c}{c} = \dfrac{(3 \times 10^{-10} \text{ mole/cm}^3)(10^3 \text{ mmoles/mole})(10^3 \text{ cm}^3/\text{liter})}{30 \text{ mmoles/liter}} = 10^{-5}$.

The correct answer is B.

*18.91. Set Up: The change in electrical energy stored in the cell can be calculated from $U = \frac{1}{2}CV^2$. The specific capacitance of the cell is $1 \text{ } \mu\text{F/cm}^2$. Assume that the egg cell is spherical with a diameter of $100 \text{ } \mu\text{m} = 10^{-2}$ cm as before.

Solve: The capacitance of the cell is $4\pi r^2 (1 \text{ } \mu\text{F/cm}^2) = 4\pi (10^{-2} \text{ cm})^2 (1 \text{ } \mu\text{F/cm}^2) = 0.00126 \text{ } \mu\text{F} = 1.26$ nF. The change in electrical energy needed to return the cell from +30 mV back to −70 mV is

$$\Delta U = \frac{1}{2}C(V_f^2 - V_i^2) = \frac{1}{2}(1.26 \times 10^{-9} \text{ F})[(-70 \times 10^{-3} \text{ V})^2 - (+30 \times 10^{-3} \text{ V})^2] = 2.5 \times 10^{-12} \text{ J} \approx 3 \text{ pJ}.$$

The correct answer is D.

CURRENT, RESISTANCE, AND DIRECT-CURRENT CIRCUITS

Problems 3, 5, 9, 13, 15, 19, 21, 25, 27, 29, 33, 35, 39, 43, 47, 49, 53, 55, 57, 61, 63, 65, 69, 73, 75, 77, 81, 83, 87, 89, 93, 95

Solutions to Problems

*19.3. Set Up: The number of ions that enter gives the charge that enters the axon in the specified time. $I = \dfrac{\Delta Q}{\Delta t}$.

Solve: $\Delta Q = (5.6 \times 10^{11} \text{ ions})(1.60 \times 10^{-19} \text{ C/ion}) = 9.0 \times 10^{-8} \text{ C}$

$$I = \frac{\Delta Q}{\Delta t} = \frac{9.0 \times 10^{-8} \text{ C}}{10 \times 10^{-3} \text{ s}} = 9.0 \, \mu\text{A}$$

*19.5. Set Up: The volume of the cylindrical wire is given by $V = \pi r^2 L$, where $r = (2.05 \text{ mm})/2 = \underline{1.025} \text{ mm}$ and $L = 25.0 \text{ cm}$. The current passing a given point in the wire is $I = \dfrac{\Delta Q}{\Delta t}$. We assume that the density of *free electrons* in copper is 8.5×10^{28} electrons/m^3. We assume that all of the electrons travel with the drift velocity v_d (see section 19.1).

Solve: **(a)** The number of free electrons in the wire is
$\pi(\underline{1.025} \times 10^{-3} \text{ m})^2(0.250 \text{ m})(8.5 \times 10^{28} \text{ electrons/m}^3) = 7.0 \times 10^{22}$.

(b) Imagine that you are standing near the end of the wire where the current exits. In a time equal to $t = \dfrac{L}{v_\text{d}} = \dfrac{0.25 \text{ m}}{v_\text{d}}$ all of the free electrons in this section of wire will exit past you. Thus,
$\Delta Q = (7.0 \times 10^{22} \text{ electrons})(1.60 \times 10^{-19} \text{ C/electron}) = 1.1 \times 10^4 \text{ C}$ pass in a time t. The resulting current is $I = 1.55 \text{ A} = \dfrac{\Delta Q}{\Delta t} = \dfrac{1.1 \times 10^4 \text{ C}}{(0.25 \text{ m})/v_\text{d}}$. Solving for the drift velocity we obtain

$v_\text{d} = (1.55 \text{ A})(0.25 \text{ m})/(1.1 \times 10^4 \text{ C}) = 3.5 \times 10^{-5} \text{ m/s} = 0.035 \text{ mm/s}$.

Reflect: The charges inside of a conductor that can move are known as *free charges*. The current in a conductor is carried by its free charges.

*19.9. Set Up: $R = \dfrac{\rho L}{A}$. The length of the wire in the spring is the circumference πd of each coil times the number of coils.

Solve: $L = (75)\pi d = (75)\pi(3.50 \times 10^{-2} \text{ m}) = 8.25 \text{ m}.$

$$A = \pi r^2 = \pi d^2/4 = \pi(3.25 \times 10^{-3} \text{ m})^2/4 = 8.30 \times 10^{-6} \text{ m}^2.$$

$$\rho = \frac{RA}{L} = \frac{(1.74 \, \Omega)(8.30 \times 10^{-6} \text{ m}^2)}{8.25 \text{ m}} = 1.75 \times 10^{-6} \, \Omega \cdot \text{m}.$$

Reflect: The value of ρ we calculated is about a factor of 100 times larger than ρ for copper. The metal of the spring is not a very good conductor.

*19.13. Set Up: $R = \dfrac{\rho L}{A}$. $L_{new} = 3L$. The volume of the wire remains the same when it is stretched.

Solve: Volume $= LA$ so $LA = L_{new} A_{new}$. $A_{new} = \dfrac{L}{L_{new}} A = \dfrac{A}{3}$.

$$R_{new} = \frac{\rho L_{new}}{A_{new}} = \frac{\rho(3L)}{A/3} = 9\frac{\rho L}{A} = 9R.$$

Reflect: When the length increases the resistance increases, and when the area decreases the resistance increases.

*19.15. Set Up: First use Ohm's law to find the resistance at 20.0°C; then calculate the resistivity from the resistance. Finally use the dependence of resistance on temperature to calculate the temperature coefficient of resistance. Ohm's law is $R = V/I$, $R = \rho L/A$, $R = R_0[1 + \alpha(T - T_0)]$, and the radius is one-half the diameter.

Solve: (a) At 20.0°C, $R = V/I = (15.0 \text{ V})/(18.5 \text{ A}) = 0.811 \, \Omega$. Using $R = \rho L/A$ and solving for ρ gives $\rho = RA/L = R\pi(D/2)^2/L = (0.811 \, \Omega)\pi[(0.00500 \text{ m})/2]^2/(1.50 \text{ m}) = 1.06 \times 10^{-5} \, \Omega \cdot \text{m}$.

(b) At 92.0°C, $R = V/I = (15.0 \text{ V})/(17.2 \text{ A}) = 0.872 \, \Omega$. Using $R = R_0[1 + \alpha(T - T_0)]$ with T_0 taken as 20.0°C, we have $0.872 \, \Omega = (0.811 \, \Omega)[1 + \alpha(92.0°C - 20.0°C)]$. This gives $\alpha = 0.00105 \, (°C)^{-1}$

Reflect: The results are typical of ordinary metals.

*19.19. Set Up: Apply $R_T = R_0(1 + \alpha(T - T_0))$. Since $V = IR$ and V is the same, $\dfrac{R_T}{R_{20}} = \dfrac{I_{20}}{I_T}$. For tungsten, $\alpha = 4.5 \times 10^{-3} \, (\text{C}°)^{-1}$.

Solve: The ratio of the current at 20°C to that at the higher temperature is $(0.860 \text{ A})/(0.220 \text{ A}) = 3.909$.

$\dfrac{R_T}{R_{20}} = 1 + \alpha(T - T_0) = 3.909$, where $T_0 = 20°C$.

$$T = T_0 + \frac{R_T/R_{20} - 1}{\alpha} = 20°C + \frac{3.909 - 1}{4.5 \times 10^{-3} \, (\text{C}°)^{-1}} = 666°C.$$

Reflect: As the temperature increases, the resistance increases and for constant applied voltage the current decreases. The resistance increases by nearly a factor of four.

*19.21. Set Up: $V = IR$, where I is the current through the battery and V is the potential difference applied to the person.

Solve: $V = IR = (5.0 \times 10^{-3} \text{ A})(1000 \, \Omega) = 5.0 \text{ V}$. This is well within the range of common household voltages.

*19.25. Set Up: $R = \dfrac{\rho L}{A}$. $V = IR$ so $I = \dfrac{V}{R} = \dfrac{VA}{\rho L} = \dfrac{V\pi(d/2)^2}{\rho L} = \dfrac{V\pi d^2}{4\rho L}$.

Solve: (a) $L_{new} = 2L$. $I_{new} = \dfrac{V\pi d^2}{4\rho L_{new}} = \dfrac{V\pi d^2}{4\rho(2L)} = \dfrac{1}{2}I$

(b) $d_{new} = 2d$. $I_{new} = \dfrac{V\pi d_{new}^2}{4\rho L} = \dfrac{V\pi(2d)^2}{4\rho L} = 4I$

(c) $L_{new} = 2L$; $d_{new} = 2d$. $I_{new} = \dfrac{V\pi d_{new}^2}{4\rho L_{new}} = \dfrac{V\pi(2d)^2}{4\rho(2L)} = 2I$

Reflect: I increases when R decreases. R decreases when L decreases or d increases.

*19.27. Set Up: When the switch is open there is no current and the terminal voltage of the battery equals its emf, $\mathcal{E}$. When the switch is closed, current I flows and the terminal voltage V of the battery is $V = \mathcal{E} - Ir$. The current I is the same at all points of the circuit.

Solve: $\mathcal{E} = 3.08$ V. $V = \mathcal{E} - Ir$, with $V = 2.97$ V and $I = 1.65$ A, so $r = \dfrac{V - \mathcal{E}}{I} = \dfrac{3.08 \text{ V} - 2.97 \text{ V}}{1.65 \text{ A}} = 0.067\,\Omega$.

$V - IR = 0$ and $R = \dfrac{V}{I} = \dfrac{2.97 \text{ V}}{1.65 \text{ A}} = 1.80\,\Omega$.

Reflect: When current flows through the battery, the terminal voltage is less than the emf because of the voltage across the internal resistance.

*19.29. Set Up: The terminal voltage of the battery is given by $V_{ab} = \mathcal{E} - Ir$. The terminal voltage also equals the voltage across the resistor, so $V_{ab} = IR$. We have two unknowns $(\mathcal{E}, r)$ so we need two equations.

Solve: (a) With the 1500 M Ω resistor, $I = \dfrac{V_{ab}}{R} = \dfrac{2.50 \text{ V}}{1500 \times 10^6\,\Omega} = 1.67 \times 10^{-9}$ A.

$$V_{ab} = \mathcal{E} - Ir \text{ gives } 2.50 \text{ V} = \mathcal{E} - (1.67 \times 10^{-9} \text{ A})r.$$

With the 5.00 Ω resistor, $I = \dfrac{V_{ab}}{R} = \dfrac{1.75 \text{ V}}{5.00\,\Omega} = 0.350$ A.

$$V_{ab} = \mathcal{E} - Ir \text{ gives } 1.75 V = \mathcal{E} - (0.350 \text{ A})r.$$

Subtracting the second equation from the first gives 0.75 V $= (0.350 \text{ A})r$ and $r = 2.14\,\Omega$.

$$\mathcal{E} = 1.75 \text{ V} + (0.350 \text{ A})r = 2.50 \text{ V}$$

(b) Now R is 7.00 Ω. $I = \dfrac{\mathcal{E}}{R + r} = \dfrac{2.50 \text{ V}}{7.00\,\Omega + 2.14\,\Omega} = 0.274$ A

$$V_{ab} = \mathcal{E} - Ir = 2.50 \text{ V} - (0.274 \text{ A})(2.14\,\Omega) = 1.91 \text{ V. Or,}$$

$$V_{ab} = IR = 1.91 \text{ V.}$$

Reflect: The smaller R is, the larger the current and the smaller the terminal voltage. With the 1500 M Ω resistor the current is very small and the terminal voltage differs only very little from the emf of the battery.

*19.33. Set Up: For a resistor, $P = VI$ and $V = IR$.

Solve: (a) $I = \dfrac{P}{V} = \dfrac{327 \text{ W}}{15.0 \text{ V}} = 21.8$ A

(b) $R = \dfrac{V}{I} = \dfrac{15.0 \text{ V}}{21.8 \text{ A}} = 0.688\,\Omega$

*19.35. Set Up: The voltmeter reads the terminal voltage of the battery, which is the potential difference across the appliance. The terminal voltage is less than 15.0 V because some potential is lost across the internal resistance of the battery. The equation $P = V^2/R$ gives the power dissipated by the appliance. The equation $V_{ab} = \mathcal{E} - Ir$ relates the terminal voltage of the battery to its internal resistance.

(a) Solve: $P = (11.3 \text{ V})^2 / (75.0\,\Omega) = 1.70$ W

(b) The current in the circuit can be calculated from the terminal voltage of the battery and the resistance of the attached appliance: $I = \dfrac{11.3 \text{ V}}{75.0\,\Omega} = 0.151$ A. The internal resistance of the battery, r, is given by $r = \dfrac{\mathcal{E} - V_{ab}}{I} = $

$\dfrac{15.0 \text{ V} - 11.3 \text{ V}}{0.151 \text{ A}} 24.5\,\Omega.$

Reflect: The full 15.0 V of the battery would be available only when no current (or a very small current) is flowing in the circuit. This would be the case if the appliance had a resistance much greater than 24.5 Ω.

*19.39. Set Up: $P = VI$ and energy is the product of power and time.

Solve: $P = (500 \text{ V})(80 \times 10^{-3} \text{ A}) = 40 \text{ W}.$

$$\text{Energy} = Pt = (40 \text{ W})(10 \times 10^{-3} \text{ s}) = 0.40 \text{ J}.$$

Reflect: The energy delivered depends not only on the voltage and current but also on the length of the pulse.

*19.43. Set Up: $P = I^2 R = \dfrac{V^2}{R} = VI.$ $V = IR.$ The heater consumes 540 W when $V = 120$ V. Energy $= Pt.$

Solve: (a) $P = \dfrac{V^2}{R}$ so $R = \dfrac{V^2}{P} = \dfrac{(120 \text{ V})^2}{540 \text{ W}} = 26.7 \, \Omega$

(b) $P = VI$ so $I = \dfrac{P}{V} = \dfrac{540 \text{ W}}{120 \text{ V}} = 4.50 \text{ A}$

(c) Assuming that R remains $26.7 \, \Omega$, $P = \dfrac{V^2}{R} = \dfrac{(110 \text{ V})^2}{26.7 \, \Omega} = 453 \text{ W}.$ P is smaller by a factor of $(110/120)^2.$

Reflect: (d) With the lower line voltage the current will decrease and the operating temperature will decrease. R will be less than $26.7 \, \Omega$ and the power consumed will be greater than the value calculated in part (c).

*19.47. Set Up: The resistors can be placed in any combination of series and parallel. The equivalent resistance of two resistors in series is $R_{eq} = R_1 + R_2$, which is larger than either resistance by itself. However, the equivalent resistance of two resistors in parallel is given by $R_{eq} = \dfrac{1}{\dfrac{1}{R_1} + \dfrac{1}{R_2}}$, which is always less than the smaller of the two resistances. Thus, to obtain the maximum equivalent resistance the three resistors should all be in series and to obtain the minimum equivalent resistance they should all be in parallel.

Solve: (a) For three resistors in series we have $R_{eq} = R_1 + R_2 + R_3 = 36 \, \Omega + 47 \, \Omega + 51 \, \Omega = 134 \, \Omega.$

(b) For three resistors in parallel we have $R_{eq} = \dfrac{1}{\dfrac{1}{36 \, \Omega} + \dfrac{1}{47 \, \Omega} + \dfrac{1}{51 \, \Omega}} = 15 \, \Omega.$

Reflect: In addition to these two extreme cases, we can combine the three resistors in a variety of other combinations: for example, we can form a series combination of a single resistor and a pair of resistors that are connected in parallel. Can you find all possible combinations of these three resistors?

*19.49. Set Up: For a parallel connection the full line voltage appears across each resistor. The equivalent resistance is $\dfrac{1}{R_{eq}} = \dfrac{1}{R_1} + \dfrac{1}{R_2}$ and $R_{eq} = \dfrac{R_1 R_2}{R_1 + R_2}.$

Solve: (a) $R_{eq} = \dfrac{(40.0 \, \Omega)(90.0 \, \Omega)}{40.0 \, \Omega + 90.0 \, \Omega} = 27.7 \, \Omega$

(b) $I = \dfrac{V}{R_{eq}} = \dfrac{120 \text{ V}}{27.7 \, \Omega} = 4.33 \text{ A}$

(c) For the $40.0 \, \Omega$ resistor, $I_1 = \dfrac{V}{R_1} = \dfrac{120 \text{ V}}{40.0 \, \Omega} = 3.00 \text{ A}.$ For the $90.0 \, \Omega$ resistor, $I_2 = \dfrac{V}{R_2} = \dfrac{120 \text{ V}}{90.0 \, \Omega} = 1.33 \text{ A}.$

Reflect: If either resistor is removed, the voltage and current for the other resistor is unchanged.

*19.53. Set Up: For resistors in parallel, the voltages are the same and the currents add. $\dfrac{1}{R_{eq}} = \dfrac{1}{R_1} + \dfrac{1}{R_2}$ so $R_{eq} = \dfrac{R_1 R_2}{R_1 + R_2}.$ For resistors in series, the currents are the same and the voltages add. $R_{eq} = R_1 + R_2.$

Solve: The rules for combining resistors in series and parallel lead to the sequence of equivalent circuits shown in the figure below. $R_{eq} = 3.00 \, \Omega$. In Figure (c), $I = \dfrac{48.0 \, V}{3.00 \, \Omega} = 16.0 \, A$. In Figure (b), the voltage across each resistor is 48.0 V, so $I_{12} = \dfrac{48.0 \, V}{4.00 \, \Omega} = 12.0 \, A$ and $I_{34} = \dfrac{48.0 \, V}{12.0 \, \Omega} = 4.00 \, A$. Note that $I_{12} + I_{34} = 16.0 \, A$. Then in Figure (a), $I_1 = I_2 = 12.0 \, A$ and $I_3 = I_4 = 4.00 \, A$.

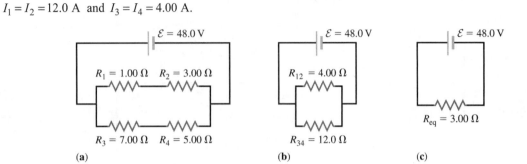

(a)   (b)   (c)

*19.55. Set Up: We need to reduce the given resistance without removing it from the circuit. We can do this by soldering a second resistor in parallel with the incorrect one. The equivalent resistance for resistors in parallel is $\dfrac{1}{R_{eq}} = \dfrac{1}{R_1} + \dfrac{1}{R_2}$.

Solve: We wish for the equivalent resistance to equal 36.5 kΩ. Thus, the unknown resistance is given by $\dfrac{1}{R_1} = \dfrac{1}{R_{eq}} - \dfrac{1}{R_2}$. This simplifies to $R_1 = \dfrac{R_{eq} R_2}{R_2 - R_{eq}} = \dfrac{(36.5 \, k\Omega)(69.8 \, k\Omega)}{(69.8 \, k\Omega - 36.5 \, k\Omega)} = 76.5 \, k\Omega$.

Reflect: There would probably be no easy way to solve this problem if we had soldered a resistor that was smaller than the desired value into the circuit.

*19.57. Set Up: Assume the unknown currents have the directions shown in the figure below. We have used the junction rule to write the current through the 10.0 V battery as $I_1 + I_2$. There are two unknowns, I_1 and I_2, so we will need two equations. Three possible circuit loops are shown in the figure.

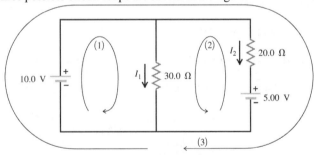

Solve: (a) Apply the loop rule to loop (1), going around the loop in the direction shown: $+10.0 \, V - (30.0 \, \Omega)I_1 = 0$ and $I_1 = 0.333 \, A$.

(b) Apply the loop rule to loop (3): $+10.0 \, V - (20.0 \, \Omega)I_2 - 5.00 \, V = 0$ and $I_2 = 0.250 \, A$.

(c) $I_1 + I_2 = 0.333 \, A + 0.250 \, A = 0.583 \, A$

Reflect: For loop (2) we get
$$+5.00 \, V + I_2(20.0 \, \Omega) - I_1(30.0 \, \Omega) = 5.00 \, V + (0.250 \, A)(20.0 \, \Omega) - (0.333 \, A)(30.0 \, \Omega) =$$
$$5.00 \, V + 5.00 \, V - 10.0 \, V = 0,$$
so that with the currents we have calculated the loop rule is satisfied for this third loop.

*19.61. Set Up: We can use the power consumption in the 5.00 Ω resistor to find the current through it. The circuit, unknown currents, and a circuit loop are all shown in the figure below.

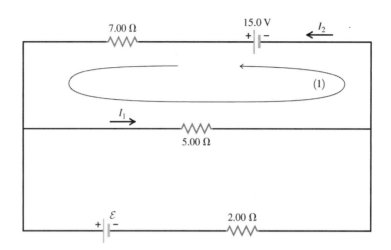

Solve: $P = I^2 R$ so $I_1 = \sqrt{\dfrac{20.0 \text{ W}}{5.00 \, \Omega}} = 2.00$ A. The loop rule for loop (1) gives

$$+15.0 \text{ V} - I_2(7.00 \, \Omega) - (2.00 \text{ A})(5.00 \, \Omega) = 0.$$

$I_2 = \dfrac{15.0 \text{ V} - 10.0 \text{ V}}{7.00 \, \Omega} = 0.714$ A. The ammeter reads 0.714 A.

*19.63. Set Up: The time constant is $\tau = RC$.

Solve: $C = \dfrac{\tau}{R} = \dfrac{2.00 \text{ s}}{500.0 \, \Omega} = 4.00 \times 10^{-3}$ F = 4.00 mF

*19.65. Set Up: The time constant is $RC = (0.895 \times 10^6 \, \Omega)(12.4 \times 10^{-6} \text{ F}) = 11.1$ s.

Solve: (a) At $t = 0$ s: $q = C\mathcal{E}(1 - e^{-t/RC}) = 0$.

At $t = 5$ s: $q = C\mathcal{E}(1 - e^{-t/RC}) = (12.4 \times 10^{-6} \text{ F})(60.0 \text{ V})(1 - e^{-(5.0 \text{ s})/(11.1 \text{ s})}) = 2.70 \times 10^{-4}$ C.

At $t = 10$ s: $q = C\mathcal{E}(1 - e^{-t/RC}) = (12.4 \times 10^{-6} \text{ F})(60.0 \text{ V})(1 - e^{-(10.0 \text{ s})/(11.1 \text{ s})}) = 4.42 \times 10^{-4}$ C.

At $t = 20$ s: $q = C\mathcal{E}(1 - e^{-t/RC}) = (12.4 \times 10^{-6} \text{ F})(60.0 \text{ V})(1 - e^{-(20.0 \text{ s})/(11.1 \text{ s})}) = 6.21 \times 10^{-4}$ C.

At $t = 100$ s: $q = C\mathcal{E}(1 - e^{-t/RC}) = (12.4 \times 10^{-6} \text{ F})(60.0 \text{ V})(1 - e^{-(100 \text{ s})/(11.1 \text{ s})}) = 7.44 \times 10^{-4}$ C.

(b) The current at time t is given by: $i = \dfrac{\mathcal{E}}{R} e^{-t/RC}$.

At $t = 0$ s: $i = \dfrac{60.0 \text{ V}}{8.95 \times 10^5 \, \Omega} e^{-0/11.1} = 6.70 \times 10^{-5}$ A.

At $t = 5$ s: $i = \dfrac{60.0 \text{ V}}{8.95 \times 10^5 \, \Omega} e^{-5/11.1} = 4.27 \times 10^{-5}$ A.

At $t = 10$ s: $i = \dfrac{60.0 \text{ V}}{8.95 \times 10^5 \, \Omega} e^{-10/11.1} = 2.72 \times 10^{-5}$ A.

At $t = 20$ s: $i = \dfrac{60.0 \text{ V}}{8.95 \times 10^5 \, \Omega} e^{-20/11.1} = 1.11 \times 10^{-5}$ A.

At $t = 100$ s: $i = \dfrac{60.0 \text{ V}}{8.95 \times 10^5 \ \Omega} e^{-100/11.1} = 8.20 \times 10^{-9} \text{ A}.$

(c) The graphs of $q(t)$ and $i(t)$ are given in Figures (a) and (b) below.

Reflect: The charge on the capacitor increases in time as the current decreases.

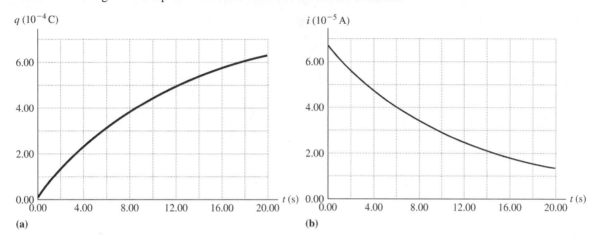

(a)   (b)

*19.69. Set Up: $q = C/V$. After a long time the current has dropped to zero, there is no potential drop across the resistor and the full battery voltage is applied to the capacitor network. Just after the switch is closed the charge and voltage across each capacitor is zero and the battery voltage equals the voltage drop across the resistor. The time constant is $\tau = RC$, where C is the equivalent capacitance of the network. $i = I_0 e^{-t/RC}$.

Solve: (a) The 20.0 pF, 30.0 pF, and 40.0 pF capacitors are in series and their equivalent capacitance C_s is given by

$$\frac{1}{C_s} = \frac{1}{20.0 \text{ pF}} + \frac{1}{30.0 \text{ pF}} + \frac{1}{40.0 \text{ pF}}.$$

$C_s = 9.23 \ \mu F$. After a long time the voltage across the 10.0 pF capacitor is 50.0 V. The charge on this capacitor is $q_{10} = C_{10} V_{10} = (10.0 \text{ pF})(50.0 \text{ V}) = 500 \text{ pC}$. The voltage across C_s is 50.0 V and

$$q_5 = C_s V_5 = (9.23 \text{ pF})(50.0 \text{ V}) = 461 \text{ pC}.$$

This is the charge on each of the capacitors in series, so $q_{20} = q_{30} = q_{40} = 461 \text{ pC}.$

(b) $V_{10} = 50.0 \text{ V}. \quad V_{20} = \dfrac{q_{20}}{C_{20}} = \dfrac{461 \text{ pC}}{20.0 \text{ pF}} = 23.1 \text{ V}. \quad V_{30} = \dfrac{q_{30}}{C_{30}} = \dfrac{461 \text{ pC}}{30.0 \text{ pF}} = 15.4 \text{ V}.$

$$V_{40} = \frac{q_{40}}{C_{40}} = \frac{461 \text{ pC}}{40.0 \text{ pF}} = 11.5 \text{ V}.$$

Note that $V_{20} + V_{30} + V_{40} = 50.0 \text{ V}$, as it should.

(c) $I = \dfrac{\mathcal{E}}{R} = \dfrac{50.0 \text{ V}}{20.0 \ \Omega} = 2.50 \text{ A}$

(d) The 10.0 pF capacitor and C_s are in parallel so their equivalent is $C_{eq} = 10.0 \text{ pF} + 9.23 \text{ pF} = 19.2 \text{ pF}.$
$\tau = RC = (20.0 \ \Omega)(19.2 \text{ pF}) = 384 \text{ ps}$

Reflect: The charges on the capacitors start at zero and increase to their final values. The current starts at its maximum value and then decays to zero.

*19.73. Set Up: The ohmmeter reads the equivalent resistance between points a and b. Replace series and parallel combinations by their equivalent. For resistors in parallel we have $\dfrac{1}{R_{eq}} = \dfrac{1}{R_1} + \dfrac{1}{R_2}$. For resistors in series we have $R_{eq} = R_1 + R_2$.

Solve: Circuit (a): The $75.0\,\Omega$ and $40.0\,\Omega$ resistors are in parallel and have equivalent resistance $26.09\,\Omega$. The $25.0\,\Omega$ and $50.0\,\Omega$ resistors are in parallel and have an equivalent resistance of $16.67\,\Omega$. The equivalent network is given in Figure (a) below. $\dfrac{1}{R_{eq}} = \dfrac{1}{100.0\,\Omega} + \dfrac{1}{23.05\,\Omega}$, so $R_{eq} = 18.7\,\Omega$.

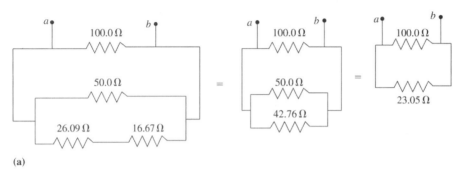

(a)

Circuit (b): The $30.0\,\Omega$ and $45.0\,\Omega$ resistors are in parallel and have equivalent resistance $18.0\,\Omega$. The equivalent network is given in Figure (b) below. $\dfrac{1}{R_{eq}} = \dfrac{1}{10.0\,\Omega} + \dfrac{1}{30.3\,\Omega}$, so $R_{eq} = 7.5\,\Omega$.

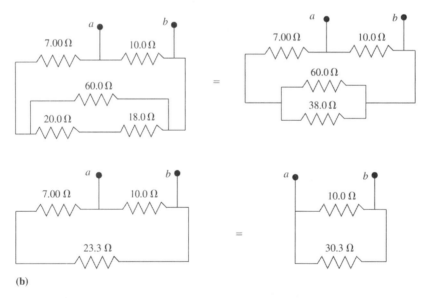

(b)

Reflect: In circuit (a) the resistance along one path between a and b is $100.0\,\Omega$, but that is not the equivalent resistance between these points. A similar comment can be made about circuit (b).

*19.75. Set Up: $R_T = R_0(1 + \alpha[T - T_0])$. $R = \dfrac{V}{I}$. $P = VI$. When the temperature increases the resistance increases and the current decreases.

Solve: (a) $\dfrac{V}{I_T} = \dfrac{V}{I_0}(1 + \alpha[T - T_0])$. $I_0 = I_T(1 + \alpha[T - T_0])$.

$$T - T_0 = \frac{I_0 - I_T}{\alpha I_T} = \frac{1.35 \text{ A} - 1.23 \text{ A}}{(1.23 \text{ A})(4.5 \times 10^{-4}(\text{C}°)^{-1})} = 217 \text{ C}°.$$
$$T = 20°\text{C} + 217 \text{ C}° = 237°\text{C}$$

(b)   (i) $P = VI = (120 \text{ V})(1.35 \text{ A}) = 162 \text{ W}$

(ii) $P = (120 \text{ V})(1.23 \text{ A}) = 148 \text{ W}$

*19.77. Set Up: The circuit is sketched in the figure below. r_{total} is the combined internal resistance of both batteries. The power delivered to the bulb is $I^2 R$. Energy $= Pt$.

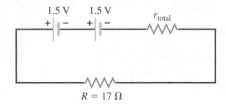

1.5 V   1.5 V r_{total}

$R = 17 \, \Omega$

Solve: (a) $r_{total} = 0$. The loop rule gives

$$1.5 \text{ V} + 1.5 \text{ V} - I(17 \, \Omega) = 0. \quad I = 0.1765 \text{ A}.$$
$$P = I^2 R = (0.1765 \text{ A})^2 (17 \, \Omega) = 0.530 \text{ W}.$$

This is also $(3.0 \text{ V})(0.1765 \text{ A})$.

(b) Energy $= (0.530 \text{ W})(5.0 \text{ h})(3600 \text{ s/h}) = 9540 \text{ J}$

(c) $P = \dfrac{0.530 \text{ W}}{2} = 0.265 \text{ W}.$ $P = I^2 R$ so $I = \sqrt{\dfrac{P}{R}} = \sqrt{\dfrac{0.265 \text{ W}}{17 \, \Omega}} = 0.125 \text{ A}.$

The loop rule gives $1.5 \text{ V} + 1.5 \text{ V} - IR - I r_{total}$. $r_{total} = \dfrac{3.0 \text{ V} - (0.125 \text{ A})(17 \, \Omega)}{0.125 \text{ A}} = 7.0 \, \Omega.$

Reflect: When the power to the bulb has decreased to half its initial value, the total internal resistance of the two batteries is nearly half the resistance of the bulb. Compared to a single battery, using two identical batteries in series doubles the emf but also doubles the total internal resistance.

*19.81. Set Up: The network is sketched in the figure below. If the current in R_1 is I, it is $I/2$ for R_2 and R_3. $P = I^2 R$. Since the current is greatest in R_1, that resistor dissipates the most power.

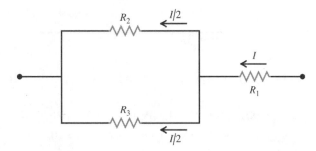

R_2 $I/2$

I

R_1

R_3

$I/2$

Solve: $P_1 = 32.0 \text{ W}$ so $I = \sqrt{\dfrac{P}{R}} = \sqrt{\dfrac{32.0 \text{ W}}{2.00 \, \Omega}} = 4.00 \text{ A}.$ The total power dissipated therefore is

$$(4.00 \text{ A})^2 (2.00 \, \Omega) + (2.00 \text{ A})^2 (2.00 \, \Omega) + (2.00 \text{ A})^2 (2.00 \, \Omega) = 48.0 \text{ W}.$$

Reflect: R_1 is dissipating its maximum of 32.0 W but R_2 and R_3 are dissipating only 8.0 W each.

*19.83. Set Up: Since they are in parallel, the voltage across each device is 120 V. $P = VI$. The total current drawn from the outlet is the sum of the individual currents.

Solve: (a) Toaster: $I = \dfrac{P}{V} = \dfrac{1800 \text{ W}}{120 \text{ V}} = 15.0 \text{ A}$. Frypan: $I = \dfrac{1400 \text{ W}}{120 \text{ V}} = 11.7 \text{ A}$. Lamp: $I = \dfrac{75 \text{ W}}{120 \text{ V}} = 0.625 \text{ A}$.

(b) $I = 15.0 \text{ A} + 11.7 \text{ A} + 0.625 \text{ A} = 27.3 \text{ A}$. This is greater than 20 A and the circuit breaker will blow.

*19.87. Set Up: The current delivered by the power plant can be calculated from its power and the known voltage: $I = \dfrac{P}{V}$. The power dissipated by the wires can be calculated from the current and resistance of the wires: $P = I^2 R$.

Solve: (a) With a voltage of 220 V the current is $I = \dfrac{150 \times 10^3 \text{ W}}{220 \text{ V}} = \underline{681}.8 \text{ A}$. With this current the power dissipated by the wire is $P = I^2 R = (\underline{681}.8 \text{ A})^2 (0.25 \,\Omega) = \underline{116} \times 10^3 \text{ W}$. Thus, the percentage of the total power dissipated by the wire is $\dfrac{\underline{116} \text{ kW}}{150 \text{ kW}} \cdot 100\% = 77\%$.

(b) We can repeat the steps above—or notice that 22 kV is 100 times the previous voltage. Since current is inversely proportional to voltage and the power dissipated is proportional to current squared the power dissipated will now be $\left(\dfrac{1}{100}\right)^2 \cdot 77\% = 0.0077\%$.

Reflect: Power is most efficiently transferred at relatively low current and high voltage.

*19.89. Set Up: There is no current in the middle branch because there is not a complete conducting path for that branch. There is only a single current in the circuit, as shown in the figure below. To find $V_{ab} = V_a - V_b$, start at point b and travel to point a.

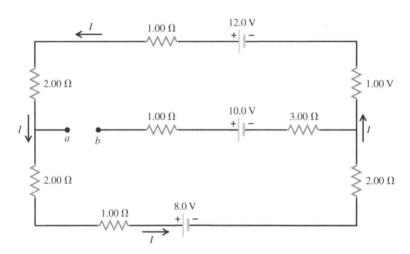

Solve: $12.0 \text{ V} - 8.0 \text{ V} - I(2.00 \,\Omega + 2.00 \,\Omega + 1.00 \,\Omega + 2.00 \,\Omega + 1.00 \,\Omega + 1.00 \,\Omega) = 0$. $I = 0.444 \text{ A}$. Going from a to b through the 12.0 V battery gives

$$V_a + I(2.00 \,\Omega) + I(1.00 \,\Omega) - 12.0 \text{ V} + I(1.00 \,\Omega) + 10.0 \text{ V} = V_b.$$

$$V_a - V_b = 12.0 \text{ V} - 10.0 \text{ V} - (0.444 \text{ A})(4.00 \,\Omega) = +0.22 \text{ V}.$$

Or, going from a to b through the 8.0 V battery gives

$$V_a - I(2.00 \,\Omega) - I(1.00 \,\Omega) - 8.0 \text{ V} - I(2.00 \,\Omega) + 10.0 \text{ V} = V_b \text{ and}$$

$$V_a - V_b = -2.0 \text{ V} + (0.444 \text{ A})(5.00 \,\Omega) = +0.22 \text{ V}.$$

The voltmeter reads 0.22 V. $V_a - V_b$ is positive, so point a is at higher potential.

Reflect: Since there is no current in the middle branch, the two resistances in that branch could be removed without affecting the potential difference V_{ab}. The 10.0 V emf doesn't affect the current but does affect V_{ab}

Solutions to Passage Problems

*19.93. **Set Up:** We know that resistance can be calculated from the dimensions of the conductor according to $R = \rho \dfrac{L}{A}$. We are told that $\rho = 100 \, \Omega \cdot \text{cm} = 1 \, \Omega \cdot \text{m}$. *Conductance* is defined to be the reciprocal of resistance and so it

is given by $\dfrac{1}{R} = \dfrac{A}{\rho L}$, and it is measured in Ω^{-1}, which is known as a siemens (S) (so we have $\text{S} = \Omega^{-1}$).

Solve: Based on our model the resistance of the channel is $R = \rho \dfrac{L}{A} = \dfrac{(1 \, \Omega \cdot \text{m})(0.3 \times 10^{-9})}{\pi(0.3 \times 10^{-9})^2} \approx 10^9 \, \Omega$. Thus, the

conductance is $\dfrac{1}{R} = \dfrac{1}{10^9 \, \Omega} = 1 \, \text{nS}$. The correct answer is B.

*19.95. **Set Up:** The time constant for a single channel is $\tau = RC$, where R is the resistance of the channel and C is its capacitance. Assuming the results of the previous problem we have that the channel density is 10^{10} channels/cm^2 and the resistance of a single channel is $10^{11} \, \Omega$. We are told that the *specific capacitance* is $1 \, \mu\text{F/cm}^2$.

Solve: The capacitance of a single channel is $\dfrac{1 \, \mu\text{F/cm}^2}{10^{10} \, \text{channels/cm}^2} = 10^{-10} \, \mu\text{F}$. Thus, the time constant of the channel

must be $\tau = RC = (10^{11} \, \Omega)(10^{-10} \, \mu\text{F}) = 10 \, \mu\text{s}$. The correct answer is B.

MAGNETIC FIELD AND MAGNETIC FORCES

Problems 1, 3, 9, 11, 13, 17, 19, 23, 25, 29, 33, 35, 37, 41, 45, 47, 51, 53, 55, 59, 63, 67, 69, 71, 75, 77, 81, 83, 87, 91, 93

Solutions to Problems

*20.1. Set Up: The force $\vec{F}$ on the particle is in the direction of the deflection of the particle. The directions of $\vec{v}$, $\vec{B}$ and $\vec{F}$ are shown in the figure below. Apply the right-hand rule to the directions of $\vec{v}$ and $\vec{B}$. See if your thumb is in the direction of $\vec{F}$, or opposite to that direction. Use $F = |q|vB\sin\phi$ with $\phi = 90°$ to calculate F.

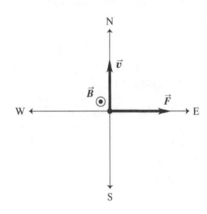

Solve: (a) When you apply the right-hand rule to $\vec{v}$ and $\vec{B}$, your thumb points east. $\vec{F}$ is in this direction, so the charge is positive.

(b) $F = |q|vB\sin\phi = (8.50 \times 10^{-6} \text{ C})(4.75 \times 10^3 \text{ m/s})(1.25 \text{ T})\sin 90° = 0.0505 \text{ N}$

*20.3. Set Up: The directions of $\vec{v}$ and $\vec{B}$ are shown in the figure below. A proton has charge $e = +1.60 \times 10^{-19}$ C. $F = |q|vB\sin\phi$. In part (a), $\phi = 55.0°$. An electron has charge $-e$.

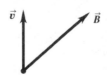

Solve: **(a)** The right-hand rule says $\vec{F}$ is into the page in the figure above.

$$F = |q|vB\sin\phi = (1.60\times10^{-19}\text{ C})(3.60\times10^3\text{ m/s})(0.750\text{ T})\sin55.0° = 3.54\times10^{-16}\text{ N}.$$

(b) F is maximum when $\phi = 90°$, when $\vec{v}$ is perpendicular to $\vec{B}$. $F_{max} = |q|vB = 4.32\times10^{-16}$ N. F is minimum when $\phi = 0°$ or 180°, when $\vec{v}$ is either parallel or antiparallel to $\vec{B}$. $F_{min} = 0$.

(c) $|q|$ is the same for an electron and a proton, so $F = 3.45\times10^{-16}$ N, the same as for a proton. Since the proton and electron have charges of opposite sign, the forces on them are in opposite directions. The force on the electron is directed out of the page in the figure above.

Reflect: Only the component of $\vec{v}$ perpendicular to $\vec{B}$ contributes to the magnetic force. Therefore, this force is zero when the charged particle moves along the direction of $\vec{B}$ and the force is maximum when the particle moves in a direction perpendicular to $\vec{B}$. When the sign of the charge changes, the force reverses direction. And, even though the force magnitude is the same for the electron and proton, the effect of the force (the acceleration) for the electron would be much greater, because of its smaller mass.

*20.9. Set Up: Take $\vec{v}$ and $\vec{B}$ as shown in the textbook. The direction of the magnetic force at each point is given by the right-hand rule. We may use $F = |q|vB\sin\phi$ to find the magnitude of the force and $a = F/m$ to find the acceleration of the charge. According to the figure $\phi = 45°$.

Solve: According to the right-hand rule, $\vec{F}$ is in the $-z$ direction. The acceleration of the charge is given by $a = |q|vB\sin\phi/m = (1.22\times10^{-8}\text{ C})(3.00\times10^4\text{ m/s})(1.25\text{ T})\sin45°/(1.81\times10^{-3}\text{ kg}) = 1.79\times10^{-1}\text{ m/s}^2$.

Reflect: For a given velocity and magnetic field strength, the maximum acceleration of the charge occurs when the magnetic field is perpendicular to the velocity of the charge. In this case, the maximum acceleration would be 0.253 m/s². On the other hand, if the magnetic field were parallel to the velocity of the charge, the charge would have an acceleration of zero.

*20.11. Set Up: The relation between v, B, and E is given by Eq. (20.3). $\vec{E}$ points from the positive plate to the negative plate, so is directed upward in the figure. The electric field is related to the potential difference V between the plates by $V = Ed$, where d is the separation between the plates.

Solve: **(a)** $E = \dfrac{V}{d} = \dfrac{150\text{ V}}{4.50\times10^{-2}\text{ m}} = 3.33\times10^3\text{ V/m}$. $B = \dfrac{E}{v} = \dfrac{3.33\times10^3\text{ V/m}}{3.25\times10^3\text{ m/s}} = 1.02\text{ T}$.

(b) The forces and fields are shown in the figure below. Assume a positive charge; the same field directions also work for a negative charge. $\vec{E}$ is upward so the electric force $\vec{F}_E$ is upward. In order for $\vec{F}_B$ to be downward, to oppose $\vec{F}_E$, $\vec{B}$ must be out of the page, as shown.

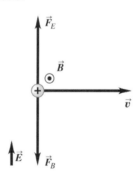

Reflect: Ions with speed $v = E/B$ are undeflected no matter what is the sign or magnitude of their charge. If v is not equal to E/B, then either F_B or F_E is larger than the other, there is a net force on the charge and the charge does not travel in a straight line.

*20.13. Set Up: Eq. (20.4) says $R = \dfrac{mv}{|q|B}$. A proton has mass $m = 1.67 \times 10^{-27}$ kg and charge

$q = +e = 1.60 \times 10^{-19}$ C.

Solve: $v = \dfrac{R|q|B}{m} = \dfrac{(0.0613 \text{ m})(1.60 \times 10^{-19} \text{ C})(0.250 \text{ T})}{1.67 \times 10^{-27} \text{ kg}} = 1.47 \times 10^{6}$ m/s.

*20.17. Set Up: $R = \dfrac{mv}{|q|B}$.

Solve: $RB = \dfrac{mv}{|q|} = \text{constant}$. $R_1 B_1 = R_2 B_2$ and $R_2 = R_1 \left(\dfrac{B_1}{B_2} \right) = R \left(\dfrac{B_1}{3B_1} \right) = R/3$.

Reflect: The radius of the circular path decreases when B increases.

*20.19. Set Up: In the acceleration process the electric potential energy decrease, $|q|V$, equals the increase in kinetic energy, $\frac{1}{2}mv^2$. Use this to find the speed v of the protons after they have been accelerated. In the magnetic field,

$$R = \dfrac{mv}{|q|B}.$$

Protons have charge $+e$ and mass 1.67×10^{-27} kg. Electrons have charge $-e$ and mass 9.11×10^{-31} kg.

Solve: (a) $|q|V = \frac{1}{2}mv^2$. $v = \sqrt{\dfrac{2|q|V}{m}} = \sqrt{\dfrac{2(1.60 \times 10^{-19} \text{ C})(0.745 \times 10^3 \text{ V})}{1.67 \times 10^{-27} \text{ kg}}} = 3.78 \times 10^5$ m/s. $R = \dfrac{mv}{|q|B}$ and

$$B = \dfrac{mv}{|q|R} = \dfrac{(1.67 \times 10^{-27} \text{ kg})(3.78 \times 10^5 \text{ m/s})}{(1.60 \times 10^{-19} \text{ C})(0.875 \text{ m})} = 4.51 \times 10^{-3} \text{ T}.$$

(b) $B = \dfrac{mv}{|q|R}$. In $\dfrac{mv}{|q|R}$ only m changes. $\dfrac{B}{m} = \dfrac{v}{|q|R} = \text{constant}$ and $\dfrac{B_p}{m_p} = \dfrac{B_e}{m_e}$.

$$B_e = \left(\dfrac{m_e}{m_p} \right) B_p = \left(\dfrac{9.11 \times 10^{-31} \text{ kg}}{1.67 \times 10^{-27} \text{ kg}} \right)(4.51 \times 10^{-3} \text{ T}) = 2.46 \times 10^{-6} \text{ T}.$$

Reflect: For lighter particles a smaller force and hence smaller B is needed to produce the same path. To produce electrons of the same speed as the protons a smaller accelerating voltage would be required.

*20.23. Set Up: In a velocity selector, $E = vB$ (Section 20.2). For motion in a circular arc in a magnetic field B', $R = \dfrac{mv}{|q|B'}$ (Section 20.3). The ions have charge $+e$.

Solve: (a) $v = \dfrac{E}{B} = \dfrac{155 \text{ V/m}}{0.0315 \text{ T}} = 4.92 \times 10^3$ m/s

(b) $m = \dfrac{R|q|B'}{v} = \dfrac{(0.175 \text{ m})(1.60 \times 10^{-19} \text{ C})(0.0175 \text{ T})}{4.92 \times 10^3 \text{ m/s}} = 9.96 \times 10^{-26}$ kg

Reflect: Ions with larger $m/|q|$ move in a path of larger radius.

*20.25. Set Up: $|q|$, v, and B are the same as in Problem 20.24. For $m = 1.99 \times 10^{-26}$ kg (^{12}C), $R_{12} = 12.5$ cm. The separation of the isotopes at the detector is $2(R_{15} - R_{14})$.

Solve: $R = \dfrac{mv}{|q|B}$. $\dfrac{R}{m} = \dfrac{v}{|q|B} = \text{constant}$. $\dfrac{R_{14}}{m_{14}} = \dfrac{R_{12}}{m_{12}}$ and

$$R_{14} = R_{12} \left(\frac{m_{14}}{m_{12}} \right) = (12.5 \text{ cm}) \left(\frac{2.32 \times 10^{-26} \text{ kg}}{1.99 \times 10^{-26} \text{ kg}} \right) = 14.6 \text{ cm}.$$

$$R_{15} = R_{12} \left(\frac{m_{15}}{m_{12}} \right) = (12.5 \text{ cm}) \left(\frac{2.49 \times 10^{-26} \text{ kg}}{1.99 \times 10^{-26} \text{ kg}} \right) = 15.6 \text{ cm}.$$

The separation of the isotopes at the detector is $2(R_{15} - R_{14}) = 2(15.6 \text{ cm} - 14.6 \text{ cm}) = 2.0 \text{ cm}.$

Reflect: The separation is large enough to be easily detectable. Since the diameter of the ion path is large, about 30 cm, the uniform magnetic field within the instrument must extend over a large area.

*20.29. Set Up: $F = IlB \sin \phi$. The direction of $\vec{F}$ is determined by applying the right-hand rule to the directions of I and $\vec{B}$. 1 gauss $= 10^{-4}$ T.

Solve: (a) The directions of I and $\vec{B}$ are sketched in Figure (a) below. $\phi = 90°$ and

$$F = (1.5 \text{ A})(2.5 \text{ m})(0.55 \times 10^{-4} \text{ T}) = 2.1 \times 10^{-4} \text{ N}.$$

The right-hand rule says that $\vec{F}$ is directed out of the page, so is upward.

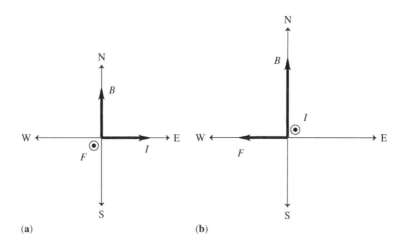

(a)   (b)

(b) The directions of I and $\vec{B}$ are sketched in Figure (b) above. $\phi = 90°$ and $F = 2.1 \times 10^{-4}$ N. $\vec{F}$ is directed east to west.

(c) $\vec{B}$ and the direction of the current are antiparallel. $\phi = 180°$ and $F = 0$.

(d) The magnetic force of 2.1×10^{-4} N is not large enough to cause significant effects.

Reflect: The magnetic force is a maximum when the directions of I and $\vec{B}$ are perpendicular and it is zero when the current and magnetic field are either parallel or antiparallel.

*20.33. Set Up: Label the three segments in the field as a, b, and c. Let x be the length of segment a. Segment b has length 0.300 m and segment c has length $0.600 \text{ cm} - x$. Figure (a) below shows the direction of the force on each segment. For each segment, $\phi = 90°$. The total force on the wire is the vector sum of the forces on each segment.

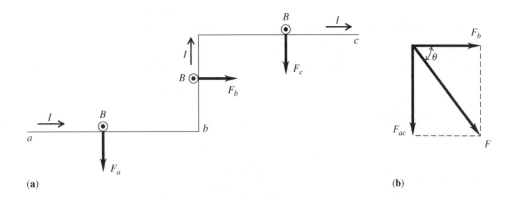

(a)   (b)

Solve: $F_a = IlB = (4.50 \text{ A})x(0.240 \text{ T})$. $F_c = (4.50 \text{ A})(0.600 \text{ m} - x)(0.240 \text{ T})$. Since $\vec{F}_a$ and $\vec{F}_c$ are in the same direction their vector sum has magnitude $F_{ac} = F_a + F_c = (4.50 \text{ A})(0.600 \text{ m})(0.240 \text{ T}) = 0.648 \text{ N}$ and is directed toward the bottom of the page in Figure (a). $F_b = (4.50 \text{ A})(0.300 \text{ m})(0.240 \text{ T}) = 0.324 \text{ N}$ and is directed to the right. The vector addition diagram for $\vec{F}_{ac}$ and $\vec{F}_b$ is given in Figure (b) above.

$$F = \sqrt{F_{ac}^2 + F_b^2} = \sqrt{(0.648 \text{ N})^2 + (0.324 \text{ N})^2} = 0.724 \text{ N}. \quad \tan\theta = \frac{F_{ac}}{F_b} = \frac{0.648 \text{ N}}{0.324 \text{ N}} \text{ and}$$

$$\theta = 63.4°.$$

The net force has magnitude 0.724 N and its direction is specified by $\theta = 63.4°$ in Figure (b).

Reflect: All three current segments are perpendicular to the magnetic field, so $\phi = 90°$ for each in the force equation. The direction of the force on a segment depends on the direction of the current for that segment.

*20.35. Set Up: $\tau = IAB \sin\phi$. Since the plane of the loop is parallel to the field, the field is perpendicular to the normal to the loop and $\phi = 90°$. The magnetic moment of the loop is $\mu = IA$.

Solve: (a) $\tau = IAB = (6.2 \text{ A})(0.050 \text{ m})(0.080 \text{ m})(0.19 \text{ T}) = 4.7 \times 10^{-3} \text{ N} \cdot \text{m}$

(b) $\mu = IA = (6.2 \text{ A})(0.050 \text{ m})(0.080 \text{ m}) = 0.025 \text{ A} \cdot \text{m}^2$

Reflect: The torque is a maximum when the field is in the plane of the loop and $\phi = 90°$.

*20.37. Set Up: $\tau = IAB \sin\phi$. The coil as viewed along the axis of rotation is shown in Figure (a) below for its original position and in Figure (b) after it has rotated $30.0°$.

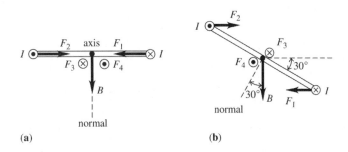

(a)   (b)

Solve: (a) The forces on each side of the coil are shown in Figure (a). $\vec{F}_1 + \vec{F}_2 = 0$ and $\vec{F}_3 + \vec{F}_4 = 0$. The net force on the coil is zero. $\phi = 0°$ and $\sin\phi = 0$, so $\tau = 0$. The forces on the coil produce no torque.

(b) The net force is still zero. $\phi = 30.0°$ and the net torque is

$$\tau = (1)(1.40 \text{ A})(0.220 \text{ m})(0.350 \text{ m})(1.50 \text{ T})\sin 30.0° = 0.0808 \text{ N} \cdot \text{m}.$$

The net torque is clockwise in Figure (b) above and is directed so as to increase the angle ϕ.

Reflect: For any current loop in a uniform magnetic field the net force on the loop is zero. The torque on the loop depends on the orientation of the plane of the loop relative to the magnetic field direction.

*20.41. Set Up: For a long straight wire, $B = \dfrac{\mu_0 I}{2\pi r}$. The magnetic field of the earth is of the order of 10^{-4} T.

$\mu_0 = 4\pi \times 10^{-7}$ T·m/A.

Solve: $B = \dfrac{(4\pi \times 10^{-7}\ \text{T·m/A})(10\ \text{A})}{2\pi(0.050\ \text{m})} = 4.0 \times 10^{-5}$ T. This field is the same order of magnitude as the earth's field.

*20.45. Set Up: For a long, thin, straight wire we have $B = \dfrac{\mu_0 I}{2\pi r}$.

Solve: $B = \dfrac{\mu_0 I}{2\pi r} = \dfrac{(4\pi \times 10^{-7}\ \text{T·m/A})(0.65\ \text{A})}{2\pi(0.0030\ \text{m})} = 4.3 \times 10^{-5}$ T.

Reflect: $B = \dfrac{\mu_0 I}{2\pi r}$ applies precisely only to an infinitely long straight wire.

*20.47. Set Up: $B = \dfrac{\mu_0 I}{2\pi r}$. $B_1 = 8.0\ \mu\text{T}$ when $r_1 = 2.0$ cm.

Solve: (a) $Br = \dfrac{\mu_0 I}{2\pi} = \text{constant}$ and $B_1 r_1 = B_2 r_2$. $B_2 = B_1\left(\dfrac{r_1}{r_2}\right) = (8.0\ \mu\text{T})\left(\dfrac{2.0\ \text{cm}}{4.0\ \text{cm}}\right) = 4.0\ \mu\text{T}$.

(b) $r_2 = r_1\left(\dfrac{B_1}{B_2}\right) = (2.0\ \text{cm})\left(\dfrac{8.0\ \mu\text{T}}{1.0\ \mu\text{T}}\right) = 16.0$ cm.

(c) $\dfrac{B}{I} = \dfrac{\mu_0}{2\pi r} = \text{constant}$ and $\dfrac{B_1}{I_1} = \dfrac{B_2}{I_2}$. $B_2 = B_1\left(\dfrac{I_2}{I_1}\right) = (8.0\ \mu\text{T})\left(\dfrac{2I}{I}\right) = 16.0\ \mu\text{T}$

Reflect: B is proportional to I and to $1/r$. When I is doubled, B is doubled. When r is doubled, B is halved.

*20.51. Set Up: $B = \dfrac{\mu_0 I}{2\pi r}$. The direction of $\vec{B}$ is given by the right-hand rule in Section 20.7. Call the wires a and b, as indicated in the figure below. The magnetic fields of each wire at points P_1 and P_2 are also shown in the figure.

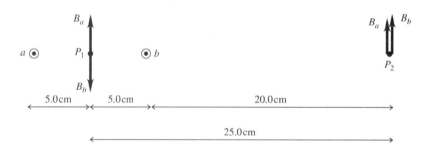

Solve: (a) At P_1, $B_a = B_b$ and the two fields are in opposite directions, so the net field is zero.

(b) $B_a = \dfrac{\mu_0 I}{2\pi r_a}$. $B_b = \dfrac{\mu_0 I}{2\pi r_b}$. $\vec{B}_a$ and $\vec{B}_b$ are in the same direction so

$$B = B_a + B_b = \dfrac{\mu_0 I}{2\pi}\left(\dfrac{1}{r_a} + \dfrac{1}{r_b}\right) = \dfrac{(4\pi \times 10^{-7}\ \text{T·m/A})(4.00\ \text{A})}{2\pi}\left[\dfrac{1}{0.300\ \text{m}} + \dfrac{1}{0.200\ \text{m}}\right] = 6.67 \times 10^{-6}\ \text{T}$$

$\vec{B}$ has magnitude $6.67\ \mu\text{T}$ and is directed toward the top of the page.

Reflect: At points directly to the left of both wires the net field is directed toward the bottom of the page.

*20.53. Set Up: $F = \dfrac{\mu_0 l l I'}{2\pi r}$. $\mu_0 = 4\pi \times 10^{-7}$ N/A^2. Parallel conductors carrying currents in opposite directions repel each other. Parallel conductors carrying currents in the same direction attract each other.

Solve: $F = \dfrac{(4\pi \times 10^{-7}\text{ N/A}^2)(15\text{ m})(25\text{ A})(75\text{ A})}{2\pi(0.35\text{ m})} = 0.0161$ N. Since the currents are in the same direction the force is attractive.

Reflect: The currents are large but the force per meter on each wire is very small.

*20.55. Set Up: Measurements on a typical extension cord give that the centers of the two wires are separated by about 4 mm. $F = \dfrac{\mu_0 l I I'}{2\pi r}$. $4\pi \times 10^{-7}$ N/A^2.

Solve: $F = \dfrac{(4\pi \times 10^{-7}\text{ N/A}^2)(2.0\text{ m})(5.0\text{ A})^2}{2\pi(4.0 \times 10^{-3}\text{ m})} = 2.5 \times 10^{-3}$ N. Since the currents are in opposite directions the wires repel.

*20.59. Set Up: The magnetic field at the center of N circular loops is $B = \dfrac{N\mu_0 I}{2R}$.

Solve: $N = \dfrac{2RB}{\mu_0 I} = \dfrac{2(6.00 \times 10^{-2}\text{ m})(6.39 \times 10^{-4}\text{ T})}{(4\pi \times 10^{-7}\text{ T}\cdot\text{m/A})(2.50\text{ A})} = 24.4$. Therefore, 24 turns are required.

*20.63. Set Up: The magnetic field at the center of a circular loop is $B = \dfrac{\mu_0 I}{2R}$. By symmetry each segment of the loop that has length Δl contributes equally to the field, so the field at the center of a semicircle is $\frac{1}{2}$ that of a full loop. Since the straight sections produce no field at P, the field at P is $B = \dfrac{\mu_0 I}{4R}$.

Solve: $B = \dfrac{\mu_0 I}{4R}$. The direction of $\vec{B}$ is given by the right-hand rule: $\vec{B}$ is directed into the page.

Reflect: For a quarter-circle section of wire the magnetic field at its center of curvature is $B = \dfrac{\mu_0 I}{8R}$.

*20.67. Set Up: The magnetic field at the center of a circular loop is $B_{\text{loop}} = \dfrac{\mu_0 I}{2R}$. The magnetic field at the center of a solenoid is $B_{\text{solenoid}} = \mu_0 n I$, where $n = \dfrac{N}{L}$ is the number of turns per meter.

Solve: (a) $B_{\text{loop}} = \dfrac{\mu_0 I}{2R} = \dfrac{(4\pi \times 10^{-7}\text{ T}\cdot\text{m/A})(2.00\text{ A})}{2(0.050\text{ m})} = 2.51 \times 10^{-5}$ T. **(b)** $n = \dfrac{N}{L} = \dfrac{1000}{5.00\text{ m}} = 200$ m^{-1}.

$B_{\text{solenoid}} = \mu_0 n I = (4\pi \times 10^{-7}\text{ T}\cdot\text{m/A})(200\text{ m}^{-1})(2.00\text{ A}) = 5.03 \times 10^{-4}$ T. $B_{\text{solenoid}} = 20B_{\text{loop}}$. The field at the center of a circular loop depends on the radius of the loop. The field at the center of a solenoid depends on the length of the solenoid, not on its radius.

Reflect: Eq. (20.14) for the field at the center of a solenoid is only correct for a very long solenoid, one whose length L is much greater than its radius R. We cannot consider the limit that L gets small and expect the expression for the solenoid to go over to the expression for N circular loops.

*20.69. Set Up: The magnetic field at the center of N circular loops is $B = \dfrac{\mu_0 N I}{2R}$. The magnetic field of the earth is given to be 0.55 G $= 5.5 \times 10^{-5}$ T. The number of turns can be determined from the length of the wire and the diameter of the coil: $N = L/(\pi D)$

Solve: The number of turns is $N = L/(\pi D) = \dfrac{25 \text{ m}}{\pi(0.44 \text{ m})} = 18$. The required current is $I = \dfrac{2RB}{\mu_0 N} =$

$\dfrac{2(0.22 \text{ m})(5.5 \times 10^{-5} \text{ T})}{18(4\pi \times 10^{-7} \text{ T} \cdot \text{m/A})} = 1.1 \text{ A}$.

Reflect: For a fixed current, the magnetic field, $B = \dfrac{\mu_0 NI}{2R}$, is determined by the ratio $\dfrac{N}{R}$. If a given length of wire is used to construct the coil, we know that halving R would double the number of turns (N) that we can make — and so the B at the center of the coil would be quadrupled.

*20.71. Set Up: The Biot and Savart law is $\Delta B = \dfrac{\mu_0}{4\pi} \dfrac{I \Delta l \sin\theta}{r^2}$. The wire and the vector $\vec{r}$ from each short segment to point P are shown in the figure below. For segment A, $\tan\theta_A = \dfrac{5.00 \text{ cm}}{7.00 \text{ cm}}$ and $\theta_A = 35.5°$. $r_A = 8.60$ cm. For segment C, $\theta_C = 90°$ and $r_C = 5.00$ cm.

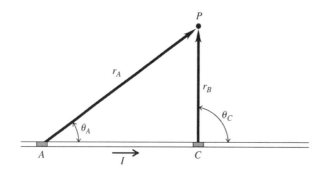

Solve: (a) $\Delta B = \left(\dfrac{4\pi \times 10^{-7} \text{ T} \cdot \text{m/A}}{4\pi} \right) \dfrac{(10.0 \text{ A})(2.00 \times 10^{-3} \text{ m}) \sin 35.5°}{(8.60 \times 10^{-2} \text{ m})^2} = 1.57 \times 10^{-7} \text{ T}$.

The right-hand rule says that $\Delta \vec{B}$ is directed out of the page.

(b) $\Delta B = \left(\dfrac{4\pi \times 10^{-7} \text{ T} \cdot \text{m/A}}{4\pi} \right) \dfrac{(10.0 \text{ A})(2.00 \times 10^{-3} \text{ m}) \sin 90°}{(5.00 \times 10^{-2} \text{ m})^2} = 8.00 \times 10^{-7} \text{ T}$.

The right-hand rule says that $\Delta \vec{B}$ is directed out of the page.

Reflect: Both segments produce magnetic field at P that is in the same direction. At P the field due to segment C is larger than that due to segment A.

*20.75. Set Up: $\mu = K_m \mu_0$, with $\mu_0 = 4\pi \times 10^{-7}$ T·m/A. The magnetic field in the material is a factor of K_m greater than it is in vacuum.

Solve: (a) $\mu = (1.00026)(4\pi \times 10^{-7} \text{ T} \cdot \text{m/A}) = 1.257 \times 10^{-6} \text{ T} \cdot \text{m/A}$

(b) $B_{\text{inside}} = K_m B_{\text{external}} = (1.00026)(1.3500 \text{ T}) = 1.3504 \text{ T}$

Reflect: The magnetic field inside the paramagnetic material is larger than the external field, but only slightly larger.

*20.77. Set Up: Use $v_y{}^2 = v_{0y}{}^2 + 2a_y y$ to calculate v_y. With $+y$ downward, $v_{0y} = 0$, $a_y = +9.80$ m/s^2 and $y = 125$ m. The direction of the force is given by the right-hand rule and the magnitude is given by $F = |q|vB\sin\phi$. The charge of the ball is $-(4.00 \times 10^8)e$.

Solve: The directions of $\vec{v}$ and $\vec{B}$ are shown in the figure below, where the downward direction is into the page. The right-hand rule (rhr) direction is north, but since the charge is negative the force is opposite the rhr direction and is to the south. At the bottom of the shaft the speed of the ball is

$$v = \sqrt{2g(125 \text{ m})} = \sqrt{2(9.80 \text{ m/s}^2)(125 \text{ m})} = 49.5 \text{ m/s}.$$

$$|q| = (4.00 \times 10^8)(1.60 \times 10^{-19} \text{ C}) = 6.40 \times 10^{-11} \text{ C}.$$

$$\phi = 90°, \text{ so } F = |q|vB = (6.40 \times 10^{-11} \text{ C})(49.5 \text{ m/s})(0.250 \text{ T}) = 7.92 \times 10^{-10} \text{ N}.$$

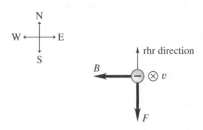

Reflect: The magnetic force is much less than the gravity force on the ball.

*20.81. Set Up: The ion has charge $q = +e$. After being accelerated through a potential difference V the ion has kinetic energy qV. The acceleration in the circular path is v^2/R.

Solve: $K = qV = +eV$. $\frac{1}{2}mv^2 = eV$ and

$$v = \sqrt{\frac{2eV}{m}} = \sqrt{\frac{2(1.60 \times 10^{-19} \text{ C})(220 \text{ V})}{1.16 \times 10^{-26} \text{ kg}}} = 7.79 \times 10^4 \text{ m/s}. \quad F_B = |q|vB\sin\phi. \quad \phi = 90°.$$

$\vec{F} = m\vec{a}$ gives $|q|vB = m\dfrac{v^2}{R}$.

$$R = \frac{mv}{|q|B} = \frac{(1.16 \times 10^{-26} \text{ kg})(7.79 \times 10^4 \text{ m/s})}{(1.60 \times 10^{-19} \text{ C})(0.723 \text{ T})} = 7.81 \times 10^{-3} \text{ m} = 7.81 \text{ mm}.$$

*20.83. Set Up: The direction of the current is shown in the figure below. The direction of $\vec{B}$ is given by the right-hand rule. $B = \dfrac{\mu_0 I}{2\pi r}$.

Solve: (a) Below the line the magnetic field is toward the east.

$$B = \frac{(4\pi \times 10^{-7} \text{ T} \cdot \text{m/A})(800 \text{ A})}{2\pi(5.50 \text{ m})} = 2.91 \times 10^{-5} \text{ T}.$$

(b) The field due to the current is the same order of magnitude as the earth's field so it is problem.

*20.87. Set Up: The current in the bar is horizontal toward the bottom of the figure, so the magnetic force on it is vertically upwards (out of the page). The net force on the bar is equal to the magnetic force minus the gravitational force, so Newton's second law gives the acceleration. The bar is in parallel with the 10.0-Ω resistor, so we must use circuit analysis to find the initial current through it. First find the current. The equivalent resistance across the battery is 30.0 Ω, so the total current is 4.00 A, half of which goes through the bar. Applying Newton's second law to the bar gives $\sum F = ma = F_B - mg = iLB - mg$.

Solve: Solving for the acceleration gives

$$a = \frac{iLB - mg}{m} = \frac{(2.0\text{A})(1.50\text{ m})(1.60\text{ T}) - 3.00\text{ N}}{(3.00\text{ N}/9.80\text{ m/s}^2)} = 5.88\text{ m/s}^2.$$

The direction is upward.

Reflect: Once the bar is free of the conducting wires, its acceleration will become 9.8 m/s^2 downward since only gravity will be acting on it.

*20.91. Set Up: Turning the charged loop creates a current, and the external magnetic field exerts a torque on that current. The current is $I = q/T = q/(1/f) = qf = q(\omega/2\pi) = q\omega/2\pi$. The torque is $\tau = \mu B \sin\phi$.

Solve: In this case, $\phi = 90°$ and $\mu = AB$, giving $\tau = IAB$. Combining the results for the torque and current and

using $A = \pi r^2$ gives $\tau = \left(\dfrac{q\omega}{2\pi}\right)\pi r^2 B = \frac{1}{2}q\omega r^2 B$

Reflect: Any moving charge creates a current—so turning the loop creates a current, which in turn causes a magnetic field.

Solutions to Passage Problems

*20.93. Set Up: The torque on a magnetic dipole is given by $\tau = \mu B \sin\phi$. The magnetic moment of the proton is 1.4×10^{-26} J/T $= 1.4 \times 10^{-26}$ N·m/T.

Solve: Using the given values we have $\tau = \mu B \sin\phi = (1.4 \times 10^{-26}\text{ N·m/T})(2\text{ T})\sin 90° = 2.8 \times 10^{-26}$ N·m. Thus, the correct answer is C.

ELECTROMAGNETIC INDUCTION

Problems 3, 5, 9, 11, 13, 19, 21, 23, 27, 31, 33, 37, 39, 41, 45, 47, 49, 53, 55, 59, 61, 63

Solutions to Problems

*21.3. Set Up: The total flux through the bottle is zero because it is a closed surface. The total flux through the bottle is the flux through the plastic plus the flux through the open cap, so the sum of these must be zero. $\Phi_{\text{plastic}} + \Phi_{\text{cap}} = 0$.

$$\Phi_{\text{plastic}} = -\Phi_{\text{cap}} = -B\,A\cos\varphi = -B(\pi r^2)\cos\varphi$$

Solve: Substituting the numbers gives $\Phi_{\text{plastic}} = -(1.75\ T)\pi(0.0125\ \text{m})^2 \cos 25° = -7.8 \times 10^{-4}\ \text{Wb}$

Reflect: It would be impossible to calculate the flux through the plastic directly because of the complicated shape of the bottle, but with a little thought we can find this flux through a simple indirect calculation.

*21.5. Set Up: $\mathcal{E} = N\left|\dfrac{\Delta\Phi_B}{\Delta t}\right|$. $\Phi_B = BA\cos\phi$. $\phi = 0°$, A is constant and B is changing.

Solve: $\mathcal{E} = N\left|\dfrac{\Delta\Phi_B}{\Delta t}\right| = NA\left|\dfrac{\Delta B}{\Delta t}\right|$. $I = \dfrac{\mathcal{E}}{R} = \dfrac{NA|\Delta B/\Delta t|}{R}$.

$$\left|\dfrac{\Delta B}{\Delta t}\right| = \dfrac{RI}{NA} = \dfrac{(40.0\ \Omega)(0.150\ \text{A})}{(200)\pi(3.00 \times 10^{-2}\ \text{m})^2} = 10.6\ \text{T/s}.$$

*21.9. Set Up: Since the field is uniform, $\Phi_B = BA\cos\phi$. Since the plane of the circuit is perpendicular to the field, $\phi = 0°$ and $\cos\phi = 1$. $\mathcal{E} = \left|\dfrac{\Delta\Phi_B}{\Delta t}\right|$. A is constant and B is changing, so $\mathcal{E} = A\left|\dfrac{\Delta B}{\Delta t}\right|$ and $\left|\dfrac{\Delta B}{\Delta t}\right| = 0.25\ \text{T/s}$. The current in the circuit is given by Ohm's law, $\mathcal{E} = IR$.

Solve: $\mathcal{E} = (0.300\ \text{m})(0.600\ \text{m})(0.25\ \text{T/s}) = 0.045\ \text{V}$. $I = \dfrac{\mathcal{E}}{R} = \dfrac{0.045\ \text{V}}{15\ \Omega} = 3.0\ \text{mA}$.

Reflect: The induced emf depends on the rate of change of the flux so it depends on the rate at which B is changing. It does not depend on the initial magnitude of the field.

*21.11. Set Up: The maximum emf is produced when the opposite sides of the coil are moving perpendicular to the field with speed $v = \omega(L/2)$. The maximum emf induced in each of the two sides on one turn is BLv. The maximum total emf is $2NBLv = NBL^2\omega$.

Solve: (a) $\mathcal{E}_{max} = NBL^2\omega$. $\omega = \dfrac{\mathcal{E}_{max}}{NBL^2} = \dfrac{(20.0\times10^{-3}\text{ V})}{(15)(0.0250\text{ T})(0.120\text{ m})^2} = 3.70$ rad/s.

(b) The emf changes polarity as the coil rotates and the average emf is zero.

Reflect: The maximum emf is proportional to the angular velocity of the coil. As ω increases, the flux through the coil changes more rapidly.

*21.13. Set Up: The field of the induced current is directed to oppose the change in flux.

Solve: (a) The field is into the page and is increasing so the flux is increasing. The field of the induced current is out of the page. To produce field out of the page the induced current is counterclockwise.

(b) The field is into the page and is decreasing so the flux is decreasing. The field of the induced current is into the page. To produce field into the page the induced current is clockwise.

(c) The field is constant so the flux is constant and there is no induced emf and no induced current.

*21.19. Set Up: By Lenz's law, the induced current flows to oppose the flux change that caused it.

Solve: The magnetic field is outward through the round coil and is decreasing, so the magnetic field due to the induced current must also point outward to oppose this decrease. Therefore the induced current is counterclockwise.

Reflect: Careful! Lenz's law does not say that the induced current flows to oppose the magnetic flux. Instead it says that the current flows to oppose the *change* in flux.

*21.21. Set Up: The bar and magnetic field are shown in the figure below. $V_{ab} = vBL$. To determine which end is at higher potential consider the direction of the magnetic force on a positive charge in the moving rod. Accumulation of positive charge produces higher potential.

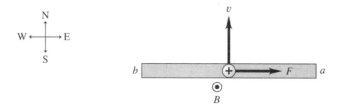

Solve: (a) $V_{ab} = vBL = (11.5\text{ m/s})(1.22\text{ T})(0.150\text{ m}) = 2.10$ V.

(b) The magnetic force on a positive charge in the bar is eastward. Positive charge accumulates at end a of the bar and that end (the east end) is at higher potential.

(c) If the bar is moving east to west the force is to the north. The force is perpendicular to the bar and there is no charge accumulation at its ends; the potential difference between its ends is zero.

Reflect: If the bar moves parallel to the magnetic field there is no magnetic force and no potential difference is produced.

*21.23. Set Up: There will be a motional emf between the underbody and roof of your car due to the motion of your car through the earth's magnetic field. Follow the derivation of Equation 21.7—except that the velocity of the car is not perpendicular to the magnetic field. Thus, the magnetic force on charges within the body of your car is given by $F_B = |q|vB\sin\phi$, which leads to $\mathcal{E} = EL = vBL\sin\phi$. The speed of the car is 95 km/h = (95 km/h) $(10^3$ m/km)(1 h/3600 s) = $\underline{26.39}$ m/s.

Solve: $\mathcal{E} = vBL\sin\phi = (26.39\text{ m/s})(5.5\times10^{-5}\text{ T})(1.5\text{ m})\sin35° = \underline{1.248}\times10^{-3}$ V = 1.2 mV. Using the right-hand rule we conclude that an electron in the body of the car will experience a downward force (while the car moves in the given direction). This implies that the roof of the car will be at a higher potential than the underbody.

Reflect: The maximum motional emf occurs when the velocity of the car is perpendicular to the direction of the magnetic field. For example, if the car were moving due east, the induced emf would be $\mathcal{E} = vBL\sin\phi = (26.39 \text{ m/s})$ $(5.5\times10^{-5} \text{ T})(1.5 \text{ m})\sin 90° = 2.2 \text{ mV}.$

*21.27. Set Up: Let solenoid 1 have 100 turns and solenoid 2 have 500 turns. $M = \left|\dfrac{N_2\Phi_{B2}}{i_1}\right|$. The field within the turns of solenoid 1 is $B_1 = \dfrac{\mu_0 N_1 i_1}{2\pi r}$.

Solve: $\Phi_{B2} = B_1 A = \dfrac{\mu_0 N_1 i_1 A}{2\pi r}$.

$$M = \frac{\mu_0 N_1 N_2 A}{2\pi r} = \frac{(4\pi\times10^{-7} \text{ Wb/m}\cdot\text{A})(100)(500)(4.00\times10^{-4} \text{ m}^2)}{2\pi(0.100 \text{ m})} = 4.00\times10^{-5} \text{ H}.$$

Reflect: If we interchange the roles of coils 1 and 2 we calculate the same value of M.

*21.31. Set Up: Apply Equation 21.11.

Solve: (a) $M = \dfrac{N_2\Phi_{B2}}{i_1} = \dfrac{400(0.0320 \text{ Wb})}{6.52 \text{ A}} = 1.96 \text{ H}$

(b) $M = \dfrac{N_1\Phi_{B1}}{i_2}$ so $\Phi_{B1} = \dfrac{Mi_2}{N_1} = \dfrac{(1.96 \text{ H})(2.54 \text{ A})}{700} = 7.11\times10^{-3} \text{ Wb}$

Reflect: M relates the current in one coil to the flux through the other coil. Equation 21.11 shows that M is the same for a pair of coils, no matter which one has the current and which one has the flux.

*21.33. Set Up: $\mathcal{E} = L\left|\dfrac{\Delta i}{\Delta t}\right|$

Solve: $L = \dfrac{\mathcal{E}}{|\Delta i/\Delta t|} = \dfrac{0.0160 \text{ V}}{0.0640 \text{ A/s}} = 0.250 \text{ H}$

*21.37. Set Up: Combine the two expressions for L: $L = N\Phi_B/i$ and $L = |\mathcal{E}/(\Delta i/\Delta t)|$, where Φ_B is the average flux through one turn of the solenoid.

Solve: Solving for N we have $N = |\mathcal{E}i/[\Phi_B(\Delta i/\Delta t)]| = \dfrac{(12.6\times10^{-3} \text{ V})(1.40 \text{ A})}{(0.00285 \text{ Wb})(0.0260 \text{ A/s})} = 238 \text{ turns}.$

Reflect: The induced emf depends on the time rate of change of the total flux through the solenoid.

*21.39. Set Up: A transformer transforms voltages according to $\dfrac{V_2}{V_1} = \dfrac{N_2}{N_1}$. The effective resistance of a secondary circuit of resistance R is $R_{\text{eff}} = \dfrac{R}{(N_2/N_1)^2}$. Resistance R is related to P and V by $P = \dfrac{V^2}{R}$. Conservation of energy requires $P_1 = P_2$ so $V_1 I_1 = V_2 I_2$.

Solve: (a) $V_1 = 240 \text{ V}$ and we want $V_2 = 120 \text{ V}$, so use a step-down transformer with $N_2/N_1 = \frac{1}{2}$.

(b) $P = VI$, so $I = \dfrac{P}{V} = \dfrac{1600 \text{ W}}{240 \text{ V}} = 6.67 \text{ A}.$

(c) The resistance R of the blower is $R = \dfrac{V^2}{P} = \dfrac{(120 \text{ V})^2}{1600 \text{ W}} = 9.00 \text{ }\Omega.$ The effective resistance of the blower is

$R_{\text{eff}} = \dfrac{9.00 \text{ }\Omega}{(1/2)^2} = 36.0 \text{ }\Omega.$

Reflect: $I_2 V_2 = (13.3 \text{ A})(120 \text{ V}) = 1600 \text{ W}$. Energy is provided to the primary at the same rate that it is consumed in the secondary. Step-down transformers step up resistance and the current in the primary is less than the current in the secondary.

*21.41. Set Up: $V_1 = 120 \text{ V}$. $V_2 = 13,000 \text{ V}$. $P = VI$ and $P_1 = P_2$. $\dfrac{N_1}{N_2} = \dfrac{V_1}{V_2}$.

Solve: (a) $\dfrac{N_2}{N_1} = \dfrac{V_2}{V_1} = \dfrac{13,000 \text{ V}}{120 \text{ V}} = 108$

(b) $P = V_2 I_2 = (13,000 \text{ V})(8.50 \times 10^{-3} \text{ A}) = 110 \text{ W}$

(c) $I_1 = \dfrac{P}{V_1} = \dfrac{110 \text{ W}}{120 \text{ V}} = 0.917 \text{ A}$

Reflect: Since the power supplied to the primary must equal the power delivered by the secondary, in a step-up transformer the current in the primary is greater than the current in the secondary.

*21.45. Set Up: (a) The magnetic field inside a solenoid is $B = \mu_0 n I$.

Solve: $B = \dfrac{(4\pi \times 10^{-7} \text{ T} \cdot \text{m/A})(400)(80.0 \text{ A})}{0.250 \text{ m}} = 0.161 \text{ T}$

(b) Set Up: The energy density in a magnetic field is $u = \dfrac{B^2}{2\mu_0}$.

Solve: $u = \dfrac{(0.161 \text{ T})^2}{2(4\pi \times 10^{-7} \text{ T} \cdot \text{m/A})} = 1.03 \times 10^4 \text{ J/m}^3$

(c) Set Up: The total stored energy is $U = uV$.

Solve: $U = uV = u(lA) = (1.03 \times 10^4 \text{ J/m}^3)(0.250 \text{ m})(0.500 \times 10^{-4} \text{ m}^2) = 0.129 \text{ J}$

(d) Set Up: The energy stored in an inductor is $U = \frac{1}{2} L I^2$.

Solve: Solving for L and putting in the numbers gives

$$L = \frac{2U}{I^2} = \frac{2(0.129 \text{ J})}{(80.0 \text{ A})^2} = 4.02 \times 10^{-5} \text{ H}$$

Reflect: An inductor stores its energy in the magnetic field inside of it.

*21.47. Set Up: $U = \frac{1}{2} L I^2$. $\dfrac{U}{I^2} = \frac{1}{2} L =$ constant and $\dfrac{U_1}{I_1^2} = \dfrac{U_2}{I_2^2}$.

Solve: $I_2 = I_1 \left(\dfrac{U_2}{U_1} \right)^{1/2} = I \left(\dfrac{9.0 \text{ mJ}}{3.0 \text{ mJ}} \right)^{1/2} = I\sqrt{3}$.

*21.49. Set Up: The loop rule applied to the circuit gives $\mathcal{E} - iR - L\dfrac{\Delta i}{\Delta t} = 0$. $L\dfrac{\Delta i}{\Delta t} = v_L$, the voltage across the inductor. The current as a function of time is given by $i = \dfrac{\mathcal{E}}{R}(1 - e^{-(R/L)t})$. At $t = 0$, $i = 0$. $i = i_{\max} = \dfrac{\mathcal{E}}{R}$ at $t \to \infty$.

Solve: (a) At $t = 0$, $i = 0$ and $\dfrac{\Delta i}{\Delta t} = \dfrac{\mathcal{E}}{L} = \dfrac{6.00 \text{ V}}{2.50 \text{ H}} = 2.40 \text{ A/s}$.

(b) At $t = 0$, $\dfrac{\Delta i}{\Delta t} = 2.40 \text{ A/s}$ and $v_L = L\dfrac{\Delta i}{\Delta t} = (2.50 \text{ H})(2.40 \text{ A/s}) = 6.00 \text{ V}$. Initially, $i = 0$, the voltage across the resistor is zero, and the full battery emf appears across the inductor.

(c) The time constant is $\tau = \dfrac{L}{R} = \dfrac{2.50\ \text{H}}{8.00\ \Omega} = 0.313\ \text{s}$. When $t = 0.313\ \text{s}$,

$$i = \dfrac{\mathcal{E}}{R}(1 - e^{-1}) = \dfrac{6.00\ \text{V}}{8.00\ \Omega}(1 - e^{-1}) = 0.474\ \text{A}.$$

(d) When $t \to \infty$, $i = \dfrac{\mathcal{E}}{R} = \dfrac{6.00\ \text{V}}{8.00\ \Omega} = 0.750\ \text{A}.$

Reflect: Initially $i = 0$ and the full battery voltage is across the inductor. After a long time, the full battery voltage is across the resistor.

*21.53. Set Up: The current as a function of time is given by $i = \dfrac{\mathcal{E}}{R}(1 - e^{-(R/L)t})$. The energy stored in the inductor

is $U = \frac{1}{2}Li^2$. U reaches $\frac{1}{2}$ its maximum value when i is $\dfrac{1}{\sqrt{2}}$ times its maximum value.

Solve: (a) The maximum current is $\dfrac{\mathcal{E}}{R}$. $i = \dfrac{1}{2}\dfrac{\mathcal{E}}{R}$ gives $\dfrac{1}{2} = 1 - e^{-(R/L)t}$ and $e^{-(R/L)t} = \dfrac{1}{2}$. $-\left(\dfrac{R}{L}\right)t = \ln\left(\dfrac{1}{2}\right).$

$$t = -\dfrac{L\ln\left(\frac{1}{2}\right)}{R} = -\dfrac{(1.50 \times 10^{-3}\ \text{H})\ln\left(\frac{1}{2}\right)}{0.750 \times 10^3\ \Omega} = 1.39\ \mu\text{s}.$$

(b) $\dfrac{1}{\sqrt{2}} = 1 - e^{-(R/L)t}$ and $t = -\dfrac{L\ln(1 - 1/\sqrt{2})}{R} = -\dfrac{(1.50 \times 10^{-3}\ \text{H})\ln(1 - 1/\sqrt{2})}{0.750 \times 10^3\ \Omega} = 2.46\ \mu\text{s}.$

*21.55. Set Up: The energy stored in a capacitor is $U_C = \frac{1}{2}Cv^2$. The energy stored in an inductor is $U_L = \frac{1}{2}Li^2$. Energy conservation requires that the total stored energy be constant. The current is a maximum when the charge on the capacitor is zero.

Solve: (a) Initially $v = 16.0\ \text{V}$ and $i = 0$. $U_L = 0$ and

$$U_C = \tfrac{1}{2}Cv^2 = \tfrac{1}{2}(5.00 \times 10^{-6}\ \text{F})(16.0\ \text{V})^2 = 6.40 \times 10^{-4}\ \text{J}.$$

The total energy stored is $0.640\ \text{mJ}$.

(b) The current is maximum when $q = 0$ and $U_C = 0$. $U_C + U_L = 6.40 \times 10^{-4}\ \text{J}$ so $U_L = 6.40 \times 10^{-4}\ \text{J}$.

$$\tfrac{1}{2}Li_{\text{max}}^2 = 6.40 \times 10^{-4}\ \text{J} \text{ and } i_{\text{max}} = \sqrt{\dfrac{2(6.40 \times 10^{-4}\ \text{J})}{3.75 \times 10^{-3}\ \text{H}}} = 0.584\ \text{A}.$$

*21.59. Set Up: At $t = 0$ the angle ϕ between the normal to the loop and the magnetic field is zero. $\phi = \omega t$. $\Phi_B = BA\cos\phi$. From the discussion of generators in Section 21.3, the rate of change of Φ_B is $-\omega BA\sin\omega t$. The induced emf is $\omega AB\sin\omega t$.

Solve: (a) $\Phi_B = BA\cos\omega t$. The graph of Φ_B versus t is sketched in Figure **(a)** below.

(b) The graph of $-\omega BA\sin\omega t$ versus t is sketched in Figure **(b)** below.

(c) The graph of $\mathcal{E} = \omega AB\sin\omega t$ versus t is sketched in Figure **(c)** below.

(d) Doubling ω halves the period of $\mathcal{E}$ and doubles its amplitude. The graph of $\mathcal{E} = 2\omega AB\sin(2\omega t)$ versus t is sketched in Figure **(d)** below.

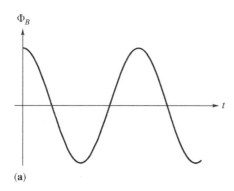

(a)

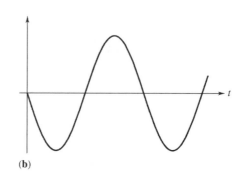

(b)

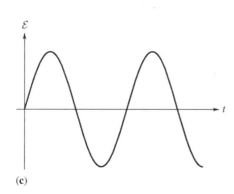

(c)

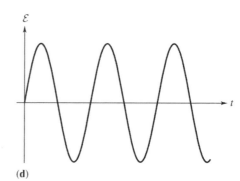

(d)

*21.61. Set Up: Use $U_C = \frac{1}{2}CV_C^2$ (energy stored in a capacitor) to solve for C. Then use Equation 21.30 and $\omega = 2\pi f$ to solve for the L that gives the desired current oscillation frequency.

Solve: $V_C = 12.0$ V; $U_C = \frac{1}{2}CV_C^2$ so $C = 2U_C/V_C^2 = 2(0.0160 \text{ J})/(12.0 \text{ V})^2 = 222 \, \mu\text{F}$

We know that $f = \dfrac{1}{2\pi\sqrt{LC}}$ so $L = \dfrac{1}{(2\pi f)^2 C}$. Setting $f = 3500$ Hz gives $L = 9.31 \, \mu\text{H}$

Reflect: Note that f is in Hz and ω is in rad/s; we must be careful not to confuse the two.

*21.63. Set Up: The emf in the inductor is $\mathcal{E} = L\left|\dfrac{\Delta i}{\Delta t}\right|$. $\mathcal{E}$ is limited by the loop rule, so the rate of change of the current is limited and the current in the inductor can't jump suddenly from one value to another.

Solve: (a) The current in the inductor is zero when the switch is open and is still zero immediately after the switch is closed. The current through the $30.0 \, \Omega$ resistor equals the current through the inductor and therefore is zero. The voltage across the $20.0 \, \Omega$ resistor must equal the battery voltage, so the current through this resistor is $\dfrac{60.0 \text{ V}}{20.0 \, \Omega} = 3.00$ A.

(b) After a long time the current is no longer changing and the voltage across the inductor is zero. The voltage across each resistor is 60.0 V, so the current through the $20.0 \, \Omega$ resistor is 3.00 A and the current through the $30.0 \, \Omega$ resistor is 2.00 A.

(c) The current through the inductor just after the switch is opened is the same as it was just before the switch was opened, 2.00 A. With the switch open there is only one current loop and the current is 2.00 A through each resistor.

ALTERNATING CURRENT

Problems 3, 5, 11, 13, 17, 19, 21, 25, 27, 31, 33, 37, 39, 43, 45

Solutions to Problems

*22.3. Set Up: The reactance of capacitors and inductors depends on the angular frequency at which they are operated, as well as their capacitance or inductance. The reactances are $X_C = 1/\omega C$ and $X_L = \omega L$.

Solve: (a) Equating the reactances gives $\omega L = \dfrac{1}{\omega C} \Rightarrow \omega = \dfrac{1}{\sqrt{LC}}$

(b) Using the numerical values we get $\omega = \dfrac{1}{\sqrt{LC}} = \dfrac{1}{\sqrt{(5.00 \text{ mH})(3.50\mu\text{F})}} = 7560 \text{ rad/s}$

$X_C = X_L = \omega L = (7560 \text{ rad/s})(5.00 \text{ mH}) = 37.8 \text{ }\Omega$

Reflect: At other angular frequencies, the two reactances could be very different.

*22.5. Set Up: We have $V_L = I\omega L$, where ω is the angular frequency, in rad/s and $f = \dfrac{\omega}{2\pi}$ is the frequency in Hz.

Solve: $V_L = I\omega L$ so $f = \dfrac{V_L}{2\omega I L} = \dfrac{(12.0 \text{ V})}{2\pi(2.60 \times 10^{-3} \text{ A})(4.50 \times 10^{-4} \text{ H})} = 1.63 \times 10^6 \text{ Hz}.$

Reflect: When f is increased, I decreases.

*22.11. Set Up: Use Equation 22.18: $V^2 = V_R^2 + (V_L - V_C)^2$

Solve: $V = \sqrt{(30.0 \text{ V})^2 + (50.0 \text{ V} - 90.0 \text{ V})^2} = 50.0 \text{ V}$

Reflect: Note that the voltage amplitudes do not simply add to give 170.0 V for the source voltage.

*22.13. Set Up: $Z = \sqrt{R^2 + (X_L - X_C)^2}$. $X_L = \omega L$. $X_C = \dfrac{1}{\omega C}$. $I = \dfrac{V}{Z}$. $\tan\phi = \dfrac{X_L - X_C}{R}$.

Solve: (a) *1000 rad/s:* $X_L = \omega L = (1000 \text{ rad/s})(0.900 \text{ H}) = 900 \text{ }\Omega.$

$$X_C = \dfrac{1}{\omega C} = \dfrac{1}{(1000 \text{ rad/s})(2.00 \times 10^{-6} \text{ F})} = 500 \text{ }\Omega.$$

$$Z = \sqrt{(200 \text{ }\Omega)^2 + (900 \text{ }\Omega - 500 \text{ }\Omega)^2} = 447 \text{ }\Omega.$$

750 rad/s: $X_L = \omega L = (750 \text{ rad/s})(0.900 \text{ H}) = 675 \, \Omega.$

$$X_C = \frac{1}{\omega C} = \frac{1}{(750 \text{ rad/s})(2.00 \times 10^{-6} \text{ F})} = 667 \, \Omega.$$

$$Z = \sqrt{(200 \, \Omega)^2 + (675 \, \Omega - 667 \, \Omega)^2} = 200 \, \Omega.$$

500 rad/s: $X_L = \omega L = (500 \text{ rad/s})(0.900 \text{ H}) = 450 \, \Omega.$

$$X_C = \frac{1}{\omega C} = \frac{1}{(500 \text{ rad/s})(2.00 \times 10^{-6} \text{ F})} = 1000 \, \Omega.$$

$$Z = \sqrt{(200 \, \Omega)^2 + (450 \, \Omega - 1000 \, \Omega)^2} = 585 \, \Omega.$$

(b) $I = \dfrac{V}{Z}.$ Z decreases and then increases. I increases as ω varies from 1000 rad/s to around 750 rad/s and then I decreases as ω continues to decrease.

(c) $\tan\phi = \dfrac{X_L - X_C}{R} = \dfrac{900 \, \Omega - 500 \, \Omega}{200 \, \Omega} = 2.00$ and $\phi = +63.4°.$

(d) The phasor diagram is sketched in the figure below

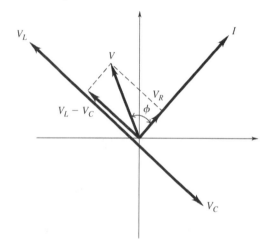

*22.17. Set Up: From Problem 22.14, $R = 200 \, \Omega.$ For a resistor, $P_R = I_{\text{rms}}^2 R.$ For the source, $P_{\text{source}} = V_{\text{rms}} I_{\text{rms}} \cos\phi.$ $\cos\phi = \dfrac{R}{Z}$ and $\dfrac{V_{\text{rms}}}{Z} = I_{\text{rms}}.$

Solve: (a) $P_R = I_{\text{rms}}^2 R = (0.250 \text{ A})^2 (200 \, \Omega) = 12.5 \text{ W}$

(b) $P_{\text{source}} = V_{\text{rms}} I_{\text{rms}} \left(\dfrac{R}{Z}\right) = I_{\text{rms}}^2 R = 12.5 \text{ W}$

(c) Energy is stored and released in the capacitor and inductor but not dissipated. The average rate of dissipation of energy for each is zero.

*22.19. Set Up: The resonance angular frequency is $\omega_0 = \dfrac{1}{\sqrt{LC}}.$ The period, the time duration of one cycle, is $T = \dfrac{2\pi}{\omega_0}.$

Solve: (a) $\omega_0 = \dfrac{1}{\sqrt{(5.0 \times 10^{-3} \text{ H})(2.0 \times 10^{-6} \text{ C})}} = 1.0 \times 10^4 \text{ rad/s}$

(b) $T = \dfrac{2\pi}{\omega_0} = \dfrac{2\pi}{1.0 \times 10^4 \text{ rad/s}} = 6.28 \times 10^{-4} \text{ s}$

*22.21. Set Up: At resonance $X_L = X_C, \phi = 0$ and $Z = R$. $R = 150\,\Omega, L = 0.750$ H, $C = 0.0180\,\mu$F, $V = 150$ V

Solve: (a) At the resonance frequency $X_L = X_C$ and from $\tan\phi = \dfrac{X_L - X_C}{R}$ we have that $\phi = 0°$ and the power

factor is $\cos\phi = 1.00$.

(b) $P_{\text{av}} = \frac{1}{2}VI\cos\phi$ (Equation 22.29)

At the resonance frequency $Z = R$, so $I = \dfrac{V}{Z} = \dfrac{V}{R}$

$$P_{\text{av}} = \frac{1}{2}V\left(\frac{V}{R}\right)\cos\phi = \frac{1}{2}\frac{V^2}{R} = \frac{1}{2}\frac{(150\text{ V})^2}{150\,\Omega} = 75.0\text{ W}$$

(c) Evaluate: When C and f are changed but the circuit is kept on resonance, nothing changes in $P_{\text{av}} = V^2/(2R)$, so the average power is unchanged: $P_{\text{av}} = 75.0$ W. The resonance frequency changes but since $Z = R$ at resonance the current doesn't change.

*22.25. Set Up: $\omega_0 = \dfrac{1}{\sqrt{LC}}$. $V = IZ$. At resonance, $Z = R$.

Solve: (a) $\omega_0 = \dfrac{1}{\sqrt{LC}} = \dfrac{1}{\sqrt{(0.200\text{ H})(80.0\times10^{-6}\text{ F})}} = 250$ rad/s

(b) $V = IZ = IR$ so $R = \dfrac{V}{I} = \dfrac{240\text{ V}}{0.600\text{ A}} = 400\,\Omega$.

(c) $X_L = \omega L = (250\text{ rad/s})(0.200\text{ H}) = 50.0\,\Omega$. $X_C = X_L = 50.0\,\Omega$.
$V_L = IX_L = (0.600\text{ A})(50.0\,\Omega) = 30.0$ V. $V_C = V_L = 30.0$ V. $V_R = IR = (0.600\text{ A})(400\,\Omega) = 240$ V.

*22.27. Set Up: In a parallel L-R-C circuit the impedance is a maximum at the resonance angular frequency of $\omega_0 = \dfrac{1}{\sqrt{LC}}$. At the resonance angular frequency, $Z = R$.

Solve: (a) $\omega_0 = \dfrac{1}{\sqrt{LC}} = \dfrac{1}{\sqrt{(11.2\times10^{-3}\text{ H})(8.50\times10^{-6}\text{ F})}} = 3.24\times10^3$ rad/s

(b) $Z = R = 125\,\Omega$

Reflect: $\omega_0 = \dfrac{1}{\sqrt{LC}}$ is the resonance angular frequency for either a series or a parallel L-R-C circuit. For the series circuit, at resonance the impedance is a minimum and the current is a maximum. For the parallel circuit, the impedance is a maximum and the current is a minimum.

*22.31. Set Up: $\tan\phi = \dfrac{X_L}{R}$, since there is no capacitance. $X_L = \omega L$.

Solve: $X_L = R\tan\phi = (48.0\,\Omega)\tan 52.3° = 62.1\,\Omega$. $L = \dfrac{X_L}{\omega} = \dfrac{X_L}{2\pi f} = \dfrac{62.1\,\Omega}{(2\pi)(80.0\text{ Hz})} = 0.124$ H

*22.33. Set Up: The capacitance of an air-filled parallel plate capacitor is $C = \dfrac{\epsilon_0 A}{d}$. The inductance of a long

solenoid was calculated in Problem 37 of Chapter 21 to be $L = \dfrac{\mu_0 AN^2}{l}$. The coil has $N = (125\text{ coils/cm})(9.00\text{ cm}) =$

1125 coils. The resonance frequency is $f_0 = \dfrac{1}{2\pi\sqrt{LC}}$. $\epsilon_0 = 8.85\times10^{-12}$ C^2/N$\cdot$m^2. $\mu_0 = 4\pi\times10^{-7}$ T$\cdot$m/A.

Solve: $C = \dfrac{\epsilon_0 A}{d} = \dfrac{(8.85\times10^{-12}\text{ C}^2/\text{N}\cdot\text{m}^2)(4.50\times10^{-2}\text{ m})^2}{8.00\times10^{-3}\text{ m}} = 2.24\times10^{-12}$ F.

$$L = \frac{\mu_0 \, AN^2}{l} = \frac{(4\pi \times 10^{-7} \text{ T} \cdot \text{m/A})\pi(0.250 \times 10^{-2} \text{ m})^2(1125)^2}{9.00 \times 10^{-2} \text{ m}} = 3.47 \times 10^{-4} \text{ H}$$

$$\omega_0 = \frac{1}{\sqrt{(3.47 \times 10^{-4} \text{ H})(2.24 \times 10^{-12} \text{ F})}} = 3.59 \times 10^7 \text{ rad/s}$$

*22.37. Set Up: The average power supplied by the source is $P = V_{rms}I_{rms}\cos\phi$. The power consumed in the resistance is $P = I_{rms}^2 R$. $\omega = 2\pi f = 2\pi(1.25 \times 10^3 \text{ Hz}) = 7.854 \times 10^3 \text{ rad/s}$. $X_L = \omega L = 157 \, \Omega$. $X_C = \frac{1}{\omega C} = 909 \, \Omega$.

Solve: (a) First, let us find the phase angle between the voltage and the current: $\tan\phi = \frac{X_L - X_C}{R} = \frac{157 \, \Omega - 909 \, \Omega}{350 \, \Omega}$

and $\phi = -65.04°$. The impedance of the circuit is $Z = \sqrt{R^2 + (X_L - X_C)^2} = \sqrt{(350 \, \Omega)^2 + (-752 \, \Omega)^2} = 830 \, \Omega$. The

average power provided by the generator is then $P = V_{rms}I_{rms}\cos(\phi) = \frac{V_{rms}^2}{Z}\cos(\phi) = \frac{(120 \text{ V})^2}{830 \, \Omega}\cos(-65.04°) = 7.32 \text{ W}$

(b) The average power dissipated by the resistor is $P_R = I_{rms}^2 R = \left(\frac{120 \text{ V}}{830 \, \Omega}\right)^2 (350 \, \Omega) = 7.32 \text{ W}$.

Reflect: Conservation of energy requires that the answers to parts (a) and (b) are equal.

*22.39. Set Up: We know R, X_C and ϕ so Eq. (22.23) tells us X_L. Use $P_{av} = I_{rms}^2 R$ from Exercise 22.38 to calculate I_{rms}. Then calculate Z and use Eq. (22.24) to calculate V_{rms} for the source. Source voltage lags current so $\phi = -54.0°$. $X_C = 350 \, \Omega$, $R = 180 \, \Omega$, $P_{av} = 140 \text{ W}$

Solve: (a) $\tan\phi = \frac{X_L - X_C}{R}$

$X_L = R\tan\phi + X_C = (180 \, \Omega)\tan(-54.0°) + 350 \, \Omega = -248 \, \Omega + 350 \, \Omega = 102 \, \Omega$

(b) $P_{av} = V_{rms}I_{rms}\cos\phi = I_{rms}^2 R$ (Exercise 22.38). $I_{rms} = \sqrt{\frac{P_{av}}{R}} = \sqrt{\frac{140 \text{ W}}{180 \, \Omega}} = 0.882 \text{ A}$

(c) $Z = \sqrt{R^2 + (X_L - X_C)^2} = \sqrt{(180 \, \Omega)^2 + (102 \, \Omega - 350 \, \Omega)^2} = 306 \, \Omega$

$V_{rms} = I_{rms}Z = (0.882 \text{ A})(306 \, \Omega) = 270 \text{ V}$.

Reflect: We could also use Equation 22.29: $P_{av} = V_{rms}I_{rms}\cos\phi$. This gives us $V_{rms} = \frac{P_{av}}{I_{rms}\cos\phi} =$

$\frac{140 \text{ W}}{(0.882 \text{ A})\cos(-54.0°)} = 270 \text{ V}$, which agrees with our previous result. The source voltage lags the current when

$X_C > X_L$, and this agrees with what we found.

*22.43. Set Up: From Problem 22.47, $R = 100.0 \, \Omega$, $C = 0.100 \, \mu\text{F}$, $L = 0.300 \text{ H}$ and $V = 240 \text{ V}$. $I = \frac{V}{Z}$.

Solve: (a) $X_L = \omega L = (400 \text{ rad/s})(0.300 \text{ H}) = 120 \, \Omega$.

$$X_C = \frac{1}{\omega C} = \frac{1}{(400 \text{ rad/s})(0.100 \times 10^{-6} \text{ F})} = 2.50 \times 10^4 \, \Omega.$$

$$Z = \sqrt{R^2 + (X_L - X_C)^2} = \sqrt{(100.0 \, \Omega)^2 + (120 \, \Omega - 2.50 \times 10^4 \, \Omega)^2} = 2.49 \times 10^4 \, \Omega.$$

$$I = \frac{V}{Z} = \frac{240 \text{ V}}{2.49 \times 10^4 \, \Omega} = 9.64 \times 10^{-3} \text{ A} = 9.64 \text{ mA}.$$

(b) $V_C = IX_C = (9.64 \times 10^{-3} \text{ A})(2.50 \times 10^4 \, \Omega) = 241 \text{ V}$

(c) $V_L = IX_L = (9.64 \times 10^{-3} \text{ A})(120 \, \Omega) = 1.16$ V

(d) $U_C = \frac{1}{2}CV_C^2 = \frac{1}{2}(0.100 \times 10^{-6} \text{ F})(241 \text{ V})^2 = 2.90 \times 10^{-3}$ J $= 2.90$ mJ.

$$U_L = \frac{1}{2}LI^2 = \frac{1}{2}(0.300 \text{ H})(9.64 \times 10^{-3} \text{ A})^2 = 1.39 \times 10^{-5} \text{ J}.$$

Solutions to Passage Problems

*22.45. Set Up: We are told that the platinum electrode behaves like an ideal capacitor in series with the resistance of the fluid, which is given by $R_A = \rho/(10a)$, where $\rho = 100 \, \Omega \cdot \text{cm} = 1 \, \Omega \cdot \text{m}$ and $d = 2a = 20 \mu m$. We know that

$X_C = \dfrac{1}{\omega C}$, where we are given $C = 10 \text{ nF} = 10^{-8}$ F and $\omega = 2\pi f = 2\pi[(5000/\pi)\text{Hz}] = 10^4$ rad/s. For an R-C circuit

we know that the impedance is given by $Z = \sqrt{R^2 + X_C^2}$.

Solve: $R_A = \rho/(10a) = (1 \, \Omega \cdot \text{m})/[10(10^{-5} \text{ m})] = 10^4 \, \Omega.$ $\quad X_C = \dfrac{1}{\omega C} = \dfrac{1}{(10^4 \text{ rad/s})(10^{-8} \text{ F})} = 10^4 \, \Omega.$ Thus we have

$Z = \sqrt{R^2 + X_C^2} = \sqrt{(10^4 \, \Omega)^2 + (10^4 \, \Omega)^2} = \sqrt{2} \times (10^4 \, \Omega).$ Thus, the correct answer is C.

ELECTROMAGNETIC WAVES

Problems 3, 5, 7, 9, 13, 15, 21, 23, 25, 27, 31, 35, 37, 41, 45, 47, 49, 55, 57, 59, 63, 67, 69, 71, 75, 77, 81, 83, 87, 91, 93

Solutions to Problems

*23.3. Set Up: The speed of light in vacuum is $c = 3.00 \times 10^8$ m/s. $d = ct$. 1 yr $= 3.156 \times 10^7$ s.

Solve: (a) $t = \dfrac{d}{c} = \dfrac{3.84 \times 10^8 \text{ m}}{3.00 \times 10^8 \text{ m/s}} = 1.28$ s

(b) $d = ct = (3.00 \times 10^8 \text{ m/s})(8.61 \text{ yr})(3.156 \times 10^7 \text{ s/yr}) = 8.15 \times 10^{16}$ m $= 8.15 \times 10^{13}$ km

Reflect: The speed of light is very large, but it can still take light a long time to travel enormous astronomical distances.

*23.5. Set Up: The wave speed is $c = 3.00 \times 10^8$ m/s. $c = f\lambda$.

Solve: (a) (i) $f = \dfrac{c}{\lambda} = \dfrac{3.00 \times 10^8 \text{ m/s}}{5.0 \times 10^3 \text{ m}} = 6.0 \times 10^4$ Hz

(ii) $f = \dfrac{3.00 \times 10^8 \text{ m/s}}{5.0 \times 10^{-6} \text{ m}} = 6.0 \times 10^{13}$ Hz

(iii) $f = \dfrac{c}{\lambda} = \dfrac{3.00 \times 10^8 \text{ m/s}}{5.0 \times 10^{-9} \text{ m}} = 6.0 \times 10^{16}$ Hz

(b) (i) $\lambda = \dfrac{c}{f} = \dfrac{3.00 \times 10^8 \text{ m/s}}{6.50 \times 10^{21} \text{ Hz}} = 4.62 \times 10^{-14}$ m $= 4.62 \times 10^{-5}$ nm

(ii) $\lambda = \dfrac{3.00 \times 10^8 \text{ m/s}}{590 \times 10^3 \text{ Hz}} = 508$ m $= 5.08 \times 10^{11}$ nm

*23.7. Set Up: For an electromagnetic wave propagating in the negative x direction, $E = -E_{max} \sin(\omega t + kx)$. $\omega = 2\pi f$ and $k = \dfrac{2\pi}{\lambda}$. $T = \dfrac{1}{f}$. $E_{max} = cB_{max}$.

Solve: (a) $E_{max} = 375$ V/m so $B_{max} = \dfrac{E_{max}}{c} = 1.25$ μT.

(b) $\omega = 5.98 \times 10^{15}$ rad/s so $f = \dfrac{\omega}{2\pi} = 9.52 \times 10^{14}$ Hz. $k = 1.99 \times 10^7$ rad/m so

$$\lambda = \frac{2\pi}{k} = 3.16 \times 10^{-7} \text{ m} = 316 \text{ nm.}$$

$$T = \frac{1}{f} = 1.05 \times 10^{-15} \text{ s.}$$

This wavelength is too short to be visible.

(c) $c = f\lambda = (9.52 \times 10^{14} \text{ Hz})(3.16 \times 10^{-7} \text{ m}) = 3.00 \times 10^8$ m/s. This is what the wave speed should be for an electromagnetic wave propagating in vacuum.

Reflect: $c = f\lambda = \left(\dfrac{\omega}{2\pi}\right)\left(\dfrac{2\pi}{k}\right) = \dfrac{\omega}{k}$ is an alternative expression for the wave speed.

*23.9. Set Up: $c = 3.00 \times 10^8$ m/s. $c = f\lambda$. $2\pi f = \omega$. $k = \dfrac{2\pi}{\lambda}$.

Solve: **(a)** $f = \dfrac{c}{\lambda}$ so f ranges from 4.29×10^{14} Hz to 7.50×10^{14} Hz.

(b) $\omega = 2\pi f$ so ω ranges from 2.70×10^{15} rad/s to 4.71×10^{15} rad/s.

(c) $k = \dfrac{2\pi}{\lambda}$ so k ranges from 8.98×10^6 rad/m to 1.57×10^7 rad/m.

*23.13. Set Up: Equations 23.3 apply for motion parallel to the $+x$-axis when the electric and magnetic field are oriented as shown in Figure 23.5 in the text. Equation 23.4 relates E_{max} to B_{max}. We know that $f = 6.10 \times 10^{15}$ Hz and $B_{max} = 5.80 \times 10^{-4}$ T.

Solve: **(a)** We are given $B_{max} = 5.80 \times 10^{-4}$ T so we can calculate

$E_{max} = cB_{max} = (3.00 \times 10^8$ m/s$)(5.80 \times 10^{-4}$ T$) = 1.74 \times 10^5$ V/m.

We are told that the magnetic field is parallel to the y-axis. By rotating the electromagnetic wave shown in Figure 23.5 of the textbook to align its magnetic field with the $+y$-axis, we see that the electric field will then be parallel to the $-z$-axis.

(b) We first calculate ω and k: $\omega = 2\pi f = 2\pi(6.10 \times 10^{14}$ Hz$) = 3.83 \times 10^{15}$ rad/s, and

$$k = \frac{2\pi}{\lambda} = \frac{2\pi f}{c} = \frac{\omega}{c} = \frac{3.83 \times 10^{15} \text{ rad/s}}{3.00 \times 10^8 \text{ m/s}} = 1.28 \times 10^7 \text{ rad/m.}$$

Note that Equations 23.3 describes motion along the $+x$-axis so that E has a positive value when B has a positive value. However, as explained in part (a), when B is aligned with the y-axis, E and B must have opposite signs. Thus,

$$B = B_{max} \sin(\omega t - kx) = (5.80 \times 10^{-4} \text{ T})\sin[(3.83 \times 10^{15} \text{ rad/s})t - (1.28 \times 10^7 \text{ rad/m})x]$$

$$E = -E_{max} \sin(\omega t - kx) = -(1.74 \times 10^5 \text{ V/m})\sin[(3.83 \times 10^{15} \text{ rad/s})t - (1.28 \times 10^7 \text{ rad/m})x].$$

Reflect: $\vec{E}$ and $\vec{B}$ are perpendicular and oscillate in synchronization.

*23.15. Set Up: Note that Equations 23.5 are the equations for an electromagnetic wave moving parallel to the $-x$-axis with a magnetic field that is parallel to the z-axis, as shown in Figure 23.6 in the textbook. The relation between E_{max} and B_{max} is $E_{max} = cB_{max}$.

Solve: **(a)** The phase of the wave is given by $kx + \omega t$, so the wave is traveling in the $-x$ direction.

(b) We know that $f = \dfrac{c}{\lambda}$, where $\lambda = \dfrac{2\pi}{k}$. Thus we have $f = \dfrac{kc}{2\pi} = \dfrac{(1.38 \times 10^4 \text{ rad/m})(3.0 \times 10^8 \text{ m/s})}{2\pi} = 6.59 \times 10^{11}$ Hz.

(c) First we find $E_{max} = cB_{max} = (3.00 \times 10^8$ m/s$)(8.25 \times 10^{-9}$ T$) = 2.48$ V/m. By rotating the electromagnetic wave shown in Figure 23.6 in the textbook so that its magnetic field is aligned with the $+y$-axis we see that the electric field will now be aligned with the $+z$-axis. Thus, E and B must have the same sign: modify Equations 23.5 to read

$$E = cB_{max} \sin(\omega t + kx) = (2.48 \text{ V/m}) \sin((4.14 \times 10^{12} \text{ rad/s})t + (1.38 \times 10^4 \text{ rad/m})x).$$

Reflect: $\vec{E}$ and $\vec{B}$ have the same phase and are in perpendicular directions.

*23.21. Set Up: The radiation pressure is $\dfrac{I}{c}$ for a totally absorbing surface and $\dfrac{2I}{c}$ for a totally reflecting surface.

Solve: (a) $\dfrac{I}{c} = \dfrac{6.00 \text{ W/m}^2}{3.00 \times 10^8 \text{ m/s}} = 2.00 \times 10^{-8} \text{ Pa}$

(b) $\dfrac{2I}{c} = 2(2.00 \times 10^{-8} \text{ Pa}) = 4.00 \times 10^{-8} \text{ Pa}$

Reflect: For the same intensity of light, the radiation pressure is twice as great for a totally reflecting surface versus a totally absorbing surface. For ordinary light intensities, the radiation pressure is very small.

*23.23. Set Up: $I = \dfrac{P}{A}$ and $I = \frac{1}{2}\epsilon_0 c E_{max}{}^2$.

Solve: $P = \frac{1}{2}A\epsilon_0 c E_{max}{}^2$. $\dfrac{E_{max,150}{}^2}{E_{max,75}{}^2} = \dfrac{150 \text{ W}}{75 \text{ W}}$ and $\dfrac{E_{max,150}}{E_{max,75}} = \sqrt{2}$.

Reflect: The power output of the source is proportional to the square of the electric field amplitude in the emitted waves.

*23.25. Set Up: We have $c = f\lambda$, $E_{max} = cB_{max}$ and $I = E_{max}B_{max}/2\mu_0$.

Solve: (a) $f = \dfrac{c}{\lambda} = \dfrac{3.00 \times 10^8 \text{ m/s}}{0.354 \text{ m}} = 8.47 \times 10^8 \text{ Hz}.$

(b) $B_{max} = \dfrac{E_{max}}{c} = \dfrac{0.0540 \text{ V/m}}{3.00 \times 10^8 \text{ m/s}} = 1.80 \times 10^{-10} \text{ T}.$

(c) $I = S_{av} = \dfrac{E_{max}B_{max}}{2\mu_0} = \dfrac{(0.0540 \text{ V/m})(1.80 \times 10^{-10} \text{ T})}{2\mu_0} = 3.87 \times 10^{-6} \text{ W/m}^2.$

Reflect: Alternatively, $I = \frac{1}{2}\epsilon_0 c E_{max}{}^2$.

*23.27. Set Up: For reflection, $\theta_r = \theta_a$. The angles of incidence and reflection at each reflection are shown in the figure below. For the rays to be perpendicular when they cross, $\alpha = 90°$.

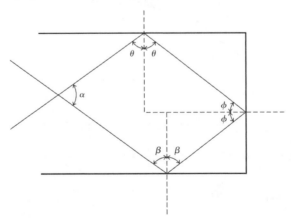

Solve: (a) $\theta + \phi = 90°$ and $\beta + \phi = 90°$, so $\beta = \theta$. $\dfrac{\alpha}{2} + \beta = 90°$ and $\alpha = 180° - 2\theta$.

(b) $\theta = \frac{1}{2}(180° - \alpha) = \frac{1}{2}(180° - 90°) = 45°$.

Reflect: As $\theta \to 0°$, $\alpha \to 180°$. This corresponds to the incident and reflected rays traveling in nearly the same direction. As $\theta \to 90°$, $\alpha \to 0°$. This corresponds to the incident and reflected rays traveling in nearly opposite directions.

*23.31. Set Up: Use the distance and time to find the speed of light in the plastic. We know that $n = \dfrac{c}{v}$, where $c = 3.00 \times 10^8$ m/s.

Solve: We have $v = \dfrac{d}{t} = \dfrac{2.50 \text{ m}}{11.5 \times 10^{-9}\text{s}} = 2.17 \times 10^8$ m/s; thus, $n = \dfrac{c}{v} = \dfrac{3.00 \times 10^8 \text{ m/s}}{2.17 \times 10^8 \text{ m/s}} = 1.38$.

Reflect: In air light travels this same distance in $\dfrac{2.50 \text{ m}}{3.00 \times 10^8 \text{ m/s}} = 8.3$ ns.

*23.35. Set Up: $\lambda = \dfrac{\lambda_0}{n}$. From Table 23.1, $n_{\text{water}} = 1.333$ and $n_{\text{benzene}} = 1.501$.

Solve: (a) $\lambda_{\text{water}} n_{\text{water}} = \lambda_{\text{benzene}} n_{\text{benzene}} = \lambda_0$. $\lambda_{\text{benzene}} = \lambda_{\text{water}} \left(\dfrac{n_{\text{water}}}{n_{\text{benzene}}} \right) = (438 \text{ nm}) \left(\dfrac{1.333}{1.501} \right) = 389$ nm.

(b) $\lambda_0 = \lambda_{\text{water}} n_{\text{water}} = (438 \text{ nm})(1.333) = 584$ nm

Reflect: λ is smallest in benzene, since n is largest for benzene.

*23.37. Set Up: The path of the light ray as it travels through the sheet of glass is sketched below. Assume that the index of refraction of air is n_1 and the index of refraction of the glass is n_2.

Solve: First note that the transmitted light ray forms a transversal between the two parallel faces so that $\theta_2 = \theta_3$. Next apply Snell's law at point A to obtain $n_1 \sin \theta_1 = n_2 \sin \theta_2$, and substitute $\theta_2 = \theta_3$ to obtain $n_1 \sin \theta_1 = n_2 \sin \theta_3$. Finally apply Snell's law at point B to obtain $n_2 \sin \theta_3 = n_1 \sin \theta_4$. Combining these last two equations we obtain $n_1 \sin \theta_1 = n_2 \sin \theta_3 = n_1 \sin \theta_4$, which gives $n_1 \sin \theta_1 = n_1 \sin \theta_4$. Thus we conclude that $\theta_1 = \theta_4$. The transmitted ray emerges parallel to the incident ray, and it is only slightly displaced if the sheet is thin.

Reflect: Since we did not need to specify values for n_1 or n_2, this result would also be true if the glass sheet were submerged in water.

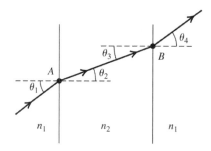

*23.41. Set Up: Snell's law is $n_a \sin \theta_a = n_b \sin \theta_b$. Light is incident in material a and refracts into material b.

Solve: (a) $n_a = 1.25$, $n_b = 1.66$, and $\theta_a = 51.0°$. $\sin \theta_b = \left(\dfrac{n_a}{n_b} \right) \sin \theta_a = \left(\dfrac{1.25}{1.66} \right) \sin 51.0° = 0.585$ and $\theta_b = 35.8°$.

(b) Now $n_a = 1.66$, $n_b = 1.25$, and $\theta_a = 35.8°$. $\sin \theta_b = \left(\dfrac{n_a}{n_b} \right) \sin \theta_a = \left(\dfrac{1.66}{1.25} \right) \sin 35.8° = 0.777$ and $\theta_b = 51.0°$.

Reflect: Reflected rays are also reversible.

*23.45. Set Up: $n_a \sin \theta_a = n_b \sin \theta_b$. The light is in diamond and encounters an interface with air, so $n_a = 2.42$ and $n_b = 1.00$. The largest θ_a is when $\theta_b = 90°$.

Solve: $(2.42) \sin \theta_a = (1.00) \sin 90°$. $\sin \theta_a = \dfrac{1}{2.42}$ and $\theta_a = 24.4°$.

*23.47. Set Up: The largest angle of incidence for which any light refracts into the air is the critical angle for water → air. The figure below shows a ray incident at the critical angle and therefore at the edge of the ring of light. The radius of this circle is r and $d = 10.0$ m is the distance from the ring to the surface of the water.

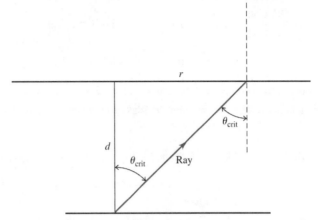

Solve: From the figure, $r = d \tan\theta_{\text{crit}}$. θ_{crit} is calculated from $n_a \sin\theta_a = n_b \sin\theta_b$ with $n_a = 1.333$, $\theta_a = \theta_{\text{crit}}$,

$n_b = 1.00$ and $\theta_b = 90°$. $\sin\theta_{\text{crit}} = \dfrac{(1.00)\sin 90°}{1.333}$ and $\theta_{\text{crit}} = 48.6°$. $r = (10.0 \text{ m}) \tan 48.6° = 11.3$ m.

$A = \pi r^2 = \pi (11.3 \text{ m})^2 = 401 \text{ m}^2$.

Reflect: When the incident angle in the water is larger than the critical angle, no light refracts into the air.

*23.49. Set Up: The ray has an angle of incidence of $0°$ at the first surface of the glass, so it enters the glass without being bent, as shown in the figure below. If no light refracts out of the glass at the glass to air interface, then the incident angle at that interface is θ_{crit}. The figure shows that $\alpha + \theta_{\text{crit}} = 90°$.

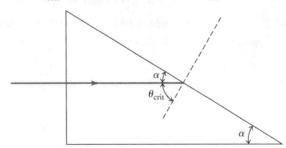

Solve: (a) For the glass-air interface $\theta_a = \theta_{\text{crit}}$, $n_a = 1.52$, $n_b = 1.00$ and $\theta_b = 1.00$. $n_a \sin\theta_a = n_b \sin\theta_b$ gives

$\sin\theta_{\text{crit}} = \dfrac{(1.00)(\sin 90°)}{1.52}$ and $\theta_{\text{crit}} = 41.1°$. $\alpha = 90° - \theta_{\text{crit}} = 48.9°$.

(b) Now the second interface is glass → water and $n_b = 1.333$. $n_a \sin\theta_a = n_b \sin\theta_b$ gives $\sin\theta_{\text{crit}} = \dfrac{(1.333)(\sin 90°)}{1.52}$

and $\theta_{\text{crit}} = 61.3°$. $\alpha = 90° - \theta_{\text{crit}} = 28.7°$.

Reflect: The critical angle increases when the air is replaced by water and rays are bent as they refract out of the glass.

*23.55. Set Up: $\lambda = \dfrac{\lambda_0}{n}$ and $v = \dfrac{c}{n}$. $n_a \sin\theta_a = n_b \sin\theta_b$.

Solve: (a) red: $n = 1.62$; violet: $n = 1.67$

(b) $\lambda_{\text{red}} = \dfrac{700 \text{ nm}}{1.62} = 432 \text{ nm}$; $\lambda_{\text{violet}} = \dfrac{400 \text{ nm}}{1.67} = 240 \text{ nm}$.

(c) $v_{red} = \dfrac{c}{n_{red}}$; $v_{violet} = \dfrac{c}{n_{violet}}$ so $\dfrac{v_{red}}{v_{violet}} = \left(\dfrac{c}{n_{red}}\right)\left(\dfrac{n_{violet}}{c}\right) = \dfrac{n_{violet}}{n_{red}} = \dfrac{1.67}{1.62} = 1.03$. The red light has a smaller n so travels faster.

(d) For red light, $\sin\theta_b = \dfrac{n_a}{n_b}\sin\theta_a = \dfrac{1.00}{1.62}\sin 65.0°$ and $\theta_b = 34.0°$. For violet light,

$$\sin\theta_b = \dfrac{n_a}{n_b}\sin\theta_\alpha = \dfrac{1.00}{1.67}\sin 65.0°$$

and $\theta_b = 32.9°$. The angle between the two rays will be $\Delta\theta = 34.0° - 32.9° = 1.1°$.

Reflect: The violet light has a larger n, slows down more in glass and is bent through a greater angle when it enters the glass.

*23.57. Set Up: For unpolarized light incident on a filter, $I = \frac{1}{2}I_0$ and the light is linearly polarized along the filter axis. For polarized light incident on a filter, $I = I_{max}(\cos\phi)^2$, where I_{max} is the intensity of the incident light, and the emerging light is linearly polarized along the filter axis.

Solve: (a) After the first filter, $I = \frac{1}{2}I_0$ and the light is polarized. After the second filter $I = \left(\frac{1}{2}I_0\right)(\cos 41.0°)^2 = 0.285 I_0$.

(b) The light is linearly polarized along the axis of the second filter.

*23.59. Set Up: When unpolarized light passes through a polarizer the intensity is reduced by a factor of $\frac{1}{2}$ and the transmitted light is polarized along the axis of the polarizer. When polarized light of intensity I_{max} is incident on a polarizer, the transmitted intensity is $I = I_{max}\cos^2\phi$, where ϕ is the angle between the polarization direction of the incident light and the axis of the filter.

Solve: (a) At point A the intensity is $I_0/2$ and the light is polarized along the vertical direction. At point B the intensity is $(I_0/2)(\cos 60°)^2 = 0.125 I_0$, and the light is polarized along the axis of the second polarizer. At point C the intenisty is $(0.125 I_0)(\cos 30°)^2 = 0.0938 I_0$.

(b) Now for the last filter $\phi = 90°$ and $I = 0$.

Reflect: Adding the middle filter increases the transmitted intensity.

*23.63. Set Up: From Malus's law, the intensity of the emerging light is proportional to the *square* of the cosine of the angle between the polarizing axes of the two filters. If the angle between the two axes is θ, the intensity of the emerging light is $I = I_{max}\cos^2\theta$.

Solve: At the angle θ we have $I = I_{max}\cos^2\theta$, and at the new angle α we have $\frac{1}{2}I = I_{max}\cos^2\alpha$. Taking the ratio of the intensities gives $\dfrac{I_{max}\cos^2\alpha}{I_{max}\cos^2\theta} = \dfrac{\frac{1}{2}I}{I}$, which gives us $\cos\alpha = \dfrac{\cos\theta}{\sqrt{2}}$. Solving for α yields $\alpha = \arccos\left(\dfrac{\cos\theta}{\sqrt{2}}\right)$.

Reflect: Careful! This result is not $\cos^2\theta$.

*23.67. Set Up: The wave speed in air is $c = 3.00 \times 10^8$ m/s. $c = f\lambda$. $E_{max} = cB_{max}$. $I = \frac{1}{2}\epsilon_0 cE_{max}^2$. For a totally absorbing surface the radiation pressure is $\dfrac{I}{c}$.

Solve: (a) $f = \dfrac{c}{\lambda} = \dfrac{3.00 \times 10^8 \text{ m/s}}{3.84 \times 10^{-2} \text{ m}} = 7.81 \times 10^9$ Hz

(b) $B_{max} = \dfrac{E_{max}}{c} = \dfrac{1.35 \text{ V/m}}{3.00 \times 10^8 \text{ m/s}} = 4.50 \times 10^{-9}$ T

(c) $I = \frac{1}{2}\epsilon_0 c E_{max}^2 = \frac{1}{2}(8.854 \times 10^{-12} \ C^2/N \cdot m^2)(3.00 \times 10^8 \ m/s)(1.35 \ V/m)^2 = 2.42 \times 10^{-3} \ W/m^2$

(d) $F = (\text{pressure})A = \frac{IA}{c} = \frac{(2.42 \times 10^{-3} \ W/m^2)(0.240 \ m^2)}{3.00 \times 10^8 \ m/s} = 1.94 \times 10^{-12} \ N$

Reflect: The intensity depends only on the amplitudes of the electric and magnetic fields and is independent of the wavelength of the light.

*23.69. Set Up: Energy $= Pt$. For absorption the radiation pressure is $\frac{I}{c}$, where $I = \frac{P}{A}$. $\lambda = \frac{\lambda_0}{n}$. $I = \frac{1}{2}\epsilon_0 c E_{max}^2$

and $E_{max} = c B_{max}$.

Solve: (a) Energy $= Pt = (250 \times 10^{-3} \ W)(1.50 \times 10^{-3} \ s) = 3.75 \times 10^{-4} \ J = 0.375 \ mJ$.

(b) $I = \frac{P}{A} = \frac{250 \times 10^{-3} \ W}{\pi(255 \times 10^{-6} \ m)^2} = 1.22 \times 10^6 \ W/m^2$. The average pressure is

$$\frac{I}{c} = \frac{1.22 \times 10^6 \ W/m^2}{3.00 \times 10^8 \ m/s} = 4.08 \times 10^{-3} \ Pa.$$

(c) $\lambda = \frac{\lambda_0}{n} = \frac{810 \ nm}{1.34} = 604 \ nm$. $f = \frac{v}{\lambda} = \frac{c}{\lambda_0} = \frac{3.00 \times 10^8 \ m/s}{810 \times 10^{-9} \ m} = 3.70 \times 10^{14} \ Hz$; f is the same in the air and in the

vitreus humor.

(d) $E = \sqrt{\frac{2I}{\epsilon_0 c}} = \sqrt{\frac{2(1.22 \times 10^6 \ W/m^2)}{(8.85 \times 10^{-12} \ C^2/N \cdot m^2)(3.00 \times 10^8 \ m/s)}} = 3.03 \times 10^4 \ V/m.$

$$B_{max} = \frac{E_{max}}{c} = 1.01 \times 10^{-4} \ T.$$

*23.71. Set Up: The minimum intensity required for detection can be determined from $I = \frac{1}{2}\epsilon_0 c E_{max}^2$, where

$E_{max} = 10.0 \ mV/m = 10^{-2} \ V/m$.

Solve: Assuming that the intensity of the transmission is distributed uniformly over a sphere of radius r centered on the station, the total power output of the station must be $P = 4\pi r^2 I$. Solving for r we obtain

$$r = \sqrt{\frac{P}{4\pi I}} = \sqrt{\frac{P}{4\pi\left(\frac{1}{2}\epsilon_0 c E_{max}^2\right)}} = \sqrt{\frac{P}{2\pi\epsilon_0 c E_{max}^2}} = \sqrt{\frac{50.0 \times 10^3 \ W}{2\pi\epsilon_0 c(10^{-2} \ V/m)^2}} = 1.73 \times 10^5 \ m, \text{ which is 173 km.}$$

Reflect: For a given power output, the intensity varies according to $I = \frac{P}{4\pi r^2}$, which is the inverse-square law for electromagnetic radiation.

*23.75. Set Up: $n_a \sin\theta_a = n_b \sin\theta_b$. $n_a = 1.00$. $n_b = 1.80$. $\theta_b = \theta_a/2$. $\sin a = 2\sin(a/2)\cos(a/2)$.

Solve: Snell's law gives $(1.00)\sin\theta_a = (1.80)\sin(\theta_a/2)$. $\sin\theta_a = 2\sin(\theta_a/2)\cos(\theta_a/2)$, so

$$2\sin(\theta_a/2)\cos(\theta_a/2) = (1.80)\sin(\theta_a/2).$$

$\cos(\theta_a/2) = 0.900$. $\theta_a/2 = 25.84°$ and $\theta_a = 51.7°$.

Reflect: When the angle of incidence increases the angle of refraction increases. The angle of refraction is smaller than the angle of incidence when $n_b > n_a$, which is the case here.

*23.77. Set Up: As the light crosses the glass-air interface along AB, it is refracted and obeys Snell's law. Snell's law is $n_a \sin\theta_a = n_b \sin\theta_b$, and we know that $n = 1.000$ for air. At point B the angle of the prism is $30.0°$.

Solve: Apply Snell's law at AB. The prism angle at A is $60.0°$, so for the upper ray, the angle of incidence at AB is $60.0° + 12.0° = 72.0°$. Using this value gives $n_1 \sin 60.0° = \sin 72.0°$ and $n_1 = 1.10$. For the lower ray, the angle of incidence at AB is $60.0° + 12.0° + 8.50° = 80.5°$, giving $n_2 \sin 60.0° = \sin 80.5°$ and $n_2 = 1.14$.

Reflect: The lower ray is deflected more than the upper ray because that wavelength has a slightly greater index of refraction than the upper ray.

*23.81. Set Up: The rays are incident on the prism in the normal direction so they do not change direction as they enter the prism. Apply Snell's law to the refraction of the rays as they exit the prism. The rays are sketched in the figure below. The angle between the two rays after they emerge from the prism is 2γ.

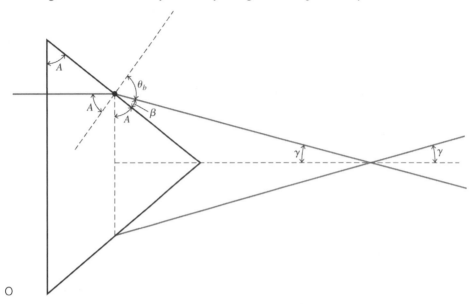

Solve: The angle of incidence is $A = 25.0°$. $n_a \sin\theta_a = n_b \sin\theta_b$ gives $(1.66)\sin(25.0°) = (1.00)\sin\theta_b$. $\theta_b = 44.55°$. $\beta = 90° - \theta_b = 45.45°$. $\gamma + \beta + A = 90°$, so $\gamma = 90° - A - \beta = 90° - 25.0° - 45.45° = 19.55°$. The angle between the emerging rays is $2\gamma = 39.1°$.

Reflect: If $A \to 0°$, $\beta \to 90°$ and $\gamma \to 0°$. The rays remain parallel if the prism is replaced by a rectangular slab of glass.

*23.83. Set Up: The path of the ray is sketched in the figure below. The problem asks us to calculate θ'_b.

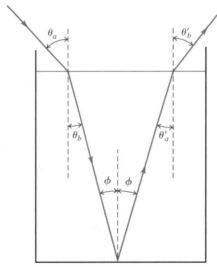

Solve: Apply Snell's law to the air $\to$ liquid refraction. $(1.00)\sin(42.5°) = (1.63)\sin\theta_b$ and $\theta_b = 24.5°$. $\theta_b = \phi$ and $\phi = \theta'_a$, so $\theta'_a = \theta_b = 24.5°$. Snell's law applied to the liquid $\to$ air refraction gives $(1.63)\sin(24.5°) = (1.00)\sin\theta'_b$ and $\theta'_b = 42.5°$.

Reflect: The light emerges from the liquid at the same angle from the normal as it entered the liquid.

*23.87. Set Up: Let the light initially be in the material with refractive index n_a and let the final slab have refractive index n_b. In part (a) let the middle slab have refractive index n_1. Apply Snell's law to each refraction.

Solve: (a) 1st interface: $n_a \sin\theta_a = n_1 \sin\theta_1$. 2nd interface: $n_1 \sin\theta_1 = n_b \sin\theta_b$. Combining the two equations gives $n_a \sin\theta_a = n_b \sin\theta_b$. This is the equation that would apply if the middle slab were absent.

(b) For N slabs, $n_a \sin\theta_a = n_1 \sin\theta_1$, $n_1 \sin\theta_1 = n_2 \sin\theta_2, \ldots, n_{N-2} \sin\theta_{N-2} = n_b \sin\theta_b$. Combining all these equations gives $n_a \sin\theta_a = n_b \sin\theta_b$. The final direction of travel depends on the angle of incidence in the first slab and the refractive indices of the first and last slabs.

*23.91. Set Up: No light enters the gas because total internal reflection must have occurred at the water-gas interface. At the minimum value of S, the light strikes the water-gas interface at the critical angle. We apply Snell's law, $n_a \sin\theta_a = n_b \sin\theta_b$, at that surface.

Solve: (a) In the water, $\theta = \dfrac{S}{R} = (1.09 \text{ m})/(1.10 \text{ m}) = 0.991 \text{ rad} = 56.77°$. This is the critical angle. So, using the refractive index for water from Table 33.1, we get $n = (1.333) \sin 56.77° = 1.12$

(b) (i) The laser beam stays in the water all the time, so

$$t = 2R/v = 2R/\left(\dfrac{c}{n_{\text{water}}}\right) = \dfrac{Dn_{\text{water}}}{c} = (2.20 \text{ m})(1.333)/(3.00 \times 10^8 \text{ m/s}) = 9.78 \text{ ns}$$

(ii) The beam is in the water half the time and in the gas the other half of the time.

$$t_{\text{gas}} = \dfrac{Rn_{\text{gas}}}{c} = (1.10 \text{ m})(1.12)/(3.00 \times 10^8 \text{ m/s}) = 4.09 \text{ ns}$$

The total time is $4.09 \text{ ns} + (9.78 \text{ ns})/2 = 8.98 \text{ ns}$

Reflect: The gas must be under considerable pressure to have a refractive index as high as 1.12.

Solutions to Passage Problems

*23.93. Set Up: The figure below shows the incident light ray as it passes through the oil and then through the glass slab. Note that the light ray forms a transversal through the parallel faces of the oil layer and then through the parallel faces of the glass slab. Let α, β, and γ be the exit angles into the oil, glass, and air as shown below.

Solve: At the air-oil interface we apply Snell's law to obtain $1.00\sin30° = 1.8\sin\alpha$. Next, at the oil-glass interface we apply Snell's law to obtain $1.8\sin\alpha = 1.5\sin\beta$. Finally, at the glass-air interface we apply Snell's law to obtain $1.5\sin\beta = 1.00\sin\gamma$. Equating these three expressions gives $1.00\sin30° = 1.8\sin\alpha = 1.5\sin\beta = 1.00\sin\gamma$. This implies that $1.00\sin30° = 1.00\sin\gamma$ so we now know that $\gamma = 30°$. Note that this result does not depend on the index of refraction of the air, the oil or the glass provided that the ray does not experience total internal reflection at the oil-glass interface. The correct answer is C.

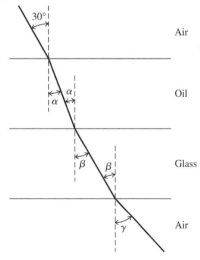

GEOMETRIC OPTICS

Problems 3, 5, 9, 11, 13, 15, 19, 23, 27, 29, 33, 37, 39, 43, 45, 49, 53, 55, 57, 61, 63, 65, 69

Solutions to Problems

*24.3. Set Up: The two mirrors 1 and 2 and the object are shown in Figure (a) below. For each mirror the image is the same distance behind the mirror as the object is in front of the mirror.

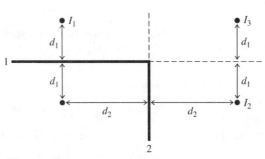

(a)

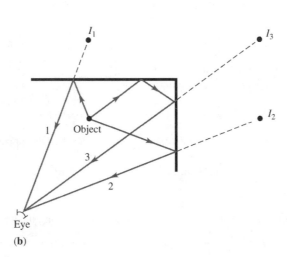

(b)

Solve: (a) Mirror 1 forms image I_1 and mirror 2 forms image I_2. The image I_1 also serves as an object for mirror 2 and image I_3 is formed and is located as shown. Image I_2 serves as an object for mirror 1 and the image is also at the location of I_3. Three images are formed: I_1, I_2, and I_3.

(b) Figure (b) above shows a ray from the object to the eye for each image. For images I_1 and I_2 there is only one reflection, from one mirror or the other. For I_3 the ray has two reflections, off one mirror and then off the other.

Reflect: All three images are virtual, erect, and the same height as the object.

*24.5. Set Up: To find the location of the image use $\dfrac{1}{s}+\dfrac{1}{s'}=\dfrac{2}{R}$. For a concave mirror $R>0$. To find the height of the image use $m=\dfrac{y'}{y}=-\dfrac{s'}{s}$.

Solve: (a) $s=15.0$ cm. $\dfrac{1}{s'}=\dfrac{2}{R}-\dfrac{1}{s}=\dfrac{2s-R}{Rs}$, so

$$s'=\frac{Rs}{2s-R}=\frac{(10.0\text{ cm})(15.0\text{ cm})}{30.0\text{ cm}-10.0\text{ cm}}=7.5\text{ cm}.$$
$$m=-\frac{s'}{s}=-\frac{7.5\text{ cm}}{15.0\text{ cm}}=-\frac{1}{2}$$

and $y'=my=-\frac{1}{2}(8.00\text{ mm})=-4.00$ mm. The image is 7.5 cm in front of the mirror. The image is 4.0 mm tall and is inverted.

(b) $s=10.0$ cm. $s'=\dfrac{(10.0\text{ cm})(10.0\text{ cm})}{20.0\text{ cm}-10.0\text{ cm}}=10.0$ cm. $m=-\dfrac{10.0\text{ cm}}{10.0\text{ cm}}=-1.00$. $y'=my=-8.00$ mm. The image is 10.0 cm in front of the mirror, at the location of the object. The image is 8.00 mm tall and is inverted.

(c) $s=2.50$ cm. $s'=\dfrac{(10.0\text{ cm})(2.50\text{ cm})}{5.00\text{ cm}-10.0\text{ cm}}=-5.00$ cm. $m=-\dfrac{-5.00\text{ cm}}{2.50\text{ cm}}=+2.00$. $y'=my=16.0$ mm. The image is 5.00 cm behind the mirror. The image is 16.0 mm tall and is erect.

(d) $s=1000.0$ cm. $s'=\dfrac{(10.0\text{ cm})(1000.0\text{ cm})}{2000.0\text{ cm}-10.0\text{ cm}}=+5.00$ cm. $m=-\dfrac{5.00\text{ cm}}{1000.0\text{ cm}}=-5.00\times10^{-3}$. $y'=my=-0.040$ mm. The image is 5.00 cm in front of the mirror. The image is 0.040 mm tall and is inverted.

Reflect: From $s'=\dfrac{Rs}{2s-R}$ we see that the image is real if $s>R/2$ and virtual if $s<R/2$. Real images are in front of the mirror and are inverted. Virtual images are behind the mirror and are erect.

*24.9. Set Up: For a convex mirror, $R<0$, so $R=-18.0$ cm and $f=\dfrac{R}{2}=-9.00$ cm.

Solve: (a) $\dfrac{1}{s}+\dfrac{1}{s'}=\dfrac{1}{f}$. $s'=\dfrac{sf}{s-f}=\dfrac{(1300\text{ cm})(-9.00\text{ cm})}{1300\text{ cm}-(-9.00\text{ cm})}=-8.94$ cm.

$m=-\dfrac{s'}{s}=-\dfrac{-8.94\text{ cm}}{1300\text{ cm}}=6.88\times10^{-3}$. $|y'|=|m|y=(6.88\times10^{-3})(1.5\text{ m})=0.0103\text{ m}=1.03$ cm.

(b) The height of the image is much less than the height of the car, so the car appears to be farther away than its actual distance.

Reflect: Problem 24.11 shows that the image formed by a convex mirror is always virtual and smaller than the object.

*24.11. Set Up: The shell behaves as a spherical mirror. The equation relating the object and image distances to the focal length of a spherical mirror is $\dfrac{1}{s}+\dfrac{1}{s'}=\dfrac{1}{f}$, and its magnification is given by $m=-\dfrac{s'}{s}$.

Solve: $\dfrac{1}{s} + \dfrac{1}{s'} = \dfrac{1}{f} \Rightarrow \dfrac{1}{s} = \dfrac{2}{-18.0 \text{ cm}} - \dfrac{1}{-6.00 \text{ cm}} \Rightarrow s = 18.0 \text{ cm}$ from the vertex. $m = -\dfrac{s'}{s} = -\dfrac{-6.00 \text{ cm}}{18.0 \text{ cm}} = \dfrac{1}{3} \Rightarrow$

$y' = \dfrac{1}{3}(1.5 \text{ cm}) = 0.50 \text{ cm}.$ The image is 0.50 cm tall, erect, and virtual.

Reflect: Since the magnification is less than one, the image is smaller than the object.

*24.13. Set Up: For a concave mirror, $R > 0$. $R = 32.0 \text{ cm}$ and $f = \dfrac{R}{2} = 16.0 \text{ cm}.$

Solve: (a) $\dfrac{1}{s} + \dfrac{1}{s'} = \dfrac{1}{f}.$ $s' = \dfrac{sf}{s-f} = \dfrac{(12.0 \text{ cm})(16.0 \text{ cm})}{12.0 \text{ cm} - 16.0 \text{ cm}} = -48.0 \text{ cm}.$ $m = -\dfrac{s'}{s} = -\dfrac{-48.0 \text{ cm}}{12.0 \text{ cm}} = +4.00.$

(b) $s' = -48.0 \text{ cm},$ so the image is 48.0 cm to the right of the mirror. $s' < 0$ so the image is virtual.

(c) The principal-ray diagram is sketched in the figure below. The rules for principal rays apply only to paraxial rays. Principal ray 2, that travels to the mirror along a line that passes through the focus, makes a large angle with the optic axis and is not described well by the paraxial approximation. Therefore, principal ray 2 is not included in the sketch.

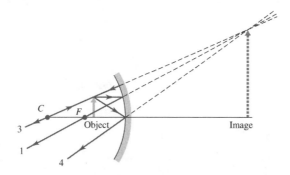

Reflect: A concave mirror forms a virtual image whenever $s < f$.

*24.15. Set Up: For a convex mirror, $R < 0$. $R = -22.0 \text{ cm}$ and $f = \dfrac{R}{2} = -11.0 \text{ cm}.$

Solve: (a) The principal-ray diagram is sketched in the figure below.

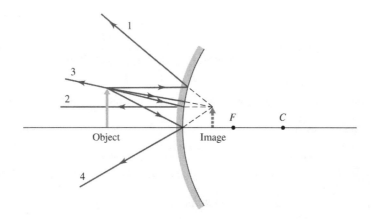

(b) $\dfrac{1}{s} + \dfrac{1}{s'} = \dfrac{1}{f}.$ $s' = \dfrac{sf}{s-f} = \dfrac{(16.5 \text{ cm})(-11.0 \text{ cm})}{16.5 \text{ cm} - (-11.0 \text{ cm})} = -6.6 \text{ cm}.$ $m = -\dfrac{s'}{s} = -\dfrac{-6.6 \text{ cm}}{16.5 \text{ cm}} = +0.400.$

$|y'| = |m|y = (0.400)(0.600 \text{ cm}) = 0.240 \text{ cm}.$ The image is 6.6 cm to the right of the mirror. It is 0.240 cm tall. $s' < 0,$ so the image is virtual. $m > 0,$ so the image is erect.

*24.19. Set Up: For a convex surface, $R > 0$. $R = +3.00$ cm. $n_a = 1.00$, $n_b = 1.60$. $\dfrac{n_a}{s} + \dfrac{n_b}{s'} = \dfrac{n_b - n_a}{R}$.

Solve: (a) $s \to \infty$. $\dfrac{n_b}{s'} = \dfrac{n_b - n_a}{R}$. $s' = \left(\dfrac{n_b}{n_b - n_a}\right) R = \left(\dfrac{1.60}{1.60 - 1.00}\right)(+3.00 \text{ cm}) = +8.00$ cm. The image is 8.00 cm to

the right of the vertex.

(b) $s = 12.0$ cm. $\dfrac{1.00}{12.0 \text{ cm}} + \dfrac{1.60}{s'} = \dfrac{1.60 - 1.00}{3.00 \text{ cm}}$. $s' = +13.7$ cm. The image is 13.7 cm to the right of the vertex.

(c) $s = 2.00$ cm. $\dfrac{1.00}{2.00 \text{ cm}} + \dfrac{1.60}{s'} = \dfrac{1.60 - 1.00}{3.00 \text{ cm}}$. $s' = -5.33$ cm. The image is 5.33 cm to the left of the vertex.

*24.23. Set Up: Light comes from the fish to the person's eye. $R = -14.0$ cm. $s = +14.0$ cm. $n_a = 1.333$ (water).

$n_b = 1.00$ (air). $\dfrac{n_a}{s} + \dfrac{n_b}{s'} = \dfrac{n_b - n_a}{R}$. $m = -\dfrac{n_a s'}{n_b s}$. The figure below shows the object and the refracting surface.

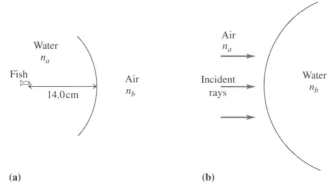

(a)   (b)

Solve: (a) $\dfrac{1.333}{14.0 \text{ cm}} + \dfrac{1.00}{s'} = \dfrac{1.00 - 1.333}{-14.0 \text{ cm}}$. $s' = -14.0$ cm. $m = -\dfrac{(1.333)(-14.0 \text{ cm})}{(1.00)(14.0 \text{ cm})} = +1.33$.

The fish's image is 14.0 cm to the left of the bowl surface so it is at the center of the bowl and the magnification is 1.33.

(b) The focal point is at the image location when $s \to \infty$. $\dfrac{n_b}{s'} = \dfrac{n_b - n_a}{R}$. $n_a = 1.00$. $n_b = 1.333$. $R = +14.0$ cm.

$\dfrac{1.333}{s'} = \dfrac{1.333 - 1.00}{14.0 \text{ cm}}$. $s' = +56.0$ cm. s' is greater than the diameter of the bowl, so the surface facing the sunlight does

not focus the sunlight to a point inside the bowl. The focal point is outside the bowl and there is no danger to the fish.
Reflect: In part (b) the rays refract when they exit the bowl back into the air so the image we calculated is not the final image.

*24.27. Set Up: Assume that the walls of the aquarium are flat and parallel so that we may use Eq. (24.13) at each

surface: $\dfrac{n_a}{s} + \dfrac{n_b}{s'} = 0$, which gives $s' = -\left(\dfrac{n_b}{n_a}\right)s$. Imagine that we are facing the tank so that we see one group of

spectators on our left and the other group on our right (with the tank between them).

Solve: First let us calculate the location of the image of the left-hand spectators that is formed by the left face of the

tank. Here we have $n_a = 1.00$, $n_b = 1.333$, and $s = 1.20$ m. We have $s' = -\left(\dfrac{n_b}{n_a}\right)s = -\left(\dfrac{1.333}{1.000}\right)(1.20 \text{ m}) = -1.60$ m,

so that this image appears 1.60 meters to the left of the left face of the tank. Next we treat this image as an object for the right face of the tank. In this case we use $s = 5.50 \text{ m} + 1.60 \text{ m} = 7.10$ m, $n_a = 1.333$, and $n_b = 1.00$. We have

$s' = -\left(\dfrac{n_b}{n_a}\right)s = -\left(\dfrac{1.00}{1.333}\right)(7.10 \text{ m}) = -5.33$ m. Thus, the second image is 5.33 meters to the left of the right face of the

tank and this is located $1.20 \text{ m} + 5.33 \text{ m} = 6.53$ meters from the spectators on the right side.

Reflect: The apparent distance between the spectators has been reduced from $5.00 \text{ m} + 2(1.20 \text{ m}) = 7.40$ meters to 6.53 meters due to the presence of the tank.

*24.29. Set Up: $m = \dfrac{y'}{y} = -\dfrac{s'}{s}$. Since the image is erect, $y' > 0$ and $m > 0$. $\dfrac{1}{s} + \dfrac{1}{s'} = \dfrac{1}{f}$.

Solve: $m = \dfrac{y'}{y} = \dfrac{1.30 \text{ cm}}{0.400 \text{ cm}} = +3.25$. $m = -\dfrac{s'}{s} = +3.25$ gives $s' = -3.25s$. $\dfrac{1}{s} + \dfrac{1}{s'} = \dfrac{1}{f}$ gives

$$\frac{1}{s} + \frac{1}{-3.25s} = \frac{1}{7.00 \text{ cm}} \quad \text{and } s = 4.85 \text{ cm.}$$

$s' = -(3.25)(4.85 \text{ cm}) = -15.8 \text{ cm}$. The object is 4.85 cm to the left of the lens. The image is 15.8 cm to the left of the lens and is virtual.

Reflect: The image is virtual because the object distance is less than the focal length.

*24.33. Set Up: $\dfrac{1}{s} + \dfrac{1}{s'} = \dfrac{1}{f}$. The type of lens determines the sign of f. $m = \dfrac{y'}{y} = -\dfrac{s'}{s}$. The sign of s' determines whether the image is real or virtual. $s = +8.00 \text{ cm}$. $s' = -3.00 \text{ cm}$. s' is negative because the image is on the same side of the lens as the object.

Solve: (a) $\dfrac{1}{f} = \dfrac{s + s'}{ss'}$ and $f = \dfrac{ss'}{s + s'} = \dfrac{(8.00 \text{ cm})(-3.00 \text{ cm})}{8.00 \text{ cm} - 3.00 \text{ cm}} = -4.80 \text{ cm}$. f is negative so the lens is diverging.

(b) $m = -\dfrac{s'}{s} = -\dfrac{-3.00 \text{ cm}}{8.00 \text{ cm}} = +0.375$. $y' = my = (0.375)(6.50 \text{ mm}) = 2.44 \text{ mm}$. $s' < 0$ and the image is virtual.

Reflect: A converging lens can also form a virtual image, if the object distance is less than the focal length. But in that case $|s'| > s$ and the image would be farther from the lens than the object is.

*24.37. Set Up: $\dfrac{1}{f} = (n-1)\left(\dfrac{1}{R_1} - \dfrac{1}{R_2}\right)$. 1 is the surface closest to the object. R is positive if the center of curvature is on the side of the lens opposite the object. For a flat surface, $R \rightarrow \infty$.

Solve: (a) L_1: $\dfrac{1}{f} = (0.5)\left(\dfrac{1}{11.5 \text{ cm}} - \dfrac{1}{-10.5 \text{ cm}}\right)$ and $f = +11.0 \text{ cm}$.

$$L_2: \frac{1}{f} = (0.5)\left(\frac{1}{10.5 \text{ cm}} - \frac{1}{-11.5 \text{ cm}}\right) \text{ and}$$
$$f = +11.0 \text{ cm.}$$

(b) L_1: $\dfrac{1}{f} = (0.5)\left(\dfrac{1}{\infty} - \dfrac{1}{-8.50 \text{ cm}}\right)$ and $f = +17.0 \text{ cm}$.

$$L_2: \frac{1}{f} = (0.5)\left(\frac{1}{8.50 \text{ cm}} - \frac{1}{\infty}\right) \text{ and}$$
$$f = +17.0 \text{ cm.}$$

(c) L_1: $\dfrac{1}{f} = (0.5)\left(\dfrac{1}{9.80 \text{ cm}} - \dfrac{1}{11.5 \text{ cm}}\right)$ and $f = +133 \text{ cm}$.

$$L_2: \frac{1}{f} = (0.5)\left(\frac{1}{-9.80 \text{ cm}} - \frac{1}{-11.5 \text{ cm}}\right) \text{ and}$$
$$f = -133 \text{ cm.}$$

(d) L_1: $\dfrac{1}{f} = (0.5)\left(\dfrac{1}{-9.20 \text{ cm}} - \dfrac{1}{\infty}\right)$ and $f = -18.4 \text{ cm}$.

$$L_2: \frac{1}{f} = (0.5)\left(\frac{1}{\infty} - \frac{1}{9.20 \text{ cm}}\right) \text{ and}$$
$$f = -18.4 \text{ cm.}$$

(e) L_1: $\dfrac{1}{f} = (0.5)\left(\dfrac{1}{-10.4 \text{ cm}} - \dfrac{1}{11.6 \text{ cm}}\right)$ and $f = -11.0$ cm.

$$L_2: \dfrac{1}{f} = (0.5)\left(\dfrac{1}{-11.6 \text{ cm}} - \dfrac{1}{10.4 \text{ cm}}\right) \text{ and}$$
$$f = -11.0 \text{ cm.}$$

*24.39. Set Up: $\dfrac{1}{f} = (n-1)\left(\dfrac{1}{R_1} - \dfrac{1}{R_2}\right)$. If R is the radius of the lens, then $R_1 = R$ and $R_2 = -R$. $\dfrac{1}{s} + \dfrac{1}{s'} = \dfrac{1}{f}$.

$m = \dfrac{y'}{y} = -\dfrac{s'}{s}$.

Solve: (a) $\dfrac{1}{f} = (n-1)\left(\dfrac{1}{R_1} - \dfrac{1}{R_2}\right) = (n-1)\left(\dfrac{1}{R} - \dfrac{1}{-R}\right) = \dfrac{2(n-1)}{R}$.

$$R = 2(n-1)f = 2(0.44)(8.0 \text{ mm}) = 7.0 \text{ mm.}$$

(b) $\dfrac{1}{s'} = \dfrac{1}{f} - \dfrac{1}{s} = \dfrac{s-f}{sf}$. $s' = \dfrac{sf}{s-f} = \dfrac{(30.0 \text{ cm})(0.80 \text{ cm})}{30.0 \text{ cm} - 0.80 \text{ cm}} = 0.82 \text{ cm} = 8.2 \text{ mm}$. The image is 8.2 mm from the lens,

on the side opposite the object. $m = -\dfrac{s'}{s} = -\dfrac{0.82 \text{ cm}}{30.0 \text{ cm}} = -0.0273$. $y' = my = (-0.0273)(16 \text{ cm}) = 0.44 \text{ cm} = 4.4 \text{ mm}$.

$s' > 0$ so the image is real. $m < 0$ so the image is inverted.

Reflect: The lens is converging and has a very short focal length. As long as the object is farther than 7.0 mm from the eye, the lens forms a real image.

*24.43. Set Up: Use the thin-lens equation to calculate the focal length of the lens: $\dfrac{1}{s} + \dfrac{1}{s'} = (n-1)\left(\dfrac{1}{R_1} - \dfrac{1}{R_2}\right)$. The

magnification of the lens is $m = -\dfrac{s'}{s}$.

Solve: (a) $\dfrac{1}{s} + \dfrac{1}{s'} = (n-1)\left(\dfrac{1}{R_1} - \dfrac{1}{R_2}\right) \Rightarrow \dfrac{1}{24.0 \text{ cm}} + \dfrac{1}{s'} = (1.52-1)\left(\dfrac{1}{-7.00 \text{ cm}} - \dfrac{1}{-4.00 \text{ cm}}\right)$

$$\Rightarrow s' = 71.2 \text{ cm, to the right of the lens.}$$

(b) $m = -\dfrac{s'}{s} = -\dfrac{71.2 \text{ cm}}{24.0 \text{ cm}} = -2.97$

Reflect: Since the magnification is negative, the image is inverted.

*24.45. Set Up: Apply $\dfrac{1}{s} + \dfrac{1}{s'} = \dfrac{1}{f}$ to each lens. $m_1 = \dfrac{y_1'}{y_1}$ and $m_2 = \dfrac{y_2'}{y_2}$.

Solve: (a) *Lens 1:* $\dfrac{1}{s} + \dfrac{1}{s'} = \dfrac{1}{f}$ gives

$$s_1' = \dfrac{s_1 f_1}{s_1 - f_1} = \dfrac{(50.0 \text{ cm})(40.0 \text{ cm})}{50.0 \text{ cm} - 40.0 \text{ cm}} = +200 \text{ cm.}$$

$$m_1 = -\dfrac{s_1'}{s_1} = -\dfrac{200 \text{ cm}}{50 \text{ cm}} = -4.00.$$

$y_1' = m_1 y_1 = (-4.00)(1.20 \text{ cm}) = -4.80 \text{ cm}$. The image I_1 is 200 cm to the right of lens 1, is 4.80 cm tall and is inverted.

(b) *Lens 2:* $y_2 = -4.80$ cm. The image I_1 is $300 \text{ cm} - 200 \text{ cm} = 100 \text{ cm}$ to the left of lens 2, so $s_2 = +100$ cm.

$$s_2' = \frac{s_2 f_2}{s_2 - f_2} = \frac{(100 \text{ cm})(60.0 \text{ cm})}{100 \text{ cm} - 60.0 \text{ cm}} = +150 \text{ cm}.$$

$$m_2 = -\frac{s_2'}{s_2} = -\frac{150 \text{ cm}}{100 \text{ cm}} = -1.50.$$

$y_2' = m_2 y_2 = (-1.50)(-4.80 \text{ cm}) = +7.20 \text{ cm}.$ The image is 150 cm to the right of the second lens, is 7.20 cm tall, and is inverted with respect to the original object.

Reflect: The overall magnification of the lens combination is $m_{\text{tot}} = m_1 m_2$.

*24.49. Set Up: $f = +14.0$ cm. $\dfrac{1}{s} + \dfrac{1}{s'} = \dfrac{1}{f}$ gives $s' = \dfrac{sf}{s - f}$. $m = \dfrac{y'}{y} = -\dfrac{s'}{s}$.

Solve: $s = 18.0$ cm

(a) $s' = \dfrac{sf}{s - f} = \dfrac{(18.0 \text{ cm})(14.0 \text{ cm})}{18.0 \text{ cm} - 14.0 \text{ cm}} = 63.0$ cm. The image is 63.0 cm to the right of the lens.

(b) $m = -\dfrac{s'}{s} = -\dfrac{63.0 \text{ cm}}{18.0 \text{ cm}} = -3.50$

(c) $s' > 0$ so the image is real.

(d) $m < 0$ so the image is inverted. The principal-ray diagram is sketched in Figure (a) below.

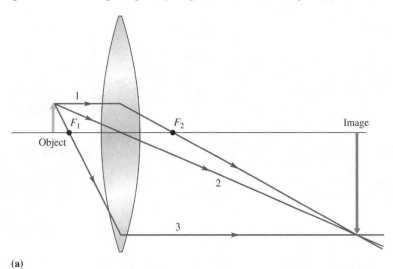

(a)

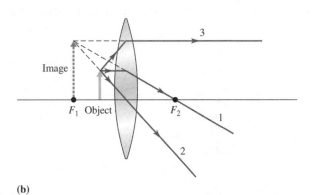

(b)

$s = 7.00$ cm

(e) $s' = \dfrac{sf}{s-f} = \dfrac{(7.00 \text{ cm})(14.0 \text{ cm})}{7.00 \text{ cm} - 14.0 \text{ cm}} = -14.0$ cm. The image is 14.0 cm to the left of the lens.

(f) $m = -\dfrac{s'}{s} = -\dfrac{-14.0 \text{ cm}}{7.00 \text{ cm}} = +2.00$

(g) $s' < 0$ so the image is virtual.

(h) $m > 0$ so the image is erect. The principal-ray diagram is sketched in Figure (b) above.

Reflect: For a converging lens, when $s > f$ the image is real and when $s < f$ the image is virtual.

*24.53. Set Up: $s = 12.0$ cm. The ray intersects the optic axis 8.0 cm to the right of the lens.

Solve: **(a)** The ray is bent toward the optic axis by the lens so the lens is converging.

(b) The ray is parallel to the optic axis and is bent so that it passes through the focal point to the right of the lens; $f = 8.0$ cm.

(c) The principal ray diagram is drawn in the figure below. The diagram shows that the image is 24.0 cm to the right of the lens.

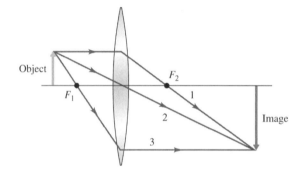

(d) $\dfrac{1}{s} + \dfrac{1}{s'} = \dfrac{1}{f}$ gives $s' = \dfrac{sf}{s-f} = \dfrac{(12.0 \text{ cm})(8.0 \text{ cm})}{12.0 \text{ cm} - 8.0 \text{ cm}} = +24.0$ cm. The calculated image position agrees with the principal ray diagram.

Reflect: The lens is converging and $s > f$, so the image is real.

*24.55. Set Up: $s = 10.0$ cm. If extended backwards the ray comes from a point on the optic axis 18.0 cm from the lens and is parallel to the optic axis after it passes through the lens.

Solve: **(a)** The ray is bent toward the optic axis by the lens so the lens is converging.

(b) The ray is parallel to the optic axis after it passes through the lens so it comes from the focal point; $f = 18.0$ cm.

(c) The principal ray diagram is drawn in the figure below. The diagram shows that the image is 22.5 cm to the left of the lens.

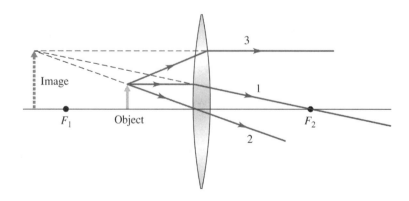

(d) $\dfrac{1}{s}+\dfrac{1}{s'}=\dfrac{1}{f}$ gives $s'=\dfrac{sf}{s-f}=\dfrac{(10.0\text{ cm})(18.0\text{ cm})}{10.0\text{ cm}-18.0\text{ cm}}=-22.5\text{ cm}$. The calculated image position agrees with the

principal ray diagram.

*24.57. Set Up: Apply $\dfrac{n_a}{s}+\dfrac{n_b}{s'}=\dfrac{n_b-n_a}{R}$ with $R\to\infty$, since we assume that the surfaces are flat. The image

formed by the first interface serves as the object for the second interface.
Solve: For the water-benzene interface to get the apparent water depth:

$$\frac{n_a}{s}+\frac{n_b}{s'}=0\Rightarrow\frac{1.33}{6.50\text{ cm}}+\frac{1.50}{s'}=0\Rightarrow s'=-7.33\text{ cm}.$$

For the benzene-air interface, to get the total apparent distance to the bottom:

$$\frac{n_a}{s}+\frac{n_b}{s'}=0\Rightarrow\frac{1.50}{(7.33\text{ cm}+2.60\text{ cm})}+\frac{1}{s'}=0\Rightarrow s'=-6.62\text{ cm}.$$

Reflect: At the water-benzene interface the light refracts into material of greater refractive index and the overall effect is that the apparent depth is greater than the actual depth.

*24.61. Set Up: We will apply the thin-lens equation Eq. (24.21): $\dfrac{1}{s}+\dfrac{1}{s'}=\dfrac{1}{f}=(n-1)\left(\dfrac{1}{R_1}-\dfrac{1}{R_2}\right)$. If the object is

on the convex side of the lens the center of curvature of each lens is on the outgoing side so the radius of curvatures are both positive: $R_1=+12.2\text{ cm}$ and $R_2=+15.4\text{ cm}$.

Solve: (a) Using the given value we have $\dfrac{1}{35.0\text{ cm}}+\dfrac{1}{s'}=(1.67-1)\left(\dfrac{1}{12.2\text{ cm}}-\dfrac{1}{15.4\text{ cm}}\right)$. Solving for s' we obtain

$s'=-58\text{ cm}$.

(b) The magnification is $m=-\dfrac{s'}{s}=-\dfrac{-58\text{ cm}}{35.0\text{ cm}}=-1.7\times$.

Reflect: If the object were on the concave side of the lens the center of curvature of each lens would be on the incoming side, and so the radius of curvatures would then both be negative and switched in order: $R_1=-15.4\text{ cm}$ and $R_2=-12.2\text{ cm}$. However, since $\left(\dfrac{1}{12.2\text{ cm}}-\dfrac{1}{15.4\text{ cm}}\right)=\left(\dfrac{1}{-15.4\text{ cm}}-\dfrac{1}{-12.2\text{ cm}}\right)$ the answer

would be the unchanged.

*24.63. Set Up: We will apply the thin-lens equation (Equation 24.21): $\dfrac{1}{f}=(n-1)\left(\dfrac{1}{R_1}-\dfrac{1}{R_2}\right)$, when $n=1.55$ and

$f=20.0\text{ cm}$.
Solve: Since $f>0$ we choose $R_1=R$ and $R_2=-R$, where R is the magnitude of the radius of curvature. Thus we

have $\dfrac{1}{f}=(n-1)\left(\dfrac{1}{R}-\dfrac{1}{-R}\right)=\dfrac{2(n-1)}{R}$. Solving for R we obtain $R=2(n-1)f=2(1.55-1)(20.0\text{ cm})=22\text{ cm}$.

Reflect: For identical convex surfaces, the relation between f and R is $f=\dfrac{1}{n-1}\cdot\dfrac{R}{2}$. This is reminiscent of the

relation for spherical mirrors, which is $f=\dfrac{R}{2}$.

*24.65. Set Up: For the second image, the image formed by the mirror serves as the object for the lens. For the mirror, $f_m=+10.0\text{ cm}$. For the lens, $f=32.0\text{ cm}$. The center of curvature of the mirror is $R=2f_m=20.0\text{ cm}$ to

the right of the mirror vertex. $\dfrac{1}{s}+\dfrac{1}{s'}=\dfrac{1}{f}$ gives $s'=\dfrac{sf}{s-f}$.

Solve: (a) The principal-ray diagrams from the two images are sketched in Figures (a) and (b) below. In Figure (b), only the image formed by the mirror is shown. This image is at the location of the candle so the principal ray diagram that shows the image formation when the image of the mirror serves as the object for the lens is analogous to that in Figure (a) and is not drawn.

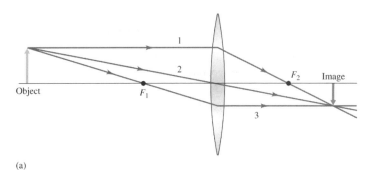

(a)

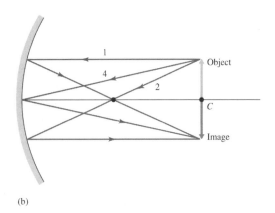

(b)

(b) Image formed by the light that passes directly through the lens: The candle is 85.0 cm to the left of the lens.

$$s' = \frac{sf}{s-f} = \frac{(85.0 \text{ cm})(32.0 \text{ cm})}{85.0 \text{ cm} - 32.0 \text{ cm}} = +51.3 \text{ cm}.$$

$$m = -\frac{s'}{s} = -\frac{51.3 \text{ cm}}{85.0 \text{ cm}} = -0.604.$$

This image is 51.3 cm to the right of the lens. $s' > 0$ so the image is real. $m < 0$ so the image is inverted. Image formed by the light that first reflects off the mirror: First consider the image formed by the mirror. The candle is 20.0 cm to the right of the mirror, so $s = +20.0$ cm.

$$s' = \frac{sf}{s-f} = \frac{(20.0 \text{ cm})(10.0 \text{ cm})}{20.0 \text{ cm} - 10.0 \text{ cm}} = 20.0 \text{ cm}.$$

$$m_1 = -\frac{s_1'}{s_1} = -\frac{20.0 \text{ cm}}{20.0 \text{ cm}} = -1.00.$$

The image formed by the mirror is at the location of the candle, so $s_2 = +85.0$ cm and $s_2' = 51.3$ cm. $m_2 = -0.604$.

$m_{\text{tot}} = m_1 m_2 = (-1.00)(-0.604) = 0.604$. The second image is 51.3 cm to the right of the lens. $s_2' > 0$, so the final image is real. $m_{\text{tot}} > 0$, so the final image is erect.

Reflect: The two images are at the same place. They are the same size. One is erect and one is inverted.

Solutions to Passage Problems

*24.69. Set Up: We will apply the given equation: $\dfrac{1}{f} = \left(\dfrac{n}{n_{liq}} - 1\right)\left(\dfrac{1}{R_1} - \dfrac{1}{R_2}\right)$.

Solve: Since the lens is concave we know that $\left(\dfrac{1}{R_1} - \dfrac{1}{R_2}\right) < 0$ and so $f < 0$ when the lens is immersed in air

$(n_{liq} = 1.00)$. If the lens is now placed in a liquid where $n_{liq} > n$, we have $\left(\dfrac{n}{n_{liq}} - 1\right) < 0$ so that

$\dfrac{1}{f} = \left(\dfrac{n}{n_{liq}} - 1\right)\left(\dfrac{1}{R_1} - \dfrac{1}{R_2}\right) > 0$. Thus, the concave lens is a converging lens when placed in a liquid with a high enough

index of refraction. The correct answer is D.

OPTICAL INSTRUMENTS

Problems 1, 7, 9, 11, 15, 19, 21, 23, 25, 29, 31, 35, 37, 39, 43, 45, 49, 51, 55, 57, 61

Solutions to Problems

*25.1. Set Up: The f-number N is $N = \dfrac{f}{D}$ and is expressed as f/N. The exposure time is inversely proportional to the aperture diameter.

Solve: (a) $f/4$ means $\dfrac{f}{D} = 4$ and $D = \dfrac{f}{4} = \dfrac{300 \text{ mm}}{4} = 75$ mm.

(b) If t is the exposure time, then $\dfrac{t_1}{t_2} = \dfrac{A_2}{A_1} = \dfrac{\frac{1}{4}\pi D_2^{\,2}}{\frac{1}{4}\pi D_1^{\,2}} = \left(\dfrac{D_2}{D_1}\right)^2$. For $f/8$, $D = \dfrac{f}{8} = 37.5$ mm.

$$t_2 = t_1 \left(\dfrac{D_1}{D_2}\right)^2 = \left(\dfrac{1}{250}\text{s}\right)\left(\dfrac{75 \text{ mm}}{37.5 \text{ mm}}\right)^2 = \dfrac{4}{250}\text{ s} = \dfrac{1}{62.5}\text{ s}$$

Reflect: Larger f-number means smaller aperture diameter and larger exposure time.

*25.7. Set Up: $\dfrac{1}{s} + \dfrac{1}{s'} = \dfrac{1}{f}$. $m = -\dfrac{s'}{s} = \dfrac{y'}{y}$. The photocells must be at the image location.

Solve: (a) $s' = \dfrac{sf}{s - f} = \dfrac{(4.25 \text{ m})(0.0750 \text{ m})}{4.25 \text{ m} - 0.0750 \text{ m}} = 0.0763 \text{ m} = 7.63$ cm

(b) $m = -\dfrac{s'}{s} = -\dfrac{0.0763 \text{ m}}{4.25 \text{ m}} = -0.0180$. $y' = my = (-0.0180)(1.85 \text{ m}) = -0.0333 \text{ m} = -3.33$ cm. The image is inverted and 3.33 cm tall. $s' > 0$ so the image is real.

(c) $\dfrac{0.0333 \text{ m}}{8.0 \times 10^{-6} \text{ m/pixel}} = 4.2 \times 10^3$ pixels

Reflect: Since the image is projected onto the photocell, it must be real. Any real image formed by a single lens is inverted.

*25.9. Set Up: $\dfrac{1}{s} + \dfrac{1}{s'} = \dfrac{1}{f}$. $m = -\dfrac{s'}{s} = \dfrac{y'}{y}$. You want $|y'| = 36$ mm. $f = 50.0$ mm. The image will be real, so $s' > 0$, $m < 0$ and $y' = -36$ mm.

Solve: (a) $m = \dfrac{y'}{y} = \dfrac{-36 \times 10^{-3} \text{ m}}{2.00 \text{ m}} = -0.0180.$ $s' = -ms = -(-0.0180)s = 0.0180s.$ Use this in $\dfrac{1}{s} + \dfrac{1}{s'} = \dfrac{1}{f}$:

$$\frac{1}{s} + \frac{1}{0.0180s} = \frac{1}{50.0 \times 10^{-3} \text{ m}}$$

and $s = 2.83$ m.

(b) $s' = 0.0180s = 5.09$ cm

*25.11. Set Up: $\dfrac{1}{s} + \dfrac{1}{s'} = \dfrac{1}{f}.$ $m = -\dfrac{s'}{s} = \dfrac{y'}{y}.$

Solve: (a) $s' = \dfrac{sf}{s-f} = \dfrac{(12.0 \text{ cm})(11.5 \text{ cm})}{12.0 \text{ cm} - 11.5 \text{ cm}} = 276 \text{ cm} = 2.76$ m

(b) $m = -\dfrac{s'}{s} = -\dfrac{276 \text{ cm}}{12.0 \text{ cm}} = -23.0.$ The dimensions of the image are $(23.0)(24 \text{ mm}) = 552 \text{ mm} = 0.552 \text{ m}$ and $(23.0)(36 \text{ mm}) = 828 \text{ mm} = 0.828 \text{ m}$

*25.15. Set Up: The magnification has magnitude $\dfrac{48.0 \text{ in.}}{4.0 \text{ in.}} = \dfrac{60.0 \text{ in.}}{5.0 \text{ in.}} = 12.$ The image is real, so $s' > 0$ and $m < 0.$ $s' = 6.0$ m. $\dfrac{1}{s} + \dfrac{1}{s'} = \dfrac{1}{f}.$ $m = -\dfrac{s'}{s}.$

Solve: (a) $s = -\dfrac{s'}{m} = -\dfrac{600 \text{ cm}}{-12} = +50 \text{ cm}.$ $f = \dfrac{ss'}{s+s'} = \dfrac{(0.50 \text{ m})(6.0 \text{ m})}{0.50 \text{ m} + 6.0 \text{ m}} = 462 \text{ mm}.$

(b) $s = 50$ cm.

Reflect: Only this particular focal length gives both an image distance of 6.0 m and the desired image size.

*25.19. Set Up: $\dfrac{1}{f} = (n-1)\left(\dfrac{1}{R_1} - \dfrac{1}{R_2}\right).$ For a double convex lens, $R_1 > 0$ and $R_2 < 0.$ To compute $\dfrac{1}{f}$ in diopters, R_1 and R_2 must be expressed in meters. Maximum power corresponds to minimum focal length.

(a) *maximum power:* $\dfrac{1}{f} = (1.43 - 1)\left(\dfrac{1}{6.0 \times 10^{-3} \text{ m}} - \dfrac{1}{-5.5 \times 10^{-3} \text{ m}}\right) = 150$ diopters

minimum power: $\dfrac{1}{f} = (1.43 - 1)\left(\dfrac{1}{10.0 \times 10^{-3} \text{ m}} - \dfrac{1}{-6.0 \times 10^{-3} \text{ m}}\right) = 115$ diopters

(b) $f_{min} = \dfrac{1}{150}$ m $= 6.67$ mm. $f_{max} = \dfrac{1}{115}$ m $= 8.70$ mm.

(c) For a distant object $s \to \infty$ and $s' = f.$ At maximum power, $f = f_{max}$ so $s' = 8.70$ mm.

(d) At minimum power, $f = f_{min} = 6.67$ mm. $\dfrac{1}{s} + \dfrac{1}{s'} = \dfrac{1}{f}$ so

$$s' = \frac{sf}{s-f} = \frac{(25 \text{ cm})(0.667 \text{ cm})}{25 \text{ cm} - 0.667 \text{ cm}} = 0.685 \text{ cm} = 6.85 \text{ mm}.$$

Reflect: Since the distance from the lens to the retina is fixed, this eye won't focus clearly on both objects.

*25.21. Set Up: For an object 25.0 cm from the eye, the corrective lens forms a virtual image at the near point of the eye. The distances from the corrective lens are $s = 23.0$ cm and $s' = -43.0$ cm.

Solve: $\dfrac{1}{s} + \dfrac{1}{s'} = \dfrac{1}{f}.$ $f = \dfrac{ss'}{s+s'} = \dfrac{(23.0 \text{ cm})(-43.0 \text{ cm})}{23.0 \text{ cm} - 43.0 \text{ cm}} = +49.4 \text{ cm}.$ The power is $\dfrac{1}{0.494 \text{ m}} = 2.02$ diopters.

Reflect: In Problem 25.20 the contact lenses have power 1.78 diopters. The power of the lenses is different for ordinary glasses versus contact lenses.

*25.23. Set Up: Apply $\dfrac{1}{s}+\dfrac{1}{s'}=\dfrac{1}{f}$. The near point is at infinity, so that is where the image must be formed for any objects that are close. The power in diopters is equal to $\dfrac{1}{f}$, with f in meters.

Solve: $\dfrac{1}{f}=\dfrac{1}{s}+\dfrac{1}{s'}=\dfrac{1}{24\text{ cm}}+\dfrac{1}{-\infty}=\dfrac{1}{0.24\text{ m}}=4.17$ diopters.

Reflect: To focus on closer objects, the power must be increased.

*25.25. Set Up: Apply the equation $\dfrac{1}{s}+\dfrac{1}{s'}=\dfrac{1}{f}$ to determine $\dfrac{1}{f}$.

Solve: We want the image of the computer screen that is formed by the student's eyeglasses to be located at the far point of her vision. Assuming that the distance between her glasses and eye is negligible this means that $s'=-17$ cm when $s=45$ cm. Thus we have $\dfrac{1}{f}=\dfrac{1}{0.45\text{ m}}+\dfrac{1}{-0.17\text{ m}}=\underline{-3.66}$ diopters.

Reflect: The eyeglass lens forms a virtual image at your far point so that your relaxed eye can comfortably view the computer screen, which is beyond your far point.

*25.29. Set Up: When the object is at the focal point of the lens and the image therefore is at infinity, then $M=\dfrac{25\text{ cm}}{f}$.

Solve: (a) $M=\dfrac{25\text{ cm}}{11.5\text{ cm}}=2.17$

(b) The object is at the focal point of the lens, so the lens is 11.5 cm from the stamp.

*25.31. Set Up: The image is not at infinity, so the equation $M=\dfrac{25\text{ cm}}{f}$ doesn't apply. The image is virtual, so $s'=-25.0$ cm.

Solve: (a) $\dfrac{1}{s}+\dfrac{1}{s'}=\dfrac{1}{f}$. $s=\dfrac{s'f}{s'-f}=\dfrac{(-25.0\text{ cm})(8.00\text{ cm})}{-25.0\text{ cm}-8.00\text{ cm}}=+6.06$ cm

(b) $m=-\dfrac{s'}{s}=-\dfrac{-25.0\text{ cm}}{6.06\text{ cm}}=+4.13.$ $|y'|=|m|y=(4.13)(1.00\text{ mm})=4.13$ mm

Reflect: The lens allows the object to be brought much closer to the eye (6.06 cm) than the near point.

*25.35. Set Up: We will assume that the given lateral magnification of the objective was calculated using the equation $|m_1|=\dfrac{s_1'}{s_1}$ (and not the approximate equation $|m_1|\approx\dfrac{s_1'}{f_1}$) and the angular magnification of the eyepiece is given by $M_2=\dfrac{(25\text{ cm})}{f_2}$. The overall magnification of the microscope is then $M=m_1M_2$. To solve this problem we must assume that the image of the objective is located very close to the focal point of the eyepiece: thus, we have $16.5\text{ cm}=s_1'+f_2$. Note that we will not use the approximation $s_1\approx f_1$ since it is not necessary given the information we have.

Solve: (a) The overall magnification of the microscope is $M=m_1M_2=(62)(10)=620\times$.

(b) We can immediately determine the focal length of the eyepiece from $f_2=\dfrac{(25\text{ cm})}{M_2}=\dfrac{(25\text{ cm})}{10}=2.5$ cm.

Since we assume that the image of the objective is located very close to the focal point of the eyepiece we have $s_1' = 16.5 \text{ cm} - 2.5 \text{ cm} = 14.0 \text{ cm}$. Thus, we obtain $s_1 = \dfrac{s_1'}{|m_1|} = \dfrac{14.0 \text{ cm}}{62} = 0.\underline{2}26 \text{ cm}$. Finally we obtain the focal length

of the objective: $f_1 = \dfrac{s_1 s_1'}{s_1 + s_1'} = \dfrac{(0.\underline{2}26 \text{ cm})(14.0 \text{ cm})}{0.\underline{2}26 \text{ cm} + 14.0 \text{ cm}} = 0.\underline{2}22 \text{ cm}$.

(c) If the object is located at $s = 0.\underline{2}26 \text{ cm}$, the image of the objective will be located at the focal point of the eyepiece—this produces a final image at infinity, which allows for comfortable viewing of a normal eye. Thus, the object should be placed approximately $0.\underline{2}26 \text{ cm} - 0.\underline{2}22 \text{ cm} = 0.004 \text{ cm}$ outside the focal point of the objective lens for comfortable viewing.

Reflect: If we assume that the magnification of the objective was calculated by using the approximate equation $|m_1| = 62 \approx \dfrac{s_1'}{f_1}$ (with $s_1' = 14.0 \text{ cm}$) we obtain slightly different answers for f_1 and s_1: $f_1 = 0.\underline{2}26 \text{ cm}$ and $s_1 = 0.\underline{2}30 \text{ cm}$. However, note that the object must still be placed roughly 0.004 cm outside the focal point of the objective for comfortable viewing.

*25.37. Set Up: $M = m_1 M_2 = \left(\dfrac{s_1'}{f_1}\right) M_2$. $s_1' = 120 \text{ mm} + f_1$.

Solve: (a) The largest M is for the smallest f_1 and largest M_2. $M_2 = 10$.

$f = 1.9 \text{ mm}$ and $s_1' = 120 \text{ mm} + 1.9 \text{ mm} = 121.9 \text{ mm}$.

$$M = \left(\frac{121.9 \text{ mm}}{1.9 \text{ mm}}\right)(10) = 640.$$

(b) The smallest M is for the largest f_1 and the smallest M_2. $M_2 = 5$,

$f_1 = 16 \text{ mm}$ and $s_1' = 120 \text{ mm} + 16 \text{ mm} = 136 \text{ mm}$.

$$M = \left(\frac{136 \text{ mm}}{16 \text{ mm}}\right)(5) = 43.$$

*25.39. Set Up: $M = -\dfrac{f_1}{f_2}$. The largest M is for the smallest f_2, where f_2 is the focal length of the eyepiece.

Solve: $M_{\max} = -\dfrac{(16.0 \text{ in.})(2.54 \text{ cm/in.})}{1.5 \text{ cm}} = -27$;

$$M_{\min} = -\frac{(16.0 \text{ in.})(2.54 \text{ cm/in.})}{8.5 \text{ cm}} = -4.8$$

Reflect: The focal length of the objective is $f_1 = 16.0 \text{ in.} = 40.6 \text{ cm}$ and $f_1 > f_2$ for each eyepiece.

*25.43. Set Up: The unaided angular size of Jupiter is $\theta = \dfrac{138{,}000 \text{ km}}{6.28 \times 10^8 \text{ km}} = 2.20 \times 10^{-4} \text{ rad} = 0.0126°$.

$M = \dfrac{\theta'}{\theta} = -\dfrac{f_1}{f_2}$. $f_1 = 19.4 \text{ m}$.

Solve: Need $M = \dfrac{0.500°}{0.0126°} = 39.7$. $f_2 = \dfrac{f_1}{|M|} = \dfrac{19.4 \text{ m}}{39.7} = 0.49 \text{ m}$.

Reflect: The angular magnification is the ratio of the angular size of the image when the optical device is used divided by the angular size of the object.

*25.45. Set Up: From Figure 23.29 in Section 23.9, for 700 nm, $n = 1.61$ and for 400 nm, $n = 1.66$. Use $\frac{1}{f} = (n-1)\left(\frac{1}{R_1} - \frac{1}{R_2}\right)$ to find f for each wavelength. Then use $\frac{1}{s} + \frac{1}{s'} = \frac{1}{f}$ to find s'. Double convex means $R_1 = 25.0$ cm and $R_2 = -35.0$ cm.

Solve: **(a)** (i) *red light:* $\frac{1}{f} = (1.61 - 1)\left(\frac{1}{0.250 \text{ m}} - \frac{1}{-0.350 \text{ m}}\right) = 23.9$ cm

(ii) *violet light:* $\frac{1}{f} = (1.66 - 1)\left(\frac{1}{0.250 \text{ m}} - \frac{1}{-0.350 \text{ m}}\right) = 22.1$ cm

(b) *red light:* $s' = \frac{sf}{s - f} = \frac{(30.0 \text{ cm})(23.9 \text{ cm})}{30.0 \text{ cm} - 23.9 \text{ cm}} = 118$ cm.

violet light: $s' = \frac{sf}{s - f} = \frac{(30.0 \text{ cm})(22.1 \text{ cm})}{30.0 \text{ cm} - 22.1 \text{ cm}} = 83.9$ cm.

*25.49. Set Up: We have $\frac{n_a}{s} + \frac{n_b}{s'} = \frac{n_b - n_a}{R}$, with $n_a = 1.00$, $n_b = 1.40$. $s = 40.0$ cm, $s' = 2.60$ cm.

Solve: $\frac{1}{40.0 \text{ cm}} + \frac{1.40}{2.60 \text{ cm}} = \frac{0.40}{R}$ and $R = 0.710$ cm.

Reflect: The cornea presents a convex surface to the object, so $R > 0$.

*25.51. Set Up: $m = -\frac{s'}{s} = \frac{y'}{y}$. $s' = 2.50$ cm. $|y'| = 5.0 \ \mu$m. The angle subtended (in radians) is height divided by distance from the eye.

Solve: **(a)** $m = -\frac{s'}{s} = -\frac{2.50 \text{ cm}}{25 \text{ cm}} = -0.10$. $y = \left|\frac{y'}{m}\right| = \frac{5.0 \ \mu\text{m}}{0.10} = 50 \ \mu$m.

(b) $\theta = \frac{y}{s} = \frac{50 \ \mu\text{m}}{25 \text{ cm}} = \frac{50 \times 10^{-6} \text{ m}}{25 \times 10^{-2} \text{ m}} = 2.0 \times 10^{-4}$ rad $= 0.0115° = 0.69$ min.

This is only a bit smaller than the typical experimental value of 1.0 min.

Reflect: The angle subtended by the object equals the angular size $\frac{|y'|}{s'} = \frac{5.0 \times 10^{-6} \text{ m}}{2.50 \times 10^{-2} \text{ m}} = 2.0 \times 10^{-4}$ rad of the image.

*25.55. Set Up: The image is the same distance behind the mirror as the object is in front of the mirror. $\frac{1}{s} + \frac{1}{s'} = \frac{1}{f}$. $m = -\frac{s'}{s} = \frac{y'}{y}$.

Solve: **(a)** $s = 1.50$ m. $s' = \frac{sf}{s - f} = \frac{(1.50 \text{ m})(19.5 \times 10^{-3} \text{ m})}{1.50 \text{ m} - 19.5 \times 10^{-3} \text{ m}} = 19.8$ mm.

(b) $m = -\frac{s'}{s} = -\frac{19.8 \times 10^{-3} \text{ m}}{1.50 \text{ m}} = -0.0132$. $|y'| = |m|y = (0.0132)(8.0 \text{ cm}) = 0.106$ cm.

Reflect: The camera focuses on the image, behind the mirror, not on the mirror surface.

*25.57. Set Up: The object for the camera lens is the image formed by the double convex lens. $\frac{1}{s} + \frac{1}{s'} = \frac{1}{f}$. $m = -\frac{s'}{s} = \frac{y'}{y}$.

Solve: **(a)** *image formed by the convex lens:*

$$s' = \frac{sf}{s - f} = \frac{(10.0 \text{ cm})(17.0 \text{ cm})}{10.0 \text{ cm} - 17.0 \text{ cm}} = -24.3 \text{ cm}.$$

The image is 24.3 cm to the left of the convex lens, so it is 24.3 cm + 5.00 cm = 29.3 cm to the left of the camera lens.

$$m = -\frac{s'}{s} = -\frac{-24.3 \text{ cm}}{10.0 \text{ cm}} = +2.43.$$

The image is $(2.43)(1.40 \text{ cm}) = 3.40 \text{ cm}$ tall.

image formed by camera lens: $s = 29.3$ cm.

$$s' = \frac{sf}{s-f} = \frac{(29.3 \text{ cm})(1.50 \text{ cm})}{29.3 \text{ cm} - 1.50 \text{ cm}} = 1.58 \text{ cm}.$$

(b) $m = -\dfrac{s'}{s} = -\dfrac{1.58 \text{ cm}}{29.3 \text{ cm}} = -0.0539.$ $y = 3.40$ cm. $|y'| = |m|y = (0.0539)(3.40 \text{ cm}) = 0.183$ cm.

(c) No. For light from the seedling to reach the camera lens it must pass through the convex lens. The camera cannot view the seedling directly.

Reflect: The overall magnification is the product $m_1 m_2$ of the magnifications of the two lenses. $m_1 m_2 = (2.43)$ $(-0.0539) = -0.131$ and the final image is 0.131 times the height of the seedling.

*25.61. Set Up: $\dfrac{1}{f} = (n-1)\left(\dfrac{1}{R_1} - \dfrac{1}{R_2}\right).$ For a double-convex lens, $R_1 > 0$ and $R_2 < 0.$ $M = -\dfrac{f_1}{f_2}.$ The length of the telescope is $f_1 + f_2$.

Solve: (a) Find the focal length of each lens. L_1: $\dfrac{1}{f} = (1.528 - 1)\left(\dfrac{1}{0.950 \text{ m}} - \dfrac{1}{-2.700 \text{ m}}\right)$ and $f = 133$ cm. L_2:

$\dfrac{1}{f} = (1.550 - 1)\left(\dfrac{1}{0.535 \text{ m}} - \dfrac{1}{-50.5 \text{ m}}\right)$ and $f = 96.3$ cm.

The largest magnification would be with L_1 as the objective and L_2 as the eyepiece.

$$M = -\frac{f_1}{f_2} = -\frac{133 \text{ cm}}{96.3 \text{ cm}} = -1.38.$$

(b) The length of the telescope would be $133 \text{ cm} + 96.3 \text{ cm} = 229$ cm.

INTERFERENCE AND DIFFRACTION

Problems 1, 5, 9, 13, 15, 19, 21, 25, 29, 31, 33, 37, 39, 43, 45, 49, 53, 55, 57, 61, 63, 67

Solutions to Problems

*26.1. Set Up: Constructive interference occurs for $r_2 - r_1 = m\lambda$, $m = 0$, ± 1, $\pm 2, \ldots$. Destructive interference occurs for $r_2 - r_1 = \left(m + \frac{1}{2}\right)\lambda$, $m = 0$, ± 1, $\pm 2, \ldots$. For this problem, $r_2 = 150$ cm and $r_1 = x$. The path taken by the person ensures that x is in the range $0 \le x \le 150$ cm.

Solve: (a) 150 cm $- x = m(34$ cm$)$. $x = 150$ cm $- m(34$ cm$)$. For $m = 0, 1, 2, 3, 4$ the values of x are 150 cm, 116 cm, 82 cm, 48 cm, 14 cm.

(b) 150 cm $- x = \left(m + \frac{1}{2}\right)(34$ cm$)$. $x = 150$ cm $- \left(m + \frac{1}{2}\right)(34$ cm$)$. For $m = 0, 1, 2, 3$ the values of x are 133 cm, 99 cm, 65 cm, 31 cm.

Reflect: When $x = 116$ cm the path difference is 150 cm $- 116$ cm $= 34$ cm, which is one wavelength. When $x = 133$ cm the path difference is 17 cm, which is one-half wavelength.

*26.5. Set Up: When she is at the midpoint between the two speakers the path difference $r_2 - r_1$ is zero. When she walks a distance d toward one speaker, r_2 increases by d and r_1 decreases by d, so the path difference changes by $2d$. Since the speakers are $180°$ out of phase, path difference $= m\lambda$, $m = 0$, ± 1, $\pm 2, \ldots$. gives destructive interference and path difference $= \left(m + \frac{1}{2}\right)\lambda$, $m = 0$, ± 1, $\pm 2, \ldots$. gives constructive interference.

Solve: $\lambda = \dfrac{v}{f} = \dfrac{340.0 \text{ m/s}}{250.0 \text{ Hz}} = 1.36$ m

(a) Path difference is zero, so the interference is destructive.

(b) Destructive interference, so the path difference equals λ. $2d = \lambda$ and $d = \dfrac{\lambda}{2} = \dfrac{1.36 \text{ m}}{2} = 68.0$ cm.

(c) Constructive interference, so the path difference equals $\lambda/2$. $2d = \dfrac{\lambda}{2}$ and $d = \dfrac{\lambda}{4} = \dfrac{1.36 \text{ m}}{4} = 34.0$ cm.

Reflect: When she moves from a point of destructive interference to a point of constructive interference the path difference changes by $\lambda/2$.

*26.9. Set Up: $\lambda = \dfrac{c}{f} = \dfrac{3.00 \times 10^8 \text{ m/s}}{6.32 \times 10^{14} \text{ Hz}} = 4.75 \times 10^{-7}$ m. Bright fringes are located at $y_m = R\dfrac{m\lambda}{d}$, when $y_m \ll R$.

Dark fringes are at $d\sin\theta = \left(m + \frac{1}{2}\right)\lambda$ and $y = R\tan\theta$. For the third bright fringe (not counting the central bright spot), $m = 3$. For the third dark fringe, $m = 2$.

Solve: (a) $d = \dfrac{m\lambda R}{y_m} = \dfrac{3(4.75\times10^{-7}\ \text{m})(0.850\ \text{m})}{0.0311\ \text{m}} = 3.89\times10^{-5}\ \text{m} = 0.0389\ \text{mm}$

(b) $\sin\theta = \left(2+\tfrac{1}{2}\right)\dfrac{\lambda}{d} = (2.5)\left(\dfrac{4.75\times10^{-7}\ \text{m}}{3.89\times10^{-5}\ \text{m}}\right) = 0.0305$ and $\theta = 1.75°$. $y = R\tan\theta = (85.0\ \text{cm})\tan 1.75° = 2.60\ \text{cm}.$

Reflect: The third dark fringe is closer to the center of the screen than the third bright fringe on one side of the central bright fringe.

*26.13. Set Up: The largest value $\sin\theta$ can have is 1.00. Bright fringes are located at angles θ given by $d\sin\theta = m\lambda$.

Solve: (a) $m = \dfrac{d\sin\theta}{\lambda}$. For $\sin\theta = 1$, $m = \dfrac{d}{\lambda} = \dfrac{0.0116\times10^{-3}\ \text{m}}{5.85\times10^{-7}\ \text{m}} = 19.8$. Therefore, the largest m for fringes on the screen is $m = 19$. There are $2(19)+1 = 39$ bright fringes, the central one and 19 above and 19 below it.

(b) The most distant fringe has $m = \pm19$.

$$\sin\theta = m\frac{\lambda}{d} = \pm19\left(\frac{5.85\times10^{-7}\ \text{m}}{0.0116\times10^{-3}\ \text{m}}\right) = \pm0.958$$

and $\theta = \pm73.3°$.

Reflect: For small θ the spacing Δy between adjacent fringes is constant but this is no longer the case for larger angles.

*26.15. Set Up: Strongly reflected means constructive interference so we will consider the interference between rays reflected from the two surfaces of the soap film. Take into account the phase difference due to the path difference of $2t$ and any phase difference due to phase changes upon reflection. As shown in the figure below there is a 180° phase change when the light is reflected from the outside surface of the bubble and no phase change when the light is reflected from the inside surface.

Solve: (a) The reflections produce a net 180° phase difference and for there to be constructive interference the path difference $2t$ must correspond to a half-integer number of wavelengths to compensate for the $\lambda/2$ shift due to the reflections. Hence the condition for constructive interference is $2t = \left(m+\dfrac{1}{2}\right)(\lambda_0/n)$ (where $m = 0, 1, 2, \ldots$). Here λ_0 is the wavelength in air and (λ_0/n) is the wavelength in the bubble, where the path difference occurs. Thus we have $\lambda_0 = \dfrac{2tn}{m+\dfrac{1}{2}} = \dfrac{2(290\ \text{nm})(1.33)}{m+\dfrac{1}{2}} = \dfrac{771.4\ \text{nm}}{m+\dfrac{1}{2}}$. For $m = 0$, $\lambda = 1543$ nm; for $m = 1$, $\lambda = 514$ nm; for $m = 2$, $\lambda = 308$ nm; ... Only 514 nm is in the visible region; the color for this wavelength is green.

(b) $\lambda_0 = \dfrac{2tn}{m+\dfrac{1}{2}} = \dfrac{2(340\ \text{nm})(1.33)}{m+\dfrac{1}{2}} = \dfrac{904.4\ \text{nm}}{m+\dfrac{1}{2}}$

For $m = 0$, $\lambda = 1809$ nm; for $m = 1$, $\lambda = 603$ nm; for $m = 2$, $\lambda = 362$ nm; ... Only 603 nm is in the visible region; the color for this wavelength is orange.

Reflect: The dominant color of the reflected light depends on the thickness of the film. If the bubble has varying thickness at different points, these points will appear to be different colors when the light reflected from the bubble is viewed.

*26.19. Set Up: Consider light reflected at the front and rear surfaces of the film. At the front surface of the film, light in air $(n = 1.00)$ reflects from the film $(n = 2.62)$ and there is a $180°$ phase shift due to the reflection. At the back surface of the film, light in the film $(n = 2.62)$ reflects from glass $(n = 1.62)$ and there is no phase shift due to reflection. Therefore, there is a net $180°$ phase difference produced by the reflections. The path difference for these two rays is $2t$, where t is the thickness of the film. The wavelength in the film is $\lambda = \dfrac{505 \text{ nm}}{2.62}$.

Solve: (a) Since the reflection produces a net $180°$ phase difference, destructive interference of the reflected light occurs when $2t = m\lambda$. $t = m\left(\dfrac{505 \text{ nm}}{2[2.62]}\right) = (96.4 \text{ nm})m$. The minimum thickness is 96.4 nm.

(b) The next three thicknesses are for $m = 2$, 3, and 4: 192 nm, 289 nm, and 386 nm.

Reflect: The minimum thickness is for $t = \lambda/2n$. Compare this to Problem 26.15, where the minimum thickness for destructive interference is $t = \lambda/4n$.

*26.21. Set Up: The fringes are produced by interference between light reflected from the top and from the bottom surfaces of the air wedge. The refractive index of glass is greater than that of air, so the waves reflected from the top surface of the air wedge have no reflection phase shift and the waves reflected from the bottom surface of the air wedge do have a half-cycle reflection phase shift. The condition for constructive interference (bright fringes) therefore is $2t = \left(m + \tfrac{1}{2}\right)\lambda$. The geometry of the air wedge is sketched in the figure below.

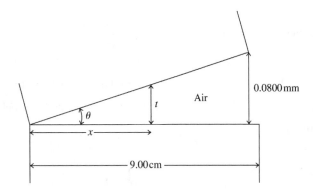

Solve: $\tan\theta = \dfrac{0.0800 \text{ mm}}{90.0 \text{ mm}} = 8.89 \times 10^{-4}$. $\tan\theta = \dfrac{t}{x}$ so $t = (8.89 \times 10^{-4})x$. $t_m = \left(m + \tfrac{1}{2}\right)\dfrac{\lambda}{2}$.

$$x_m = \left(m + \tfrac{1}{2}\right)\dfrac{\lambda}{2(8.89 \times 10^{-4})} \text{ and}$$

$$x_{m+1} = \left(m + \tfrac{3}{2}\right)\dfrac{\lambda}{2(8.89 \times 10^{-4})}.$$

The distance along the plate between adjacent fringes is

$$\Delta x = x_{m+1} - x_m = \dfrac{\lambda}{2(8.89 \times 10^{-4})} = \dfrac{656 \times 10^{-9} \text{ m}}{2(8.89 \times 10^{-4})} = 3.69 \times 10^{-4} \text{ m} = 0.369 \text{ mm}.$$

The number of fringes per cm is $\dfrac{1.00}{\Delta x} = \dfrac{1.00}{0.0369 \text{ cm}} = 27.1$ fringes/cm.

Reflect: As $t \to 0$ the interference is destructive and there is a dark fringe at the line of contact between the two plates.

*26.25. Set Up: The condition for a dark fringe is $\sin\theta = \dfrac{m\lambda}{a}$, $m = \pm1, \ \pm2, \dots$.

Solve: $\sin\theta = m\left(\dfrac{632.8 \times 10^{-9} \text{ m}}{0.00375 \times 10^{-3} \text{ m}}\right) = m(0.1687)$. $m = \pm 1$: $\theta = \pm 9.71°$. $m = \pm 2$: $\theta = \pm 19.7°$. $m = \pm 3$: $\theta = \pm 30.4°$.

$m = \pm 4$: $\theta = \pm 42.4°$. $m = \pm 5$: $\theta = \pm 57.5°$.

Reflect: There are a finite number of dark fringes because $\dfrac{m\lambda}{a} = \sin\theta$ can't be larger than 1.00. This establishes a

maximum value for m.

*26.29. Set Up: Calculate the angular positions of the minima and use $y = x\tan\theta$ to calculate the distance on the screen between them. The central bright fringe is shown in Figure (a) below and the first bright fringe on one side of the central bright fringe is shown in Figure (b) below.

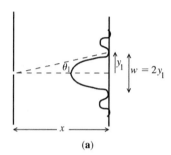

(a)

Solve: (a) The first minimum is located by $\sin\theta_1 = \dfrac{\lambda}{a} = \dfrac{633 \times 10^{-9} \text{ m}}{0.350 \times 10^{-3} \text{ m}} = 1.809 \times 10^{-3}$. Solving we obtain

$\theta_1 = 1.809 \times 10^{-3}$ rad.

We can now find the distance between the center of the central fringe and the first minimum: $y_1 = x\tan\theta_1 = (3.00 \text{ m})\tan(1.809 \times 10^{-3} \text{ rad}) = 5.427 \times 10^{-3} \text{ m}$. The width of the central bright fringe is: $w = 2y_1 = 2(5.427 \times 10^{-3} \text{ m}) = 1.09 \times 10^{-2} \text{ m} = 10.9 \text{ mm}$.

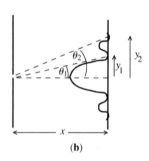

(b)

(b) The width of the first bright fringe adjacent to the central fringe is $w = y_2 - y_1$, where $y_1 = 5.427 \times 10^{-3}$ m

from part (a). To find y_2 we note that $\sin\theta_2 = \dfrac{2\lambda}{a} = 3.618 \times 10^{-3}$ so that $\theta_2 = 3.618 \times 10^{-3}$ rad. Thus,

$y_2 = x\tan\theta_2 = 1.085 \times 10^{-2}$ m and the width of the bright fringe adjacent to the central one is $w = y_2 - y_1 = 1.085 \times 10^{-2} \text{ m} - 5.427 \times 10^{-3} \text{ m} = 5.4 \text{ mm}$.

Reflect: The central bright fringe is twice as wide as the other bright fringes.

*26.31. Set Up: $v = f\lambda$ gives λ. The person hears no sound at angles corresponding to diffraction minima. The diffraction minima are located by $\sin\theta = m\lambda/a$ (where $m = \pm 1, \pm 2, \dots$).

Solve: $\lambda = v/f = (344 \text{ m/s})/(1250 \text{ Hz}) = 0.2752 \text{ m}$; $a = 1.00 \text{ m}$. $m = \pm 1$, $\theta = \pm 16.0°$; $m = \pm 2$, $\theta = \pm 33.4°$; $m = \pm 3$, $\theta = \pm 55.6°$; no solution for larger m

Reflect: $\lambda/a = 0.28$ so for the large wavelength sound waves diffraction by the doorway is a large effect. Diffraction would not be easily observable for visible light because its wavelength is much smaller and $\lambda/a \ll 1$.

*26.33. Set Up: The wavelength of light in the liquid is $\lambda = \dfrac{\lambda_{air}}{n}$. The dark fringes are located at $\sin\theta = \dfrac{m\lambda}{a}$, $m = \pm1, \pm2, \ldots$. The first dark fringes on either side of the central bright fringe correspond to $m = +1$ and $m = -1$. If a feature is located by angle θ, its distance from the center of the central maximum is $y = R\tan\theta$. In this problem θ is not small and Eq. (26.10) cannot be used.

Solve: Each dark fringe is a distance $y = 11.2$ cm from the center of the screen. $\tan\theta = \dfrac{y}{R} = \dfrac{11.2 \text{ cm}}{30.0 \text{ cm}}$ and

$\theta = 20.5°$. $\lambda = \dfrac{a\sin\theta}{1} = \dfrac{(0.00125 \times 10^{-3} \text{ m})\sin 20.5°}{1} = 438$ nm. $\lambda_{air} = n\lambda = (1.45)(438 \text{ nm}) = 635$ nm.

Reflect: Since the path difference occurs in the liquid it must be compared to the wavelength in the liquid.

*26.37. Set Up: The bright spots are located by $d\sin\theta = m\lambda$, $m = 0, \pm1, \pm2, \ldots$. The second pair correspond to

$m = \pm2$. $d = \dfrac{1}{5580 \text{ lines/cm}} = 1.79 \times 10^{-4}$ cm $= 1.79 \times 10^{-6}$ m.

Solve: **(a)** $y = R\tan\theta$. $\tan\theta = \dfrac{y}{R} = \dfrac{26.3 \text{ cm}}{42.5 \text{ cm}} = 0.6188$ and $\theta = 31.75°$.

$$\lambda = \dfrac{d\sin\theta}{m} = \dfrac{(1.79 \times 10^{-6} \text{ m})(\sin 31.75°)}{2} = 471 \text{ nm}$$

(b) The next bright spots are for $m = \pm3$. $\sin\theta = \dfrac{m\lambda}{d} = \dfrac{3(471 \times 10^{-9} \text{ m})}{1.79 \times 10^{-6} \text{ m}}$ and $\theta = 52.13°$.

$$y = R\tan\theta = (42.5 \text{ cm})(\tan 52.13°) = 54.7 \text{ cm}.$$

Reflect: $\dfrac{\lambda}{d} = \dfrac{471 \times 10^{-9} \text{ m}}{1.79 \times 10^{-6} \text{ m}} = 0.263$ and is not small. The approximation $\sin\theta \approx \tan\theta$ is not accurate.

*26.39. Set Up: The maxima are located by $d\sin\theta = m\lambda$, where $d = 1.60 \times 10^{-6}$ m for a CD and $d = 0.740 \times 10^{-6}$ m for a DVD.

Solve: (a) For a CD we have: $\theta = \arcsin\left(\dfrac{m\lambda}{d}\right) = \arcsin\left(\dfrac{m[6.328 \times 10^{-7} \text{ m}]}{1.60 \times 10^{-6} \text{ m}}\right) = \arcsin([0.396]m)$. For $m = 1$ we have

$\theta_1 = 23.3°$. For $m = 2$ we have $\theta_2 = 52.3°$. There are no other maxima.

(b) For a DVD we have: $\theta = \arcsin\left(\dfrac{m\lambda}{d}\right) = \arcsin\left(\dfrac{m[6.328 \times 10^{-7} \text{ m}]}{0.740 \times 10^{-6} \text{ m}}\right) = \arcsin([0.855]m)$. For $m = 1$ we have

$\theta_1 = 58.8°$. There are no other maxima.

Reflect: The reflective surface produces the same interference pattern as a grating with slit separation d.

*26.43. Set Up: The maxima are at angles θ given by $2d\sin\theta = m\lambda$, where $d = 0.440$ nm.

Solve: $m = 1$. $\lambda = \dfrac{2d\sin\theta}{1} = 2(0.440 \text{ nm})\sin 39.4° = 0.559$ nm.

*26.45. Set Up: Rayleigh's criterion says $\theta_{res} = 1.22\dfrac{\lambda}{D}$. $D = 7.20$ cm. $\theta_{res} = \dfrac{y}{s}$, where s is the distance of the object from the lens and $y = 4.00$ mm.

Solve: $\dfrac{y}{s} = 1.22\dfrac{\lambda}{D}$. $s = \dfrac{yD}{1.22\lambda} = \dfrac{(4.00 \times 10^{-3} \text{ m})(7.20 \times 10^{-2} \text{ m})}{1.22(550 \times 10^{-9} \text{ m})} = 429$ m.

Reflect: The focal length of the lens doesn't enter into the calculation. In practice, it is difficult to achieve resolution that is at the diffraction limit.

*26.49. Set Up: $\theta_{res} = 1.22\dfrac{\lambda}{D}$. The angular size of the object is its height divided by its distance from the eye.

Solve: $\theta_{res} = 1.22\left(\dfrac{550 \times 10^{-9}\ \text{m}}{2.00 \times 10^{-3}\ \text{m}}\right) = 3.36 \times 10^{-4}$. $\theta_{res} = \dfrac{y}{s}$ and

$$y = s\,\theta_{res} = (25.0\ \text{cm})(3.36 \times 10^{-4}) = 8.4 \times 10^{-3}\ \text{cm} = 0.084\ \text{mm}.$$

Diffraction doesn't play a significant role.

*26.53. Set Up: The interference minima are located by $d\sin\theta = \left(m + \tfrac{1}{2}\right)\lambda$. For a liquid with refractive index n,

$\lambda_{liq} = \dfrac{\lambda_{air}}{n}$.

Solve: $\dfrac{\sin\theta}{\lambda} = \dfrac{\left(m + \tfrac{1}{2}\right)}{d} = \text{constant}$, so $\dfrac{\sin\theta_{air}}{\lambda_{air}} = \dfrac{\sin\theta_{liq}}{\lambda_{liq}}$. $\dfrac{\sin\theta_{air}}{\lambda_{air}} = \dfrac{\sin\theta_{liq}}{\lambda_{air}/n}$ and $n = \dfrac{\sin\theta_{air}}{\sin\theta_{liq}} = \dfrac{\sin 35.20°}{\sin 19.46°} = 1.730$.

Reflect: In the liquid the wavelength is shorter and $\sin\theta = \left(m + \tfrac{1}{2}\right)\dfrac{\lambda}{d}$ gives a smaller θ than in air, for the same m.

*26.55. Set Up: Consider reflection from either side of the film. (a) At the front of the film, light in air $(n = 1.00)$ reflects off the film $(n = 1.45)$ and there is a 180° phase shift. At the back of the film, light in the film $(n = 1.45)$ reflects off the cornea $(n = 1.38)$ and there is no phase shift. The reflections produce a net 180° phase difference and the condition for constructive interference is $2t = \left(m + \tfrac{1}{2}\right)\lambda$, where $\lambda = \dfrac{\lambda_{air}}{n}$. $t = \left(m + \tfrac{1}{2}\right)\dfrac{\lambda_{air}}{2n}$. Minimum thickness is for $m = 0$, and is given by $t = \dfrac{\lambda_{air}}{4n} = \dfrac{600\ \text{nm}}{4(1.45)} = 103\ \text{nm}$.

(b) $\lambda_{air} = \dfrac{2nt}{m + \tfrac{1}{2}} = \dfrac{2(1.45)(103\ \text{nm})}{m + \tfrac{1}{2}} = \dfrac{299\ \text{nm}}{m + \tfrac{1}{2}}$. For $m = 0$, $\lambda_{air} = 598\ \text{nm}$. For $m = 1$, $\lambda_{air} = 199\ \text{nm}$ and all other values are smaller. No other visible wavelengths are reinforced. The condition for destructive interference is $2t = m\dfrac{\lambda_{air}}{n}$. $\lambda = \dfrac{2tn}{m} = \dfrac{299\ \text{nm}}{m}$. For $m = 1$, $\lambda_{air} = 299\ \text{nm}$ and all other values are shorter. There are no visible wavelengths for which there is destructive interference.

(c) Now both rays have a 180° phase change on reflection and the reflections don't introduce any net phase shift. The expression for constructive interference in parts (a) and (b) now gives destructive interference and the expression in (a) and (b) for destructive interference now gives constructive interference. The only visible wavelength for which there will be destructive interference is 600 nm and there are no visible wavelengths for which there will be constructive interference.

Reflect: Changing the net phase shift due to the reflections can convert the interference for a particular thickness from constructive to destructive, and vice versa.

*26.57. Set Up: The dark fringes are located by $\sin\theta = m\dfrac{\lambda}{a}$. The first dark fringes are for $m = \pm 1$. $y = R\tan\theta$ is the distance from the center of the screen. From the center to one minimum is 2.61 cm.

Solve: $\tan\theta = \dfrac{y}{R} = \dfrac{2.61\ \text{cm}}{125\ \text{cm}} = 0.02088$ and $\theta = 1.20°$. $a = \dfrac{\lambda}{\sin\theta} = \dfrac{632.8 \times 10^{-9}\ \text{m}}{\sin 1.20°} = 30.2\ \mu\text{m}$.

*26.61. Set Up: Maxima are given by $2d\sin\theta = m\lambda$, where d is the separation between crystal planes.

Solve: (a) $\theta = \arcsin\left(\dfrac{m\lambda}{2d}\right) = \arcsin\left(m\dfrac{0.125 \text{ nm}}{2(0.282 \text{ nm})}\right) = \arcsin(0.2216m)$.

For $m = 1: \theta = 12.8°$, $m = 2: \theta = 26.3°$, $m = 3: \theta = 41.7°$, and $m = 4: \theta = 62.4°$. No larger m values yield answers.

(b) If the separation $d = \dfrac{a}{\sqrt{2}}$, then $\theta = \arcsin\left(\dfrac{\sqrt{2}m\lambda}{2a}\right) = \arcsin(0.3134m)$.

So for $m = 1$: $\theta = 18.3°$, $m = 2$: $\theta = 38.8°$, and $m = 3$: $\theta = 70.1°$. No larger m values yield answers.

Reflect: In part (b), where d is smaller, the maxima for each m are at larger θ.

*26.63. Set Up: The bright fringes are located by $d\sin\theta = m\lambda$. First-order means $m = 1$. The line density is $1/d$. The largest λ appears at the largest θ. If the $\lambda_1 = 400$ nm line is at θ, then the $\lambda_2 = 700$ nm line is to be at $\theta + 30.0°$.

Solve: $\sin\theta = \dfrac{\lambda_1}{d}$. $\sin(\theta + 30.0°) = \dfrac{\lambda_2}{d}$. $\sin(\theta + 30.0°) = \sin\theta\cos 30.0° + \cos\theta\sin 30.0°$, so $(\sin\theta)(0.866) + (\cos\theta)$

$(0.500) = \dfrac{\lambda_2}{d}$. $\cos\theta = \sqrt{1 - \sin^2\theta} = \sqrt{1 - \dfrac{\lambda_1^2}{d^2}}$, so $0.866\dfrac{\lambda_1}{d} + 0.500\sqrt{1 - \dfrac{\lambda_1^2}{d^2}} = \dfrac{\lambda_2}{d}$. This gives $700 \text{ nm} - (0.866)$

$(400 \text{ nm}) = 0.500\sqrt{d^2 - (400 \text{ nm})^2}$ and $d = 812 \text{ nm} = 8.12 \times 10^{-4}$ mm. The line density is $\dfrac{1}{d} = 1230$ lines/mm.

Reflect: $2\dfrac{\lambda_2}{d} > 1$ and the second-order pattern doesn't include all of the visible spectrum. The first order is the only order that displays the entire visible spectrum.

Solutions to Passage Problems

*26.67. Set Up: If two speakers are connected to a single oscillator it is likely that some students will hear constructive interference and others will hear destructive interference.

Solve: Both speakers must still be connected since some students are still unable to hear a tone. Furthermore, if the volume were simply increased or decreased the points of destructive interference would not change. However, if the phase relationship between the speakers is changed (e.g., the polarity of the wires on one speaker is switched so that that speaker vibrates $180°$ out of phase with the other speaker) then points of destructive and constructive interference will be switched. The correct answer is C.

RELATIVITY

Problems 1, 5, 9, 11, 15, 17, 21, 23, 27, 31, 33, 37, 39, 43, 45, 49, 51, 55, 57

Solutions to Problems

*27.1. Set Up: We will work in the rest frame of the spaceship. In this rest frame the passenger sees her distance from the asteroid increase as the asteroid moves away from her at 0.9c. Simultaneously, she sees her distance from the earth decrease as the earth moves toward her at 0.9c.
Solve: The passenger receives the radio messages simultaneously when her distance from the asteroid is equal to her distance from the earth. Therefore, she deduces that the signal from the earth was sent at a time when she was farther from the earth than she was from the asteroid. Since the signal from the earth must travel a greater distance than the signal from the asteroid, the passenger deduces that the signal from the earth must have been sent before the signal from the asteroid.
Reflect: In the rest frame of an observer on Mars the two signals travel identical distances. Thus, in order to arrive simultaneously at the spaceship, the signals were sent simultaneously from Asteroid 1040A and from the earth.

*27.5. Set Up: As seen by your friend in the earth's frame of reference the round-trip distance is $d_0 = 11.5 \times 10^9$ km. As seen from your frame of reference the round-trip distance is contracted to $d = d_0\sqrt{1 - \frac{u^2}{c^2}}$. Your speed relative to the earth is $u = (45{,}000 \text{ km/h}) \cdot \left(0.2778 \frac{\text{m/s}}{\text{km/h}}\right) = 1.25 \times 10^4$ m/s and we know that $c = 3.00 \times 10^8$ m/s.

Solve: (a) As seen by your friend on the earth the trip takes $t_{\text{earth}} = \frac{d_0}{u} = \frac{11.5 \times 10^9 \text{ km}}{45{,}000 \text{ km/h}} = 2.56 \times 10^5 \text{ h} = 29.2 \text{ y.}$

(b) As seen by you the round-trip distance is $d = d_0\sqrt{1 - \frac{u^2}{c^2}}$, and the round-trip time is $t_{\text{you}} = \frac{d_0}{u}\sqrt{1 - \frac{u^2}{c^2}}$

Thus, the difference in time between your friend's watch and yours is

$$t_{\text{earth}} - t_{\text{you}} = \frac{d_0}{u}\left(1 - \sqrt{1 - \frac{u^2}{c^2}}\right) = \frac{11.5 \times 10^{12} \text{ m}}{1.25 \times 10^4 \text{ m/s}}\left(1 - \sqrt{1 - \left(\frac{1.25 \times 10^4 \text{ m/s}}{3.00 \times 10^8 \text{ ms}}\right)^2}\right) = 0.80 \text{ s.}$$

Reflect: This problem can also be solved using the time dilation formula, as explained in Example 27.2 and the subsequent discussion of the Twin Paradox.

*27.9. Set Up: A clock moving with respect to an observer appears to run more slowly than a clock at rest in the observer's frame. The clock in the spacecraft measurers the proper time Δt_0. $\Delta t = 365 \text{ days} = 8760 \text{ hours.}$

Solve: The clock on the moving spacecraft runs slow and shows the smaller elapsed time. $\Delta t_0 = \Delta t \sqrt{1 - u^2/c^2} =$
$(8760 \text{ h})\sqrt{1 - (4.80 \times 10^6 / 3.00 \times 10^8)^2} = 8758.88 \text{ h}$. The difference in elapsed times is $8760 \text{ h} - 8758.88 \text{ h} = 1.12 \text{ h}$.

*27.11. Set Up: When the meterstick is at rest with respect to you, you measure its length to be 1.000 m, and that is its proper length, l_0. $l = 0.3048 \text{ m}$.

Solve: $l = l_0 \sqrt{1 - u^2/c^2}$ gives $u = c\sqrt{1 - (l/l_0)^2} = c\sqrt{1 - (0.3048/1.00)^2} = 0.9524c = 2.86 \times 10^8 \text{ m/s}$.

*27.15. Set Up: The height measured in the earth's frame is a proper length. The lifetime measured in the muon's frame is the proper time.

Solve: (a) $l_0 = 55.0 \text{ km}$. $l = l_0 \sqrt{1 - u^2/c^2} = (55.0 \text{ km})\sqrt{1 - (0.9860)^2} = 9.17 \text{ km}$.

(b) In the muon's frame its lifetime is $2.20 \, \mu s$, so the distance it travels during its lifetime is

$$(2.20 \times 10^{-6} \text{ s})(0.9860)(3.00 \times 10^8 \text{ m/s}) = 651 \text{ m}.$$

This is $\dfrac{651 \text{ m}}{9.17 \times 10^3 \text{ m}} = 7.1\%$ of its initial height.

(c) $\Delta t = \dfrac{\Delta t_0}{\sqrt{1 - u^2/c^2}} = \dfrac{2.20 \times 10^{-6} \text{ s}}{\sqrt{1 - (0.9860)^2}} = 1.32 \times 10^{-5} \text{ s}$. The distance it travels in the earth's frame during this time is

$(1.32 \times 10^{-5} \text{ s})(0.9860)(3.00 \times 10^8 \text{ m/s}) = 3.90 \text{ km}$. This is $\dfrac{3.90 \text{ km}}{55.0 \text{ km}} = 7.1\%$ of its initial height measured in the earth's frame.

Reflect: There are two equivalent views. In the muon's frame, its distance above the surface of the earth is contracted because the earth is moving relative to the muon. In the earth's frame the lifetime of the muon is dilated due to the motion of the muon relative to the earth.

*27.17. Set Up: Let the frame attached to Arrakis be S and let the frame attached to the imperial spaceship be S'. Let the positive x direction for both frames be from the spaceship toward Arrakis. Then $v' = +0.920c$, $v = +0.360c$ and we want to solve for u.

Solve: $v' = \dfrac{v - u}{1 - uv/c^2}$ gives $u = \dfrac{v - v'}{1 - vv'/c^2} = \dfrac{0.360c - 0.920c}{1 - (0.360)(0.920)} = -0.837c$. The speed of the starship relative to Arrakis is $0.837c$. Since $u < 0$ and the direction from the spaceship toward Arrakis is positive, the spaceship is moving away from Arrakis.

Reflect: The Galilean transformation gives that the spaceship is moving away from Arrakis with a speed of $0.920c - 0.360c = 0.560c$. The Lorentz transformation gives a larger speed.

*27.21. Set Up: The rotation speed can be calculated from the radius of the star and its period: $v = \dfrac{2\pi r}{T}$. The Lorentz transformation for velocity is given by $v' = \dfrac{v - u}{1 - uv/c^2}$.

Solve: (a) $v/c = \dfrac{2\pi r}{cT} = \dfrac{2\pi(10.0 \times 10^3 \text{ m})}{(3.00 \times 10^8 \text{ m/s})(1.80 \times 10^{-3} \text{ s})} = 0.116$; thus we have $v = 0.116c$.

(b) Let the frame S be located at the center of the star, let S' be a frame located at any point on the star's equator, and let P be a point on the equator that is diametrically opposite S'. If the velocity of P relative to S is $v = +0.116c$ then the velocity of S' as seen by S must be $u = -v = -0.116c$. Thus, the velocity of P relative to S' is

$v' = \dfrac{v - u}{1 - uv/c^2} = \dfrac{2v}{1 + (v/c)^2} = \dfrac{2(0.116c)}{1 + (0.116)^2} = 0.229c$.

Reflect: If we ignore relativity, the relative velocity between these two points would be $v - u = 2v = 2(0.116c) = 0.232c,$ which is larger than the correct relativistic result as expected.

*27.23. Set Up: The momentum of a particle has magnitude $p = \dfrac{mv}{\sqrt{1 - v^2/c^2}}$. $\sqrt{1 - v^2/c^2}$ is less than 1, so $p > mv$.

Solve: (a) The problem specifies that $p = 1.010mv$. $1.010 = \dfrac{1}{\sqrt{1 - v^2/c^2}}$.

$$v = c\sqrt{1 - (1/1.010)^2} = 0.140c = 4.20 \times 10^7 \text{ m/s}.$$

(b) $p > mv$; the relativistic value is greater than the nonrelativistic value.

*27.27. Set Up: The relativistic kinetic energy is $K = \left(\dfrac{1}{\sqrt{1 - v^2/c^2}} - 1 \right) mc^2$. For an electron,

$$mc^2 = (9.11 \times 10^{-31} \text{ kg})(3.00 \times 10^8 \text{ m/s})^2 = 8.20 \times 10^{-14} \text{ J}.$$

The nonrelativistic expression for the kinetic energy is $K_{nr} = \frac{1}{2}mv^2$.

Solve: (a) $K_{nr} = \frac{1}{2}(9.11 \times 10^{-31} \text{ kg})(5.00 \times 10^7 \text{ m/s})^2 = 1.14 \times 10^{-15} \text{ J}.$

$$\frac{1}{\sqrt{1 - v^2/c^2}} = \frac{1}{\sqrt{1 - (5.00 \times 10^7 / 3.00 \times 10^8)^2}} = 1.014$$

and $K = (1.014 - 1.00)(8.20 \times 10^{-14} \text{ J}) = 1.15 \times 10^{-15} \text{ J}.$ $\dfrac{K}{K_{nr}} = 1.01.$

(b) $K_{nr} = \frac{1}{2}(9.11 \times 10^{-31} \text{ kg})(2.60 \times 10^8 \text{ m/s})^2 = 3.08 \times 10^{-14} \text{ J}.$

$$\frac{1}{\sqrt{1 - v^2/c^2}} = \frac{1}{\sqrt{1 - (2.60 \times 10^8 / 3.00 \times 10^8)^2}} = 2.00$$

and $K = (2.00 - 1.00)(8.20 \times 10^{-14} \text{ J}) = 8.20 \times 10^{-14} \text{ J}.$ $\dfrac{K}{K_{nr}} = 2.66.$

Reflect: The relativistic value of the kinetic energy, K, approaches the nonrelativistic value, K_{nr}, for small speeds. For large speeds, $K > K_{nr}$ and the ratio K/K_{nr} increases as the speed increases.

*27.31. Set Up: Use Equations 27.22 and 27.24.

Solve: (a) $E = mc^2 + K$, so $E = 4.00mc^2$ means $K = 3.00mc^2 = 4.50 \times 10^{-10} \text{ J}$

(b) $E^2 = (mc^2)^2 + (pc)^2$; $E = 4.00mc^2$, so $15.0(mc^2)^2 = (pc)^2$

$p = \sqrt{15}mc = 1.94 \times 10^{-18} \text{ kg} \cdot \text{m/s}$

(c) $E = mc^2/\sqrt{1 - v^2/c^2}$

$E = 4.00mc^2$ gives $1 - v^2/c^2 = 1/16$ and $v = \sqrt{15/16}\,c = 0.968c$

Reflect: The speed is close to c since the kinetic energy is greater than the rest energy. Nonrelativistic expressions relating E, K, p, and v will be very inaccurate.

*27.33. Set Up: The energy equivalent of mass is $E = mc^2$. $\rho = 7.86 \text{ g/cm}^3 = 7.86 \times 10^3 \text{ kg/m}^3$. For a cube, $V = L^3$.

Solve: (a) $m = \dfrac{E}{c^2} = \dfrac{1.0 \times 10^{20} \text{ J}}{(3.00 \times 10^8 \text{ m/s})^2} = 1.11 \times 10^3 \text{ kg}$

(b) $\rho = \dfrac{m}{V}$ so $V = \dfrac{m}{\rho} = \dfrac{1.11 \times 10^3 \text{ kg}}{7.86 \times 10^3 \text{ kg/m}^3} = 0.141 \text{ m}^3$. $L = V^{1/3} = 0.521 \text{ m} = 52.1 \text{ cm}$

Reflect: Particle/antiparticle annihilation has been observed in the laboratory, but only with small quantities of antimatter.

*27.37. Set Up: If an object has length l_0 when it is at rest relative to the observer making the measurement, then when it is moving at speed u its length along the direction of motion is measured by the observer to be l, where $l = l_0\sqrt{1 - u^2/c^2}$. For the markings on an empire ship to appear circular, $l = b$ when $l_0 = 1.40b$.

Solve: $b = 1.40b\sqrt{1 - u^2/c^2}$. $u = c\sqrt{1 - (1/1.40)^2} = 0.700c = 2.10 \times 10^8$ m/s.

Reflect: The length in the direction of motion is contracted but lengths perpendicular to this direction are unaffected by the motion.

*27.39. Set Up: We will use the equation for time dilation to determine the relative speed of the two frames: $\Delta t = \dfrac{\Delta t_0}{\sqrt{1 - u^2/c^2}}$. Since S is the rest frame of the event we know that $\Delta t_0 = 1.80$ s, while $\Delta t' = 2.35$ s as seen in S'.

Solve: Solving the time dilation equation for u we obtain: $\dfrac{u}{c} = \sqrt{1 - \left(\dfrac{\Delta t_0}{\Delta t'}\right)^2} = \sqrt{1 - \left(\dfrac{1.80\text{ s}}{2.35\text{ s}}\right)^2} = 0.643$. So $u = 0.643c$. For convenience, imagine that both events occur at the origin of the S frame. As seen in the S' frame the origin moves a distance $x' = u\Delta t' = (0.643c)(2.35\text{ s}) = 4.53 \times 10^8$ m during the time between the events—and this is the separation of the events as seen in the S' frame.

Reflect: It is also possible to solve this problem using the Lorentz transformation equations.

*27.43. Set Up: Let S be the lab frame and S' be the frame of the proton that is moving in the $+x$ direction, so $u = +c/2$. The reference frames and moving particles are shown in the figure below. The other proton moves in the $-x$ direction in the lab frame, so $v = -c/2$. A proton has rest mass $m_p = 1.67 \times 10^{-27}$ kg and rest energy $m_p c^2 = 938$ MeV.

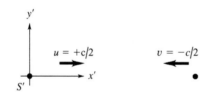

Solve: (a) $v' = \dfrac{v - u}{1 - uv/c^2} = \dfrac{-c/2 - c/2}{1 - (c/2)(-c/2)/c^2} = -\dfrac{4c}{5}$

The speed of each proton relative to the other is $\frac{4}{5}c$.

(b) In nonrelativistic mechanics the speeds just add and the speed of each relative to the other is c.

(c) $K = \dfrac{mc^2}{\sqrt{1 - v^2/c^2}} - mc^2$

(i) Relative to the lab frame each proton has speed $v = c/2$. The total kinetic energy of each proton is

$$K = \frac{938 \text{ MeV}}{\sqrt{1-\left(\frac{1}{2}\right)^2}} - (938 \text{ MeV}) = 145 \text{ MeV}.$$

(ii) In its rest frame one proton has zero speed and zero kinetic energy and the other has speed $\frac{4}{5}c$. In this frame the kinetic energy of the moving proton is

$$K = \frac{938 \text{ MeV}}{\sqrt{1-\left(\frac{4}{5}\right)^2}} - (938 \text{ MeV}) = 625 \text{ MeV}$$

(d) (i) Each proton has speed $v = c/2$ and kinetic energy $K = \frac{1}{2}mv^2 = \left(\frac{1}{2}m\right)(c/2)^2 = \frac{mc^2}{8} = \frac{938 \text{ MeV}}{8} = 117 \text{ MeV}$

(ii) One proton has speed $v = 0$ and the other has speed c. The kinetic energy of the moving proton is

$$K = \frac{1}{2}mc^2 = \frac{938 \text{ MeV}}{2} = 469 \text{ MeV}$$

Reflect: The relativistic expression for K gives a larger value than the nonrelativistic expression. The kinetic energy of the system is different in different frames.

*27.45. Set Up: The energy that enters the ice is $E = Q = mL_f$. The mass increase is the mass equivalent of this amount of energy, $m = \dfrac{E}{c^2}$.

Solve: $m = \dfrac{E}{c^2} = \dfrac{mL_f}{c^2} = \dfrac{(4.00 \text{ kg})(3.34 \times 10^5 \text{ J/kg})}{(3.00 \times 10^8 \text{ m/s})^2} = 1.48 \times 10^{-11} \text{ kg}$

*27.49. Set Up: The energy released is $E = (\Delta m)c^2$. $\Delta m = \left(\dfrac{1}{10^4}\right)(8.00 \text{ kg})$. $P_{av} = \dfrac{E}{t}$. The change in gravitational potential energy is $mg\,\Delta y$.

Solve: (a) $E = (\Delta m)c^2 = \left(\dfrac{1}{10^4}\right)(8.00 \text{ kg})(3.00 \times 10^8 \text{ m/s})^2 = 7.20 \times 10^{13} \text{ J}$

(b) $P_{av} = \dfrac{E}{t} = \dfrac{7.20 \times 10^{13} \text{ J}}{4.00 \times 10^{-6} \text{ s}} = 1.80 \times 10^{19} \text{ W}$

(c) $E = \Delta U = mg\Delta y$. $m = \dfrac{E}{g\Delta y} = \dfrac{7.20 \times 10^{13} \text{ J}}{(9.80 \text{ m/s}^2)(1.00 \times 10^3 \text{ m})} = 7.35 \times 10^9 \text{ kg}$

*27.51. Set Up: In the rest frame of the spaceship the trip takes $\Delta t_0 = 3.35$ years. As seen from the earth the trip takes $\Delta t = \dfrac{\Delta t_0}{\sqrt{1-u^2/c^2}}$. The distance to the star is 7.11 ly and the speed of light is $c = 1$ ly/y. Let u be the speed of the spaceship (in ly/y) as seen from the earth.

Solve: (a) The time for the trip as seen from the earth will be $\Delta t = \dfrac{7.11 \text{ ly}}{u} = \dfrac{3.35 \text{ y}}{\sqrt{1-u^2/c^2}}$. Solving for u we obtain

$\left(\dfrac{7.11 \text{ ly}}{3.35 \text{ y}}\right)^2 (1-u^2/c^2) = u^2$, which reduces to $u = \left(\dfrac{7.11 \text{ ly}}{3.35 \text{ y}}\right) \cdot \dfrac{1}{\sqrt{1+\left(\dfrac{7.11 \text{ ly}}{(3.35 \text{ y})(1 \text{ ly/y})}\right)^2}} = 0.905 \text{ ly/y} = 0.905c$. Thus, as

seen from the earth, the trip takes $\dfrac{7.11 \text{ ly}}{u} = \dfrac{7.11 \text{ ly}}{0.905 \text{ ly/y}} = 7.86 \text{ years.}$

(b) According to the passengers the distance is given by $x' = u\Delta t_0 = (0.905 \text{ ly/y})(3.35 \text{ y}) = 3.03 \text{ ly.}$

Reflect: The distance to the star as seen by the passengers could also be calculated by using the length contraction formula: $l = l_0\sqrt{1 - u^2/c^2}$.

Solutions to Passage Problems

*27.55. Set Up: We will use the formula for length contraction: $l = l_0\sqrt{1 - u^2/c_{alt}^2}$.

Solve: $l = l_0\sqrt{1 - u^2/c_{alt}^2} = (60\text{ m})\sqrt{1 - \left(\dfrac{180\text{ m/s}}{300\text{ m/s}}\right)^2} = 48$ m. The correct answer is C.

*27.57. Set Up: The rest energy in our universe is given by $E = mc^2$ and the rest energy in the alternate universe is given by $E_{alt} = mc_{alt}^2$.

Solve: Comparing the rest energy of the electron in our universe to the alternate universe we have: $\dfrac{E_{alt}}{E} = \dfrac{mc_{alt}^2}{mc^2}$,

which gives $E_{alt} = E \cdot \left(\dfrac{c_{alt}}{c}\right)^2 = (8.2\times10^{-14}\text{ J})(10^{-6})^2 = 8.2\times10^{-26}$ J. The correct answer is B.

PHOTONS, ELECTRONS, AND ATOMS

Problems 1, 3, 7, 11, 15, 17, 21, 23, 27, 29, 33, 37, 39, 43, 45, 47, 51, 53, 57, 59, 63, 67, 69, 71, 73

Solutions to Problems

*28.1. Set Up: $c = f\lambda$ relates frequency and wavelength and $E = hf$ relates energy and frequency for a photon. $c = 3.00 \times 10^8$ m/s. $1\,\text{eV} = 1.60 \times 10^{-16}$ J.

Solve: (a) $f = \dfrac{c}{\lambda} = \dfrac{3.00 \times 10^8\ \text{m/s}}{505 \times 10^{-9}\ \text{m}} = 5.94 \times 10^{14}$ Hz

(b) $E = hf = (6.626 \times 10^{-34}\ \text{J} \cdot \text{s})(5.94 \times 10^{14}\ \text{Hz}) = 3.94 \times 10^{-19}$ J $= 2.46$ eV

(c) $K = \frac{1}{2}mv^2$ so $v = \sqrt{\dfrac{2K}{m}} = \sqrt{\dfrac{2(3.94 \times 10^{-19}\ \text{J})}{9.5 \times 10^{-15}\ \text{kg}}} = 9.1$ mm/s

*28.3. Set Up: $P_{\text{av}} = \dfrac{\text{energy}}{t}$. $1\,\text{eV} = 1.60 \times 10^{-19}$ J. For a photon, $E = hf = \dfrac{hc}{\lambda}$. $h = 6.63 \times 10^{-34}$ J $\cdot$ s.

Solve: (a) energy $= P_{\text{av}}t = (0.600\ \text{W})(20.0 \times 10^{-3}\ \text{s}) = 1.20 \times 10^{-2}$ J $= 7.5 \times 10^{16}$ eV

(b) $E = \dfrac{hc}{\lambda} = \dfrac{(6.63 \times 10^{-34}\ \text{J} \cdot \text{s})(3.00 \times 10^8\ \text{m/s})}{652 \times 10^{-9}\ \text{m}} = 3.05 \times 10^{-19}$ J $= 1.91$ eV

(c) The number of photons is the total energy in a pulse divided by the energy of one photon:

$$\frac{1.20 \times 10^{-2}\ \text{J}}{3.05 \times 10^{-19}\ \text{J/photon}} = 3.93 \times 10^{16}\ \text{photons.}$$

Reflect: The number of photons in each pulse is very large.

*28.7. Set Up: $\frac{1}{2}mv_{\text{max}}{}^2 = hf - \phi$

Solve: $f \to f_0$, the threshold frequency, when $\frac{1}{2}mv_{\text{max}}{}^2 \to 0$. $f_0 = \phi/h$.

*28.11. Set Up: $\frac{1}{2}mv_{\text{max}}{}^2 = hf - \phi$

Solve: Use the data for 400.0 nm to calculate ϕ:

$$\phi = \frac{hc}{\lambda} - \frac{1}{2}mv_{\text{max}}{}^2 = \frac{(4.136 \times 10^{-15}\ \text{eV} \cdot \text{s})(3.00 \times 10^8\ \text{m/s})}{400.0 \times 10^{-9}\ \text{m}} - 1.10\ \text{eV} = 3.10\ \text{eV} - 1.10\ \text{eV} = 2.00\ \text{eV.}$$

Then for 300.0 nm,

$$\frac{1}{2}mv_{\text{max}}^2 = hf - \phi = \frac{hc}{\lambda} - \phi = \frac{(4.136\times10^{-15}\text{ eV}\cdot\text{s})(3.00\times10^8\text{ m/s})}{300.0\times10^{-9}\text{ m}} - 2.00\text{ eV}.$$

$$\frac{1}{2}mv_{\text{max}}^2 = 4.14\text{ eV} - 2.00\text{ eV} = 2.14\text{ eV}.$$

Reflect: When the wavelength decreases the energy of the photons increases and the photoelectrons have a larger minimum kinetic energy.

*28.15. Set Up: We will apply the equation $K_{\text{max}} = hf - \phi$. For electrons to be ejected we must have $hf - \phi \geq 0$, which gives a threshold frequency of $f_0 = \phi/h$. From the textbook we find the work function of silicon to be $\phi = 4.8\text{ eV}$ and $h = 4.14\times10^{-15}\text{ eV}\cdot\text{s}$.

Solve: (a) The threshold frequency is given by $f_0 = \phi/h$, which corresponds to a maximum wavelength of

$$\lambda_0 = \frac{c}{f_0} = \frac{ch}{\phi} = \frac{(3.00\times10^8\text{ m/s})(4.14\times10^{-15}\text{ eV}\cdot\text{s})}{4.8\text{ eV}} = 259\times10^{-9}\text{ m} = 259\text{ nm}.$$

(b) The maximum kinetic energy occurs at the shortest wavelength, which is 145 nm. The maximum kinetic energy of the electrons just after they are ejected is $K_{\text{max}} = hf - \phi = (4.14\times10^{-15}\text{ eV}\cdot\text{s})(3.00\times10^8\text{ m/s})/(145\times10^{-9}\text{ m}) - 4.8\text{ eV} = \underline{3.77}\text{ eV}$. As the electrons move to the anode they experience a reversing potential of 3.50 volts, which implies that they lose 3.50 eV of kinetic energy. Thus, the electrons reach the anode with a maximum kinetic energy of $\underline{3.77}\text{ eV} - 3.50\text{ eV} = 0.\underline{27}\text{ eV}$.

Reflect: It would take a potential of 3.77 volts to stop the electrons ejected from light with a wavelength of 145 nm.

*28.17. Set Up: Balmer's formula is $\dfrac{1}{\lambda} = R\left(\dfrac{1}{2^2} - \dfrac{1}{n^2}\right)$, where $R = 1.097\times10^7\text{ m}^{-1}$. The H_γ line corresponds to $n = 5$. $c = f\lambda$. $E = hf$.

Solve: (a) $\dfrac{1}{\lambda} = R\left(\dfrac{1}{4} - \dfrac{1}{25}\right) = \dfrac{21}{100}R.$ $\lambda = \dfrac{100}{21R} = \dfrac{100}{21(1.097\times10^7\text{ m}^{-1})} = 434\text{ nm}$

(b) $f = \dfrac{c}{\lambda} = \dfrac{3.00\times10^8\text{ m/s}}{434\times10^{-9}\text{ m}} = 6.91\times10^{14}\text{ Hz}$

(c) $E = hf = (4.136\times10^{-15}\text{ eV}\cdot\text{s})(6.91\times10^{14}\text{ Hz}) = 2.86\text{ eV}$

*28.21. Set Up: The wavelength of the photon is related to the transition energy $E_i - E_f$ of the atom by $E_i - E_f = \dfrac{hc}{\lambda}$ where $hc = 1.240\times10^{-6}\text{ eV}\cdot\text{m}$.

Solve: (a) The minimum energy to ionize an atom is when the upper state in the transition has $E = 0$, so $E_1 = -17.50\text{ eV}$. For $n = 5 \to n = 1$, $\lambda = 73.86\text{ nm}$ and $E_5 - E_1 = \dfrac{1.240\times10^{-6}\text{ eV}\cdot\text{m}}{73.86\times10^{-9}\text{ m}} = 16.79\text{ eV}$. $E_5 = -17.50\text{ eV} + 16.79\text{ eV} = -0.71\text{ eV}$. For $n = 4 \to n = 1$, $\lambda = 75.63\text{ nm}$ and $E_4 = -1.10\text{ eV}$. For $n = 3 \to n = 1$, $\lambda = 79.76\text{ nm}$ and $E_3 = -1.95\text{ eV}$. For $n = 2 \to n = 1$, $\lambda = 94.54\text{ nm}$ and $E_2 = -4.38\text{ eV}$.

(b) $E_i - E_f = E_4 - E_2 = -1.10\text{ eV} - (-4.38\text{ eV}) = 3.28\text{ eV}$ and

$$\lambda = \frac{hc}{E_i - E_f} = \frac{1.240\times10^{-6}\text{ eV}\cdot\text{m}}{3.28\text{ eV}} = 378\text{ nm}$$

Reflect: The $n = 4 \to n = 2$ transition energy is smaller than the $n = 4 \to n = 1$ transition energy so the wavelength is longer. In fact, this wavelength is longer than for any transition that ends in the $n = 1$ state.

*28.23. Set Up: Section 28.3 calculates that $r_1 = 0.5293 \times 10^{-10}$ m. $r_n = n^2 r_1$ and $E_n = -\dfrac{13.6 \text{ eV}}{n^2}$

Solve: (a) $v_1 = \dfrac{e^2}{2\epsilon_0 h} = \dfrac{(1.602 \times 10^{-19} \text{ C})^2}{2(8.854 \times 10^{-12} \text{ C}^2/\text{N} \cdot \text{m}^2)(6.626 \times 10^{-34} \text{ J} \cdot \text{s})} = 2.19 \times 10^6$ m/s

$v_n = \dfrac{v_1}{n}$ so $v_2 = 1.09 \times 10^6$ m/s and $v_3 = 7.29 \times 10^5$ m/s

(b) $r_n = n^2 r_1$, so $r_1 = 0.5293 \times 10^{-10}$ m, $r_2 = 2.117 \times 10^{-10}$ m and $r_3 = 4.764 \times 10^{-10}$ m

(c) $E_1 = -13.6$ eV, $E_2 = -3.40$ eV, $E_3 = -1.51$ eV

Reflect: As n increases, r increases, v decreases and E becomes less negative.

*28.27. Set Up: In the expression for hydrogen replace e^2 by Ze^2. For hydrogen, $E_n = -\dfrac{1}{\epsilon_0^2} \dfrac{me^4}{8n^2 h^2} = -\dfrac{13.6 \text{ eV}}{n^2}$,

so for a one-electron ion with atomic number Z, $E_n = -(13.6 \text{ eV}) \dfrac{Z^2}{n^2}$. For Be^{3+}, $Z = 4$.

Solve: (a) $E_1 = -(13.6 \text{ eV}) \dfrac{4^2}{1^2} = -218$ eV. This is a factor of 16 times the hydrogen atom value.

(b) The ionization energy is $E_\infty - E_1 = +218$ eV. This is a factor of 16 times the hydrogen atom value.

(c) $|\Delta E| = \dfrac{hc}{\lambda}$. $|\Delta E| = E_2 - E_1 = -(13.6 \text{ eV})(4^2) \left(\dfrac{1}{2^2} - \dfrac{1}{1^2} \right) = 163.2$ eV.

$$\lambda = \dfrac{hc}{|\Delta E|} = \dfrac{(4.136 \times 10^{-15} \text{ eV} \cdot \text{s})(3.00 \times 10^8 \text{ m/s})}{163.2 \text{ eV}} = 7.60 \text{ nm}.$$

This is the hydrogen value divided by 16.

(d) For the hydrogen atom, $r_{\text{hyd}} = \epsilon_0 \dfrac{n^2 h^2}{\pi m e^2}$. For a one-electron ion with atomic number Z, $r_Z = \epsilon_0 \dfrac{n^2 h^2}{\pi m Z e^2} = r_{\text{hyd}}/Z$.

The radius for Be^{3+} is $\frac{1}{4}$ times the hydrogen atom value.

Reflect: The level energies are proportional to Z^2, the wavelength for a specific transition is proportional to $1/Z^2$, and the radius of a particular orbit is proportional to $1/Z$.

*28.29. Set Up: The energy of each photon is $E = hf = \dfrac{hc}{\lambda}$. The power is the total energy per second and the total energy E_{tot} is the number of photons N times the energy E of each photon.

Solve: $P = \dfrac{E_{\text{total}}}{t} = \dfrac{Nhf}{t} = \dfrac{Nhc}{t\lambda} = \dfrac{(4.50 \times 10^{17})(6.63 \times 10^{-34} \text{ J} \cdot \text{s})(3.00 \times 10^8 \text{ m/s})}{(30 \text{ s})(645 \times 10^{-9} \text{ m})} = 4.63 \times 10^{-3} \text{ W} = 4.63 \text{ mW}.$

Reflect: Despite the very large number of photons over this short time, the average power output is relatively small.

*28.33. Set Up: $\dfrac{hc}{\lambda} = eV$, where λ is the wavelength of the x-ray and V is the accelerating voltage.

Solve: (a) $V = \dfrac{hc}{e\lambda} = \dfrac{(6.63 \times 10^{-34} \text{ J} \cdot \text{s})(3.00 \times 10^8 \text{ m/s})}{(1.60 \times 10^{-19} \text{ C})(0.150 \times 10^{-9} \text{ m})} = 8.29$ kV

(b) $\lambda = \dfrac{hc}{eV} = \dfrac{(6.63 \times 10^{-34} \text{ J} \cdot \text{s})(3.00 \times 10^8 \text{ m/s})}{(1.60 \times 10^{-19} \text{ C})(30.0 \times 10^3 \text{ V})} = 4.14 \times 10^{-11}$ m = 0.0414 nm

(c) No. A proton has the same magnitude of charge as an electron and therefore gains the same amount of kinetic energy when accelerated by the same magnitude of potential difference.

*28.37. Set Up: Apply Equation 28.19: $\lambda' - \lambda = \dfrac{h}{mc}(1 - \cos\phi) = \lambda_c(1 - \cos\phi)$. Solve for λ': $\lambda' = \lambda + \lambda_c(1 - \cos\phi)$. The largest λ' corresponds to $\phi = 180°$, so $\cos\phi = -1$.

Solve: $\lambda' = \lambda + 2\lambda_C = 0.0665 \times 10^{-9}$ m $+ 2(2.426 \times 10^{-12}$ m$) = 7.135 \times 10^{-11}$ m $= 0.0714$ nm. This wavelength occurs at a scattering angle of $\phi = 180°$.

Reflect: The incident photon transfers some of its energy and momentum to the electron from which it scatters. Since the photon loses energy its wavelength increases, $\lambda' > \lambda$.

*28.39. Set Up: An electron or a proton accelerated through a potential $V_{AC} = 4.00$ kV has a maximum kinetic energy of $K = 4.00$ keV. We will assume that this is nearly the maximum energy of the ejected photon, which is also given by $E_{photon} = hf_{max} = \dfrac{hc}{\lambda_{min}}$.

Solve: Solving for the minimum wavelength of the ejected photons we obtain

$$\lambda_{min} = \frac{hc}{E_{photon}} = \frac{(4.14 \times 10^{-15} \text{ eV} \cdot \text{s})(3.00 \times 10^8 \text{ m/s})}{4.00 \times 10^3 \text{ eV}} = 3.11 \times 10^{-10} \text{ m}$$

Reflect: In this case the energy of the ejected photons is the same for both the electron and proton beam. However, electron beams are much more easily produced and accelerated than proton beams.

*28.43. Set Up: The de Broglie wavelength is $\lambda = \dfrac{h}{p} = \dfrac{h}{mv}$. In the Bohr model, $mvr_n = n(h/2\pi)$, so $mv = nh/(2\pi r_n)$. Combine these two expressions and obtain an equation for λ in terms of n. Then $\lambda = h\left(\dfrac{2\pi r_n}{nh}\right) = \dfrac{2\pi r_n}{n}$.

Solve: **(a)** For $n = 1$, $\lambda = 2\pi r_1$ with $r_1 = a_0 = 0.529 \times 10^{-10}$ m, so $\lambda = 2\pi(0.529 \times 10^{-10}$ m$) = 3.32 \times 10^{-10}$ m
$\lambda = 2\pi r_1$; the de Broglie wavelength equals the circumference of the orbit.
(b) For $n = 4$, $\lambda = 2\pi r_4/4$.
$r_n = n^2 a_0$ so $r_4 = 16a_0$.
$\lambda = 2\pi(16a_0)/4 = 4(2\pi a_0) = 4(3.32 \times 10^{-10}$ m$) = 1.33 \times 10^{-9}$ m
$\lambda = 2\pi r_4/4$; the de Broglie wavelength is $\dfrac{1}{n} = \dfrac{1}{4}$ times the circumference of the orbit.

Reflect: As n increases the momentum of the electron increases and its de Broglie wavelength decreases. For any n, the circumference of the orbits equals an integer number of de Broglie wavelengths.

*28.45. Set Up: Both for particles with mass (electrons) and for massless particles (photons) the wavelength is related to the momentum p by $\lambda = \dfrac{h}{p}$. But for each type of particle there is a different expression that relates the energy E and momentum p. For an electron $E = \frac{1}{2}mv^2 = \dfrac{p^2}{2m}$ but for a photon $E = pc$.

Solve: *photon* $p = \dfrac{E}{c}$ and $p = \dfrac{h}{\lambda}$ so $\dfrac{h}{\lambda} = \dfrac{E}{c}$ and $\lambda = \dfrac{hc}{E} = \dfrac{1.24 \times 10^{-6} \text{ eV} \cdot \text{m}}{25 \text{ eV}} = 49.6$ nm

electron $p = \sqrt{2mE} = \sqrt{2(9.11 \times 10^{-31} \text{ kg})(25 \text{ eV})(1.60 \times 10^{-19} \text{ J/eV})} = 2.70 \times 10^{-24}$ kg·m/s

$$\lambda = \frac{h}{p} = \frac{6.63 \times 10^{-34} \text{ J} \cdot \text{s}}{2.70 \times 10^{-24} \text{ kg} \cdot \text{m/s}} = 0.245 \text{ nm}$$

Reflect: The wavelengths are quite different. For the electron $\lambda = \dfrac{h}{\sqrt{2mE}}$ and for the photon $\lambda = \dfrac{hc}{E}$ so for an

electron λ is proportional to $E^{-1/2}$ and for a photon λ is proportional to E^{-1}. It is *incorrect* to say $p = \dfrac{E}{c}$ for a

particle such as an electron that has mass; the correct relation is $p = \dfrac{\sqrt{E^2 - (mc^2)^2}}{c}$.

*28.47. Set Up: The wavelength λ of the photon is related to the transition energy E of the atom by $E = \dfrac{hc}{\lambda}$.

$\Delta E \, \Delta t \geq \dfrac{h}{2\pi}$. The minimum uncertainty in energy is $\Delta E = \dfrac{h}{2\pi \, \Delta t}$.

Solve: **(a)** The photon energy equals the transition energy of the atom, 3.50 eV.

$$\lambda = \frac{hc}{E} = \frac{(4.136 \times 10^{-15} \text{ eV} \cdot \text{s})(3.00 \times 10^{8} \text{ m/s})}{3.50 \text{ eV}} = 355 \text{ nm}.$$

(b) $\Delta E = \dfrac{6.63 \times 10^{-34} \text{ J} \cdot \text{s}}{2\pi(4.0 \times 10^{-6} \text{ s})} = 2.69 \times 10^{-29} \text{ J} = 1.6 \times 10^{-10} \text{ eV}$

*28.51. Set Up: We can use the de Broglie relation to find the momentum corresponding to the given wavelength:

$p = \dfrac{h}{\lambda}$. The required accelerating potential can be found from $eV = K = \dfrac{p^2}{2m}$.

Solve: **(a)** $eV = K = \dfrac{p^2}{2m} = \dfrac{(h/\lambda)^2}{2m}$. Solving for V we obtain

$$V = \frac{(h/\lambda)^2}{2me} = \frac{(6.63 \times 10^{-34} \text{ J} \cdot \text{s})^2}{2(6.00 \times 10^{-11} \text{ m})^2(9.11 \times 10^{-31} \text{ kg})(1.60 \times 10^{-19} \text{ C})} = 419 \text{ V}.$$

(b) We repeat our calculation for a proton—noting that only the mass changes:

$$V = \frac{(h/\lambda)^2}{2me} = \frac{(6.63 \times 10^{-34} \text{ J} \cdot \text{s})^2}{2(6.00 \times 10^{-11} \text{ m})^2(1.67 \times 10^{-27} \text{ kg})(1.60 \times 10^{-19} \text{ C})} = 0.228 \text{ V}.$$

Reflect: We could also calculate the answer to part (b) by recognizing that the voltage is reduced by the ratio of the

particle masses, $(419 \text{ V}) \dfrac{9.11 \times 10^{-31} \text{ kg}}{1.67 \times 10^{-27} \text{ kg}} = 0.229 \text{ V}.$

*28.53. Set Up: The energy of a photon is $E = hf = \dfrac{hc}{\lambda}$. 1 mole $= 6.02 \times 10^{23}$ molecules.

Solve: **(a)** The dissociation energy of one AgBr molecule is

$$E = \frac{1.00 \times 10^{5} \text{ J/mol}}{6.02 \times 10^{23} \text{ molecules/mol}} = 1.66 \times 10^{-19} \text{ J} = 1.04 \text{ eV}.$$

The photon energy is 1.04 eV.

(b) $\lambda = \dfrac{hc}{E} = \dfrac{(4.136 \times 10^{-15} \text{ eV} \cdot \text{s})(3.00 \times 10^{8} \text{ m/s})}{1.04 \text{ eV}} = 1.19 \times 10^{-6} \text{ m} = 1.19 \, \mu\text{m}$

(c) $f = \dfrac{c}{\lambda} = \dfrac{3.00 \times 10^{8} \text{ m/s}}{1.19 \times 10^{-6} \text{ m}} = 2.52 \times 10^{14} \text{ Hz}$

(d) For $f = 100 \text{ MHz} = 1.00 \times 10^{8} \text{ Hz}$. $E = hf = (4.136 \times 10^{-15} \text{ eV} \cdot \text{s})(1.00 \times 10^{8} \text{ Hz}) = 4.1 \times 10^{-7} \text{ eV}$. The radio station produces a large number of photons but they individually have far too little energy to dissociate a AgBr molecule.

Reflect: The frequency or wavelength of the radiation determines the energy of each photon. For fixed frequency, increasing the power output of the source increases the number of photons emitted per second but doesn't change the energy of each photon.

*28.57. Set Up: The H_α line in the Balmer series corresponds to the $n = 3$ to $n = 2$ transition. $E_n = -\dfrac{13.6 \text{ eV}}{n^2}$.

$\dfrac{hc}{\lambda} = \Delta E$.

Solve: (a) The atom must be given an amount of energy

$$E_3 - E_1 = -(13.6 \text{ eV})\left(\frac{1}{3^2} - \frac{1}{1^2}\right) = 12.1 \text{ eV}.$$

(b) There are three possible transitions. $n = 3 \rightarrow n = 1$: $\Delta E = 12.1 \text{ eV}$ and $\lambda = \dfrac{hc}{\Delta E} = 103 \text{ nm}$; $n = 3 \rightarrow n = 2$:

$\Delta E = -(13.6 \text{ eV})\left(\dfrac{1}{3^2} - \dfrac{1}{2^2}\right) = 1.89 \text{ eV}$ and $\lambda = 657 \text{ nm}$; $n = 2 \rightarrow n = 1$: $\Delta E = -(13.6 \text{ eV})\left(\dfrac{1}{2^2} - \dfrac{1}{1^2}\right) = 10.2 \text{ eV}$ and

$\lambda = 122 \text{ nm}$.

*28.59. Set Up: $\lambda' = \lambda + \dfrac{h}{mc}(1 - \cos\phi)$, with $\dfrac{h}{mc} = 2.426 \times 10^{-12} \text{ m}$. Conservation of energy and conservation of

momentum apply to the collision. Momentum is a vector and each of its components is conserved. A photon has

energy $\dfrac{hc}{\lambda}$ and momentum $\dfrac{h}{\lambda}$. $\phi = 180°$.

Solve: (a) $\lambda' = 0.1800 \times 10^{-9} \text{ m} + (2.426 \times 10^{-12} \text{ m})(1 - \cos 180°) = 0.1849 \text{ nm}$

(b) Energy conservation gives $\dfrac{hc}{\lambda} = \dfrac{hc}{\lambda'} + K_e$, where K_e is the final kinetic energy of the electron. $K_e =$

$hc\left(\dfrac{1}{\lambda} - \dfrac{1}{\lambda'}\right) = hc\left(\dfrac{\lambda' - \lambda}{\lambda'\lambda}\right)$.

$$K_e = (6.63 \times 10^{-34} \text{ J} \cdot \text{s})(3.00 \times 10^8 \text{ m/s})\frac{0.0049 \times 10^{-9} \text{ m}}{(0.1800 \times 10^{-9} \text{ m})(0.1849 \times 10^{-9} \text{ m})} = 2.93 \times 10^{-17} \text{ J} = 183 \text{ eV}$$

(c) $K_e = \frac{1}{2}mv^2$, so $v = \sqrt{\dfrac{2K_e}{m}} = \sqrt{\dfrac{2(2.93 \times 10^{-17} \text{ J})}{9.11 \times 10^{-31} \text{ kg}}} = 8.0 \times 10^6 \text{ m/s}$. $v \ll c$, so it is not necessary to use the

relativistic kinetic energy relationship.

Reflect: We can also find v from conservation of momentum. Let $+x$ be the direction the photon is traveling

initially. Then conservation of the x component of momentum gives $\dfrac{h}{\lambda} = -\dfrac{h}{\lambda'} + mv$.

$$v = \frac{h}{m}\left(\frac{1}{\lambda} + \frac{1}{\lambda'}\right) = \frac{h}{m}\left(\frac{\lambda + \lambda'}{\lambda\lambda'}\right) = \left(\frac{6.63 \times 10^{-34} \text{ J} \cdot \text{s}}{9.11 \times 10^{-31} \text{ kg}}\right)\left(\frac{0.1849 \times 10^{-9} \text{ m} + 0.1800 \times 10^{-9} \text{ m}}{\left[0.1849 \times 10^{-9} \text{ m}\right]\left[0.1800 \times 10^{-9} \text{ m}\right]}\right).$$

$v = 8.0 \times 10^6 \text{ m/s}$, which is the same result as computed from energy conservation.

*28.63. Set Up: For a photon, $E = \dfrac{hc}{\lambda}$. $T = 27°\text{C} = 300 \text{ K}$. $k = 1.38 \times 10^{-23} \text{ J/K}$.

Solve: $E = \frac{3}{2}(1.38 \times 10^{-23} \text{ J/K})(300 \text{ K}) = 6.21 \times 10^{-21} \text{ J}$.

$$\lambda = \frac{hc}{E} = \frac{(6.63 \times 10^{-34} \text{ J} \cdot \text{s})(3.00 \times 10^8 \text{ m/s})}{6.21 \times 10^{-21} \text{ J}} = 3.20 \times 10^{-5} \text{ m} = 32.0 \, \mu\text{m}.$$

Reflect: This photon is in the infrared region of the electromagnetic spectrum.

*28.67. Set Up: The energy of a photon is given by $E = hf = \dfrac{hc}{\lambda}$. The kinetic energy of an electron accelerated

through a potential difference ΔV is $K = \dfrac{p^2}{2m} = -q\Delta V = e\Delta V$.

Solve: (a) $E = hc/\lambda = 12$ eV

(b) Find E for an electron with $\lambda = 0.10 \times 10^{-6}$ m. $\lambda = h/p$ so $p = h/\lambda = 6.626 \times 10^{-27}$ kg·m/s.

$E = p^2/(2m) = 1.5 \times 10^{-4}$ eV. $E = e\Delta V$ so $\Delta V = 1.5 \times 10^{-4}$ V

$v = p/m = (6.626 \times 10^{-27}$ kg·m/s$)/(9.109 \times 10^{-31}$ kg$) = 7.3 \times 10^{3}$ m/s

(c) The required wavelength is the same so we need the same momentum. $E = p^2/(2m)$ but now $m = 1.673 \times 10^{-27}$ kg

so $E = 8.2 \times 10^{-8}$ eV and $|\Delta V| = 8.2 \times 10^{-8}$ V.

$v = p/m = (6.626 \times 10^{-27}$ kg·m/s$)/(1.673 \times 10^{-27}$ kg$) = 4.0$ m/s.

Reflect: The voltage required to accelerate the proton could also be found by noting that, for a given wavelength (momentum), the ratio of the required accelerating voltages is the inverse ratio of the masses: $V_p/V_e = m_e/m_p$, thus

$V_p = (1.5 \times 10^{-4}$ V$)(9.109 \times 10^{-31}$ kg$)/(1.673 \times 10^{-27}$ kg$) = 8.2 \times 10^{-8}$ V.

*28.69. Set Up: $\Delta E\, \Delta t \geq \dfrac{h}{2\pi}$. Take the minimum uncertainty product, so $\Delta E = \dfrac{h}{2\pi\, \Delta t}$, with $\Delta t = 8.4 \times 10^{-17}$ s.

$m = 264m_e$. $\Delta m = \dfrac{\Delta E}{c^2}$.

Solve: $\Delta E = \dfrac{6.63 \times 10^{-34}\text{ J·s}}{2\pi(8.4 \times 10^{-17}\text{ s})} = 1.26 \times 10^{-18}$ J.

$$\Delta m = \frac{1.26 \times 10^{-18}\text{ J}}{(3.00 \times 10^{8}\text{ m/s})^2} = 1.4 \times 10^{-35}\text{ kg.}$$

$$\frac{\Delta m}{m} = \frac{1.4 \times 10^{-35}\text{ kg}}{(264)(9.11 \times 10^{-31}\text{ kg})} = 5.8 \times 10^{-8}$$

Solutions to Passage Problems

*28.71. Set Up: The treatment dosage is $\dfrac{70\text{ Gy}}{35\text{ days}} = \dfrac{2\text{ Gy}}{\text{day}} = \dfrac{2\text{ J/kg}}{\text{day}}$ of absorbed radiation. We know that there are

10^8 cells/cm^3 and the density of the tumor is 1 g/cm^3 = 10^{-3} kg/cm^3. We are told that 1 J = 6×10^{18} eV.

Solve: The absorbed energy per cell in one day is $\dfrac{2\text{ J}}{\text{kg}} \cdot \dfrac{10^{-3}\text{ kg}}{\text{cm}^3} \cdot \dfrac{1\text{ cm}^3}{10^{8}\text{ cells}} \cdot \dfrac{6 \times 10^{18}\text{ eV}}{1\text{ J}} = 1.20 \times 10^{8}$ eV/cell. Thus, in

one day the average dosage is 120 MeV per cell. The correct answer is B.

*28.73. Set Up: We are told that each ionization in the tissue requires about 40 eV and we know (from Problem 71) that each cell receives an average of 120 MeV each day.

Solve: The number of ionizations in a day is roughly $(120 \times 10^{6}$ eV$)\dfrac{(1\text{ ionization})}{(40\text{ eV})} = 3 \times 10^{6}$ ionizations. The correct

answer is D.

ATOMS, MOLECULES, AND SOLIDS

Problems 1, 3, 7, 9, 13, 15, 21, 23, 25, 29, 33, 37, 39

Solutions to Problems

*29.1. Set Up: Equation 29.1 relates the magnitude of the orbital angular momentum L to the quantum number l:

$$L = \sqrt{l(l+1)}\,\frac{h}{2\pi}, \quad l = 0,\ 1,\ 2,\ \ldots$$

Solve: $l(l+1) = \left(2\pi\dfrac{L}{h}\right)^2 = \left(2\pi\dfrac{4.716\times10^{-34}\ \mathrm{kg\cdot m^2/s}}{6.626\times10^{-34}\ \mathrm{J\cdot s}}\right)^2 = 20.$

And then $l(l+1) = 20$ gives that $l = 4$.

Reflect: l must be integer.

*29.3. Set Up: $L = \sqrt{l(l+1)}\,\hbar$. $L_z = m_l\hbar$. $l = 0,\ 1,\ 2,\ldots,\ n-1$. $m_l = 0,\ \pm1,\ \pm2,\ldots,\ \pm l$. $\cos\theta = L_z/L$.

Solve: (a) $l = 0$: $L = 0$, $L_z = 0$. $l = 1$: $L = \sqrt{2}\,\hbar$, $L_z = \hbar,\ 0,\ -\hbar$. $l = 2$: $L = \sqrt{6}\,\hbar$, $L_z = 2\hbar,\ \hbar,\ 0,\ -\hbar,\ -2\hbar$.

(b) In each case $\cos\theta = L_z/L$. $L = 0$: θ not defined. $L = \sqrt{2}\,\hbar$: $45.0°$, $90.0°$, $135.0°$. $L = \sqrt{6}\,\hbar$: $35.3°$, $65.9°$, $90.0°$, $114.1°$, $144.7°$.

Reflect: There is no state where $\vec{L}$ is totally aligned along the z axis.

*29.7. Set Up: $m_l = 0,\ \pm1,\ \pm2,\ldots,\ \pm l$. $s = \pm\frac{1}{2}$. g means $l = 4$. $\cos\theta = L_z/L$, with $L = \sqrt{l(l+1)}\,\hbar$ and $L_z = m_l\hbar$,

Solve: (a) There are eighteen $5g$ states: $m_l = 0,\ \pm1,\ \pm2,\ \pm3,\ \pm4$, with $s = \pm\frac{1}{2}$ for each.

(b) The largest θ is for the most negative m_l. $L = 2\sqrt{5}\,\hbar$. The most negative L_z is $L_z = -4\hbar$. $\cos\theta = \dfrac{-4\hbar}{2\sqrt{5}\,\hbar}$ and $\theta = 153.4°$.

(c) The smallest θ is for the largest positive m_l, which is $m_l = +4$. $\cos\theta = \dfrac{4\hbar}{2\sqrt{5}\,\hbar}$ and $\theta = 26.6°$.

Reflect: The minimum angle between $\vec{L}$ and the z axis is for $m_l = +l$ and for that m_l, $\cos\theta = \dfrac{l}{\sqrt{l(l+1)}}$.

*29.9. Set Up: $0 \le l \le n-1$. $|m_l| \le l$. $s = \pm\frac{1}{2}$.

Solve: $n=1$, $l=0$, $m_l=0$, $s=\frac{1}{2}$. $n=1$, $l=0$, $m_l=0$, $s=-\frac{1}{2}$. $n=2$, $l=0$, $m_l=0$, $s=\frac{1}{2}$. $n=2$, $l=0$, $m_l=0$, $s=-\frac{1}{2}$. $n=2$, $l=1$, $m_l=1$, $s=\frac{1}{2}$. $n=2$, $l=1$, $m_l=1$, $s=-\frac{1}{2}$. $n=2$, $l=1$, $m_l=0$, $s=\frac{1}{2}$. $n=2$, $l=1$, $m_l=0$, $s=-\frac{1}{2}$. $n=2$, $l=1$, $m_l=-1$, $s=\frac{1}{2}$. $n=2$, $l=1$, $m_l=-1$, $s=-\frac{1}{2}$. $n=3$, $l=0$, $m_l=0$, $s=\frac{1}{2}$. $n=3$, $l=0$, $m_l=0$, $s=-\frac{1}{2}$.

*29.13. Set Up: Carbon has 6 electrons.

Solve: (a) $1s^2\,2s^2\,2p^2$

(b) The element of next larger Z with a similar electron configuration has configuration $1s^2\,2s^2\,2p^6\,3s^2\,3p^2$. $Z=14$ and the element is silicon.

Reflect: Carbon and silicon are in the same column of the periodic table.

*29.15. Set Up: Ne has 10 electrons; Ar has 18 electrons, Kr has 36 electrons.

Solve: The electron configurations are

$$\text{Ne: } 1s^2\,2s^2\,2p^6$$
$$\text{Ar: } 1s^2\,2s^2\,2p^6\,3s^2\,3p^6$$
$$\text{Kr: } 1s^2\,2s^2\,2p^6\,3s^2\,3p^6\,3d^{10}\,4s^2\,4p^6$$

These atoms have filled outer shells and are chemically inert.

Reflect: These three elements are in the same column of the periodic table.

*29.21. Set Up: The electric potential energy for two point charges is given by $U = k\dfrac{q_1 q_2}{r} = -k\dfrac{e^2}{r}$.

(a) $U = -k\dfrac{e^2}{r} = -(8.99 \times 10^9 \text{ N} \cdot \text{m}^2/\text{C}^2)\dfrac{(1.60 \times 10^{-19} \text{ C})^2}{0.29 \times 10^{-9} \text{ m}} = -7.9 \times 10^{-19} \text{ J} = -5.0 \text{ eV}$.

(b) To create a potassium bromide molecule we remove an electron from the potassium atom (add energy) and add it to the bromine atom (release of energy). In addition, the resulting ions release potential energy as they combine: Binding Energy $\approx +(4.3 \text{ eV} - 3.5 \text{ eV}) - 5.0 \text{ eV} = -4.2 \text{ eV}$.

Reflect: Our calculations assume we can treat each ion as a point charge; however, the electron charge distribution around each atom will distort as the ions approach one another—this should result in a further release of energy.

*29.23. Set Up: The density ρ is the mass per unit volume. Assume each atom occupies a volume of a^3, where a is the spacing between adjacent atoms.

Solve: There are equal numbers of K and Br atoms and the average mass is
$$\tfrac{1}{2}(6.49 \times 10^{-26} \text{ kg} + 1.33 \times 10^{-25} \text{ kg}) = 9.90 \times 10^{-26} \text{ kg}.$$

$$\rho = \frac{m}{a^3} \text{ and } a = \left(\frac{m}{\rho}\right)^{1/3} = \left(\frac{9.90 \times 10^{-26} \text{ kg}}{2.75 \times 10^3 \text{ kg/m}^3}\right)^{1/3} = 3.30 \times 10^{-10} \text{ m} = 0.330 \text{ nm}.$$

Problem 29.22 states that the spacing of adjacent atoms for NaCl is 0.282 nm. The spacing is larger for KBr, because the K and Br atoms contain more electrons and have larger atomic radii than Na and Cl.

Reflect: In a solid the atoms are closely packed together and the average spacing between atoms is approximately equal to the average diameter of the atoms.

*29.25. Set Up: The energy of a photon is given by $E_{\text{photon}} = hf = \dfrac{hc}{\lambda}$. The gap in the valence band and the conduction band is $\Delta E = 5.47$ eV.

Solve: (a) We will assume that the energy of the photon is equal to the band gap energy: $E_{\text{photon}} = \Delta E$.

Thus we have $\lambda = \dfrac{hc}{\Delta E} = \dfrac{(4.136 \times 10^{-15} \text{ eV} \cdot \text{s})(3.00 \times 10^8 \text{ m/s})}{5.47 \text{ eV}} = 2.27 \times 10^{-7}$ m $= 227$ nm, which is in the ultraviolet.

(b) Visible light lacks enough energy to excite the electrons into the conduction band, so visible light passes through the diamond unabsorbed.

Reflect: (c) Impurities can lower the gap energy, making it easier for the material to absorb shorter wavelength visible light. This allows longer wavelength visible light to pass through, giving the diamond color.

*29.29. Set Up: The ionization potential is the energy required to remove an electron from the atom and is the negative of the binding energy of the atom.

Solve: The least bound and outermost electron in the atom moves under the action of the total electric field due to the nucleus and the averaged-out electron cloud of all the electrons. The other electrons approximately screen the outermost electron from the charge of the nucleus. As electrons are removed, this screening is reduced and the outermost electron in each successive charge state is more tightly bound.

Reflect: We could describe the binding energy for a particular electron as $(13.6 \text{ eV})Z_{\text{eff}}^2/n^2$, where Z_{eff} is the effective nuclear charge of the screened nucleus. Z_{eff} increases as electrons are removed.

*29.33. Set Up: The energy gap is the energy of the maximum-wavelength photon. The energy difference is equal to the energy of the photon, so $\Delta E = hc/\lambda$.

Solve: (a) Using the photon wavelength to find the energy difference gives

$$\Delta E = hc/\lambda = (4.136 \times 10^{-15} \text{ eV} \cdot \text{s})(3.00 \times 10^8 \text{ m/s})/(1.11 \times 10^{-6} \text{ m}) = 1.12 \text{ eV}$$

(b) A wavelength of 1.11 μm $= 1110$ nm is in the infrared, shorter than that of visible light.

Reflect: Since visible photons have more than enough energy to excite electrons from the valence to the conduction band, visible light will be absorbed, which makes silicon opaque.

*29.37. Set Up: The total energy determines what shell the electron is in, which limits its angular momentum. The electron's orbital angular momentum is given by $L = \sqrt{l(l+1)}\hbar$ (where $\hbar = \dfrac{h}{2\pi}$), and its total energy in the nth shell is $E_n = -(13.6 \text{ eV})/n^2$.

Solve: (a) First find n: $E_n = -(13.6 \text{ eV})/n^2 = -0.5440$ eV which gives $n = 5$, so $l = 4, 3, 2, 1, 0$. Therefore the possible values of L are given by $L = \sqrt{l(l+1)}\hbar$, giving $L = 0, \sqrt{2}\hbar, \sqrt{6}\hbar, \sqrt{12}\hbar, \sqrt{20}\hbar$.

(b) $E_6 = -(13.6 \text{ eV})/6^2 = -0.3778$ eV. $\Delta E = E_6 - E_5 = -0.3778$ eV $- (-0.5440 \text{ eV}) = +0.1662$ eV

This must be the energy of the photon, so $\Delta E = hc/\lambda$, which gives

$\lambda = hc/\Delta E = (4.136 \times 10^{-15} \text{ eV} \cdot \text{s})(3.00 \times 10^8 \text{ m/s})/(0.1662 \text{ eV}) = 7.47 \times 10^{-6}$ m $= 7470$ nm, which is in the infrared and hence not visible.

Reflect: The electron can have any of the five possible values for its angular momentum, but it cannot have any others.

*29.39. Set Up: The shell that the electron is in limits the possibilities for its orbital angular momentum. For an electron in the n shell, its orbital angular momentum quantum number l is limited by $0 \le l < n$, and its orbital angular momentum is given by $L = \sqrt{l(l+1)}\hbar$, where $\hbar = \dfrac{h}{2\pi}$. The z-component of its angular momentum is $L_z = m_l\hbar$, where $m_l = 0, \pm 1, \ldots, \pm l$, and its spin angular momentum is $S = \sqrt{3/4}\hbar$ for all electrons. Its energy in the nth shell is $E_n = -(13.6 \text{ eV})/n^2$.

Solve: **(a)** $L = \sqrt{l(l+1)}\hbar = \sqrt{12}\hbar \Rightarrow l = 3$. Therefore the smallest that n can be is 4, so $E_n = -(13.6\text{ eV})/n^2 = -(13.6\text{ eV})/4^2 = -0.8500\text{ eV}$.

(b) For $l = 3$, $m_l = \pm 3, \pm 2, \pm 1, 0$. Since $L_z = m_l\hbar$, the largest L_z can be is $3\hbar$ and the smallest it can be is $-3\hbar$.

(c) $S = \sqrt{3/4}\hbar$ for *all* electrons.

(d) In this case, $n = 3$, so $l = 2, 1, 0$. Therefore the maximum that L can be is $L_{\text{max}} = \sqrt{2(2+1)}\hbar = \sqrt{6}\hbar$. The minimum L can be is zero when $l = 0$.

Reflect: At the quantum level, electrons in atoms can have only certain allowed values of their angular momentum.

30

NUCLEAR AND HIGH-ENERGY PHYSICS

Problems 3, 5, 9, 11, 13, 17, 19, 23, 25, 27, 31, 33, 37, 41, 43, 47, 49, 51, 55, 59, 61, 65, 67, 69

Solutions to Problems

*30.3. Set Up: The volume of a sphere is $V = \frac{4}{3}\pi R^3$. The empirical formula for the radius is $R = R_0 A^{1/3}$ where $R_0 = 1.2 \times 10^{-15}$ m. Density is $\rho = \frac{m}{V}$. $m_p = 1.67 \times 10^{-27}$ kg

Solve: (a) $V = \frac{4}{3}\pi R^3 = \frac{4}{3}\pi R_0^3 A$, so V is proportional to A.

(b) Neutrons and protons have very similar masses and A is the total number of neutrons and protons in the nucleus.

(c) $\rho = \dfrac{m_p A}{\frac{4}{3}\pi R_0^3 A} = \dfrac{3m_p}{4\pi R_0^3}$, which is independent of A. $\rho = \dfrac{3(1.67 \times 10^{-27} \text{ kg})}{4\pi(1.2 \times 10^{-15} \text{ m})^3} = 2.3 \times 10^{17}$ kg/m^3

This is 2.0×10^{13} times the density of lead and is similar to the density of a neutron star.

Reflect: The mass of an atom is close to the mass of the nucleus but the volume of the nucleus is much less than the volume of the atom, so the average density of the atom is much less than the density of the nucleus. Neutron stars are thought to be composed of nuclear matter; they are essentially one very large nucleus.

*30.5. Set Up: The constituents of the $^2_1$H neutral atom are one $^1_1$H atom and one neutron. Table 30.2 gives the atomic mass to be $M\left(^2_1\text{H}\right) = 2.014101$ u. The constituents of the $^4_2$He neutral atom are two $^1_1$H atoms and two neutrons. $M\left(^4_2\text{He}\right) = 4.002603$ u. 1 u is equivalent to 931.5 MeV. The mass defect is $\Delta M = Z m_H + N m_n - M$.

Solve: (a) $\Delta M = m_H + m_n - M = 1.007825 \text{ u} + 1.008665 \text{ u} - 2.014101 \text{ u} = 0.002389$ u.

$$E_B = (0.002389 \text{ u})(931.5 \text{ MeV/u}) = 2.23 \text{ MeV}.$$

$$E_B/A = \frac{2.23 \text{ MeV}}{2 \text{ nucleons}} = 1.11 \text{ MeV/nucleon}.$$

(b) $\Delta M = 2m_H + 2m_n - M = 2(1.007825 \text{ u} + 1.008665 \text{ u}) - 4.002603 \text{ u} = 0.03038$ u.

$$E_B = (0.03038 \text{ u})(931.5 \text{ MeV/u}) = 28.3 \text{ MeV}.$$

$$E_B/A = \frac{28.3 \text{ MeV}}{4 \text{ nucleons}} = 7.08 \text{ MeV/nucleon}.$$

(c) The binding energy per nucleon is much larger for $^4_2$He than for $^2_1$H.

Reflect: The masses in Table 30.2 are atomic masses and include the masses of the electrons that are present in the neutral atom. It is important to correctly account for the electron masses in the calculation.

*30.9. Set Up: In α-decay the nucleon number decreases by 4 and the atomic number decreases by 2. The atomic number of Po is 84. An α-particle is the nucleus of $^4_2$He, and the atomic mass of $^4_2$He is 4.00260 u. 1 u is equivalent to 931.5 MeV.

Solve: (a) $^{218}_{84}$Po $\rightarrow$ $^4_2$He + $^{214}_{82}$Pb. The daughter nucleus is lead with $A = 214$, $Z = 82$, and $N = 132$.

(b) The mass change is $m\left(^{218}_{84}\text{Po}\right) - \left[m\left(^{214}_{82}\text{Pb}\right) + m\left(^4_2\text{He}\right)\right] = 218.008973 \text{ u} - 213.999805 \text{ u} - 4.00260 \text{ u} = 0.00657 \text{ u}$

84 electrons are included on each side of the decay equation so their masses cancel. The energy released is 6.12 MeV. This is the kinetic energy of the α-particle, if recoil of the daughter nucleus is ignored.

Reflect: For a radioactive decay to be energetically possible the mass of the products must be less than the mass of the parent.

*30.11. Set Up: In the α-decay the nucleon number decreases by 4 and the atomic number decreases by 2. In β^- decay the nucleon number doesn't change and the atomic number increases by 1. The atomic number for each nucleus is identified by the chemical symbol and can be found using the periodic table in Appendix D.

Solve: (a) $^{232}_{90}$Th $\rightarrow$ $^4_2$He + $^{228}_{88}$Ra; X_1 is $^{228}_{88}$Ra; $A = 228$, $Z = 88$, $N = 140$

$^{228}_{89}$Ac $\rightarrow$ $^0_{-1}$e + $^{228}_{90}$Th; X_2 is $^{228}_{89}$Ac; $A = 228$, $Z = 89$, $N = 139$

$^{228}_{90}$Th $\rightarrow$ $^4_2$He + $^{224}_{88}$Ra; X_3 is α-particle; A = 4, Z = 2, N = 2

$^{220}_{86}$Rn $\rightarrow$ $^4_2$He + $^{216}_{84}$Po; X_4 is $^{216}_{84}$Po; $A = 216$, $Z = 84$, $N = 132$

$^{216}_{84}$Po $\rightarrow$ $^4_2$He + $^{212}_{82}$Pb; X_5 is $^{216}_{84}$Po; $A = 216$, $Z = 84$, $N = 132$

$^{212}_{82}$Pb $\rightarrow$ $^0_{-1}$e + $^{212}_{83}$Bi; X_6 is β^-; $A = 0$, $Z = -1$, $N = 0$

(b) $^{212}_{83}$Bi $\xrightarrow{\alpha}$ $^4_2$He + $^{208}_{81}$Tl and then $^{208}_{81}$Tl $\xrightarrow{\beta^-}$ $^0_{-1}$e + $^{208}_{82}$Pb or

$^{212}_{83}$Bi $\xrightarrow{\beta^-}$ $^0_{-1}$e + $^{212}_{84}$Po and then $^{212}_{84}$Po $\xrightarrow{\alpha}$ $^4_2$He + $^{208}_{82}$Pb

The end product in each case is $^{208}_{82}$Pb.

(c) The Segre chart is given in the figure below.

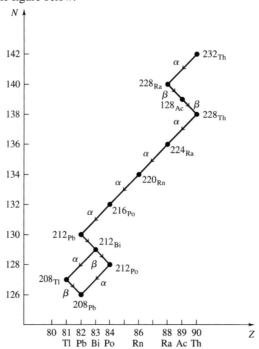

Reflect: Each decay conserves charge and nucleon number.

*30.13. Set Up: $T_{1/2} = \dfrac{\ln 2}{\lambda}$ After one half-life the activity decreases by a factor of $\frac{1}{2}$.

Solve: (a) $\lambda = \dfrac{\ln 2}{T_{1/2}} = \dfrac{\ln 2}{300.0 \text{ s}} = 2.31 \times 10^{-3} \text{ s}^{-1}$

(b) (i) After 5.00 min the decay rate is $(6.00 \times 10^3 \text{ Bq})/2 = 3.00 \times 10^3 \text{ Bq}$

(ii) After an additional 5.00 min the decay rate is $(3.00 \times 10^3 \text{ Bq})/2 = 1.50 \times 10^3 \text{ Bq}$

(iii) 25.0 min is three half-lives after 10.0 min so the decay rate is $(1.50 \times 10^3 \text{ Bq})/(2^3) = 188 \text{ Bq}$

*30.17. Set Up: The decay rate decreases by a factor of 2 in a time of one half-life.

Solve: (a) 24 d is $3T_{1/2}$ so the activity is $(375 \text{ Bq})/\left(2^3\right) = 46.9 \text{ Bq}$

(b) The activity is proportional to the number of radioactive nuclei, so the percent is $\dfrac{17.0 \text{ Bq}}{46.9 \text{ Bq}} = 36.2\%$

(c) $^{131}_{53}\text{I} \rightarrow \,_{-1}^{\;0}\text{e} + \,^{131}_{54}\text{Xe}$. The nucleus $^{131}_{54}\text{Xe}$ is produced.

Reflect: Both the activity and the number of radioactive nuclei present decrease by a factor of 2 in one half-life.

*30.19. Set Up: Let A denote the activity of the sample as a function of time: According to Equation 30.5 we have $A = \left| \dfrac{\Delta N}{\Delta t} \right| = \lambda N$, which says that the activity is proportional to N. Since N decreases according to equation 30.6, so does the activity: $A = A_0 e^{-\lambda t}$. Note that $A_0 = 7.56 \times 10^{11}$ Bq.

Solve: (a) After 92.4 minutes the activity has decreased to $A = 9.45 \times 10^{10}$ Bq. The decay constant can be found from $\lambda = \dfrac{1}{t} \ln\left(\dfrac{A_0}{A}\right) = \dfrac{1}{92.4 \text{ min}} \ln\left(\dfrac{7.56 \times 10^{11} \text{ Bq}}{9.45 \times 10^{10} \text{ Bq}}\right) = 0.0225 \, (\text{min})^{-1}$. Thus, the half-life is $T_{1/2} = \dfrac{\ln 2}{\lambda} = \dfrac{\ln 2}{0.0225 \, (\text{min})^{-1}} = 30.8 \text{ min.}$

(b) We can determine the number of radioactive nuclei initially present from the initial activity: $A_0 = \lambda N_0$, so we have $N_0 = \dfrac{A_0}{\lambda} = \dfrac{7.56 \times 10^{11} \text{ Bq}}{(0.0225 \, (\text{min})^{-1})(1 \text{ min}/60 \text{ s})} = 2.02 \times 10^{15}$ radioactive nuclei.

Reflect: Note that since 1 Bq = 1 decay/s we must convert λ from $(\text{min})^{-1}$ to s^{-1} in our answer to part (b).

*30.23. Set Up: 1 Gy = 1 J/kg and is the SI unit of absorbed dose. 1 rad = 0.010 Gy. Sv is the SI unit for equivalent dose. Equivalent dose = RBE × absorbed dose. Rem is the equivalent dose when the absorbed dose is in rad. For x-rays, RBE = 1.0. For protons, RBE = 10.
Solve: (a) 5.0 Gy, 500 rad. RBE = 1.0 so equivalent dose = absorbed dose. 5.0 Sv and 500 rem.
(b) (70.0 kg)(5.0 J/kg) = 350 J
(c) The absorbed dose and total absorbed energy are the same but the equivalent dose is 10 times larger. So the answers are: 5.0 Gy, 500 rad, 50 Sv, 5000 rem, 350 J.
Reflect: The same energy deposited by protons as x-rays is ten times greater in its biological effect.

*30.25. Set Up: We have the following conversions: 1 Gy = 1 J/kg and 1 rad = 10^{-2} Gy. Also, we have dose in rem = (RBE)(absorbed dose in rad). The RBE for slow neutrons is given to be 4.0 in this case.

Solve: (5.0 Gy)(1 rad/10^{-2} Gy) = 500 rad, dose in rem = (RBE)(absorbed dose in rad) = 4.0(500 rad) = 2000 rem, and (5.0 Gy)(1 J/kg)/Gy = 5.0 J/kg.

Reflect: The RBE is an important factor in determining the effect of a given type of radiation.

*30.27. Set Up: For x rays RBE $=1$ so the equivalent dose in Sv is the same as the absorbed dose in J/kg.

Solve: One whole-body scan delivers $(75 \text{ kg})(12 \times 10^{-3} \text{ J/kg}) = 0.90$ J. One chest x ray delivers

$$(5.0 \text{ kg})(0.20 \times 10^{-3} \text{ J/kg}) = 1.0 \times 10^{-3} \text{ J}.$$

It takes $\dfrac{0.90 \text{ J}}{1.0 \times 10^{-3} \text{ J}} = 900$ chest x rays to deliver the same total energy.

*30.31. Set Up: Since this source has a long lifetime its activity will be nearly constant: number of decays = (activity)(time). The absorbed dose is measured in grays, where $1 \text{ Gy} = 1 \text{ J/kg} = 100$ rad. Also we have: dose in rem = (RBE) (absorbed does in rad), where RBE $= 20$ for alpha particles. Finally, we know that $1 \text{ Ci} = 3.70 \times 10^{10}$ Bq, $1 \text{ y} = 3.156 \times 10^7$ s, and $1 \text{ eV} = 1.602 \times 10^{-19}$ J.

Solve: The number of alpha particles generated in 1 year is $(0.72 \times 10^{-6} \text{ Ci}) (3.7 \times 10^{10} \text{ Bq/Ci}) (3.156 \times 10^7 \text{ s}) =$ 8.41×10^{11}. The absorbed dose is $\dfrac{(8.41 \times 10^{11}) (4.0 \times 10^6 \text{ eV}) (1.602 \times 10^{-19} \text{ J/eV})}{(0.50 \text{ kg})} = 1.08 \text{ Gy} = 108$ rad. The equivalent dose is $(20)(108 \text{ rad}) = 2160$ rem.

Reflect: The equivalent does is relatively high due to the large RBE for alpha particles.

*30.33. Set Up: The reaction energy is $Q = (M_A + M_B - M_C - M_D)c^2$, where $A = {}_{2}^{4}\text{He}$, $B = {}_{3}^{7}\text{Li}$, $C = {}_{5}^{10}\text{B}$ and $D = {}_{0}^{1}\text{n}$. If neutral atom masses from Table 30.2 are used, the electron masses cancel. 1 u is equivalent to 931.5 MeV. If $Q > 0$, energy is liberated in the reaction. If $Q < 0$, energy is absorbed in the reaction.

Solve: $M_A + M_B - M_C - M_D = 4.002603 \text{ u} + 7.016005 \text{ u} - 10.012937 \text{ u} - 1.008665 \text{ u} = -2.994 \times 10^{-3}$ u $Q = (-2.994 \times 10^{-3} \text{ u})(931.5 \text{ MeV/u}) = -2.79$ MeV. The reaction absorbs 2.79 MeV.

*30.37. Set Up: 0.7% of naturally occurring uranium is the isotope ${}^{235}\text{U}$. The mass of one ${}^{235}\text{U}$ nucleus is about $235 m_p$.

Solve: (a) The number of fissions needed is $\dfrac{1.0 \times 10^{19} \text{ J}}{(200 \times 10^6 \text{ eV})(1.60 \times 10^{-19} \text{ J/eV})} = 3.13 \times 10^{29}$. The mass of ${}^{235}\text{U}$ required is $(3.13 \times 10^{29})(235 m_p) = 1.23 \times 10^5$ kg.

(b) $\dfrac{1.23 \times 10^5 \text{ kg}}{0.7 \times 10^{-2}} = 1.76 \times 10^7$ kg

Reflect: The calculation assumes 100% conversion of fission energy to electrical energy.

*30.41. Set Up: $m = \rho V$. $1 \text{ gal} = 3.788 \text{ L} = 3.788 \times 10^{-3}$ m^3. The mass of a ${}^{235}\text{U}$ nucleus is $235 m_p$.

$$1 \text{ MeV} = 1.60 \times 10^{-13} \text{ J}$$

Solve: (a) For 1 gallon, $m = \rho V = (737 \text{ kg/m}^3)(3.788 \times 10^{-3} \text{ m}^3) = 2.79 \text{ kg} = 2.79 \times 10^3$ g

$$\dfrac{1.3 \times 10^8 \text{ J/gal}}{2.79 \times 10^3 \text{ g/gal}} = 4.7 \times 10^4 \text{ J/g}$$

(b) 1 g contains $\dfrac{1.00 \times 10^{-3} \text{ kg}}{235 m_p} = 2.55 \times 10^{21}$ nuclei

$$(200 \text{ MeV/nucleus})(1.60 \times 10^{-13} \text{ J/MeV})(2.55 \times 10^{21} \text{ nuclei}) = 8.2 \times 10^{10} \text{ J/g}$$

(c) A mass of $6m_p$ produces 26.7 MeV.

$$\frac{(26.7 \text{ MeV})(1.60 \times 10^{-13} \text{ J/MeV})}{6m_p} = 4.26 \times 10^{14} \text{ J/kg} = 4.26 \times 10^{11} \text{ J/g}$$

(d) The total energy available would be $(1.99 \times 10^{30} \text{ kg})(4.7 \times 10^7 \text{ J/kg}) = 9.4 \times 10^{37} \text{ J}$

$$\text{power} = \frac{\text{energy}}{t} \quad \text{so} \quad t = \frac{\text{energy}}{\text{power}} = \frac{9.4 \times 10^{37} \text{ J}}{3.92 \times 10^{26} \text{ W}} = 2.4 \times 10^{11} \text{ s} = 7600 \text{ yr}$$

Reflect: If the mass of the sun were all proton fuel, it would contain enough fuel to last

$$(7600 \text{ yr})\left(\frac{4.3 \times 10^{11} \text{ J/g}}{4.7 \times 10^4 \text{ J/g}}\right) = 7.0 \times 10^{10} \text{ yr}.$$

*30.43. Set Up: Apply conservation of linear momentum to the collision. A photon has momentum $p = h/\lambda$, in the direction it is traveling. The energy of a photon is $E = pc = \frac{hc}{\lambda}$. All the mass of the electron and positron is converted to the total energy of the two photons, according to $E = mc^2$. The mass of an electron and of a positron is $m_e = 9.11 \times 10^{-31}$ kg.

Solve: (a) In the lab frame the initial momentum of the system is zero, since the electron and positron have equal speeds in opposite directions. According to momentum conservation, the final momentum of the system must also be zero. A photon has momentum, so the momentum of a single photon is not zero.

(b) For the two photons to have zero total momentum they must have the same magnitude of momentum and move in opposite directions. Since $E = pc$, equal p means equal E.

(c) $2E_{ph} = 2m_e c^2$ so $E_{ph} = m_e c^2$

$$E_{ph} = \frac{hc}{\lambda} \quad \text{so} \quad \frac{hc}{\lambda} = m_e c^2 \quad \text{and} \quad \lambda = \frac{h}{m_e c} = \frac{6.63 \times 10^{-34} \text{ J} \cdot \text{s}}{(9.11 \times 10^{-31} \text{ kg})(3.00 \times 10^8 \text{ m/s})} = 2.43 \text{ pm}$$

These are gamma ray photons.

Reflect: The total charge of the electron/positron system is zero and the photons have no charge, so charge is conserved in the particle-antiparticle annihilation.

*30.47. Set Up: The mass of the proton and antiproton are both equal to 938.3 MeV according to Table 30.5. The energy of a photon is given by $E = hf = \frac{hc}{\lambda}$, where $h = 4.136 \times 10^{-15}$ eV·s.

Solve: (a) The energy will be the proton rest energy, 938.3 MeV, corresponding to a frequency of

$$f = \frac{938.3 \times 10^6 \text{ eV}}{4.136 \times 10^{-15} \text{ eV} \cdot \text{s}} = 2.27 \times 10^{23} \text{ Hz} \quad \text{and a wavelength of} \quad \lambda = \frac{c}{f} = \frac{3.00 \times 10^8 \text{ m/s}}{2.27 \times 10^{23} \text{ Hz}} = 1.32 \times 10^{-15} \text{ m}.$$

(b) The energy of each photon will be $938.3 \text{ MeV} + 830 \text{ MeV} = 1768 \text{ MeV}$. Following the method our previous work this gives a frequency 4.28×10^{23} Hz and wavelength 7.02×10^{-16} m.

Reflect: Note that the frequency is directly proportional to the released energy. Thus, we have

$$(2.27 \times 10^{23} \text{ Hz})\left(\frac{1768 \text{ MeV}}{938.3 \text{ MeV}}\right) = 4.28 \times 10^{23} \text{ Hz}.$$

*30.49. Set Up: Section 30.9 says the strong interaction is 100 times as strong as the electromagnetic interaction and that the weak interaction is 10^{-9} times as strong as the strong interaction. The Coulomb force is $F_e = \frac{kq_1 q_2}{r^2}$ and the gravitational force is $F_g = G\frac{m_1 m_2}{r^2}$.

Solve: (a) $F_e = \dfrac{(9.0 \times 10^9 \text{ N} \cdot \text{m}^2/\text{C}^2)(1.60 \times 10^{-19} \text{ C})^2}{(1 \times 10^{-15} \text{ m})^2} = 200 \text{ N}$

$$F_g = \dfrac{(6.67 \times 10^{-11} \text{ N} \cdot \text{m}^2/\text{kg}^2)(1.67 \times 10^{-27} \text{ kg})^2}{(1 \times 10^{-15} \text{ m})^2} = 2 \times 10^{-34} \text{ N}$$

(b) $F_{str} \approx 100 F_e \approx 2 \times 10^4 \text{ N}$. $F_{weak} \approx 10^{-9} F_{str} \approx 2 \times 10^{-5} \text{ N}$

(c) $F_{str} > F_e > F_{weak} > F_g$

(d) $F_e \approx 1 \times 10^{36} F_g$. $F_{str} \approx 100 F_e \approx 1 \times 10^{38} F_g$. $F_{weak} \approx 10^{-9} F_{str} \approx 1 \times 10^{29} F_g$

Reflect: The gravity force is much weaker than any of the other three forces. Gravity is important only when one very massive object is involved.

*30.51. Set Up: $m_H = 1.67 \times 10^{-27}$ kg. The ideal gas law says $pV = nRT$. Normal pressure is 1.013×10^5 Pa and normal temperature is about $27 \degree C = 300$ K. 1 mole is 6.02×10^{23} atoms.

Solve: (a) $\dfrac{5.8 \times 10^{-27} \text{ kg/m}^3}{1.67 \times 10^{-27} \text{ kg/atom}} = 3.5 \text{ atoms/m}^3$

(b) $V = (4 \text{ m})(7 \text{ m})(3 \text{ m}) = 84 \text{ m}^3$ and $(3.5 \text{ atoms/m}^3)(84 \text{ m}^3) = 294 \text{ atoms}$

(c) With $p = 1.013 \times 10^5$ Pa, $V = 84 \text{ m}^3$, $T = 300$ K and the ideal gas law gives the number of moles to be

$$n = \dfrac{pV}{RT} = \dfrac{(1.013 \times 10^5 \text{ Pa})(84 \text{ m}^3)}{(8.3145 \text{ J/mol} \cdot \text{K})(300 \text{ K})} = 3.4 \times 10^3 \text{ moles}$$

$$(3.4 \times 10^3 \text{ moles})(6.02 \times 10^{23} \text{ atoms/mol}) = 2.0 \times 10^{27} \text{ atoms}$$

Reflect: The average density of the universe is very small. Interstellar space contains a very small number of atoms per cubic meter, compared to the number of atoms per cubic meter in ordinary material on the earth, such as air.

*30.55. Set Up: equivalent dose (rem) = RBE × absorbed dose (rad). 1 rad = 0.010 J/kg.

$$1 \text{ Ci} = 3.70 \times 10^{10} \text{ decays/s.}$$

Solve: (a) $15.0 \text{ Ci} = 5.55 \times 10^{11}$ decays/s. Each decay produces $1.25 \text{ MeV} = 2.00 \times 10^{-13}$ J. The ^{60}Co source releases $(5.55 \times 10^{11} \text{ decays/s})(2.00 \times 10^{-13} \text{ J}) = 0.111 \text{ J/s}$. Only half of this energy is absorbed, so the energy absorbed per second is $(0.111 \text{ J})/2 = 0.0555 \text{ J}$.

(b) The absorbed dose per second is $\dfrac{0.0555 \text{ J}}{0.500 \text{ kg}} = 0.111 \text{ J/kg} = 11.1 \text{ rad}$.

(c) The equivalent dose per second is $(0.70)(11.1 \text{ rad}) = 7.77 \text{ rem}$.

(d) $t = \dfrac{200 \text{ rem}}{7.77 \text{ rem/s}} = 25.7 \text{ s}$

Reflect: The equivalent dose is less than the absorbed dose since RBE is less than 1.00.

*30.59. Set Up: $\left| \dfrac{\Delta N}{\Delta t} \right| = \lambda N$. $\lambda = \dfrac{\ln 2}{T_{1/2}}$. $T_{1/2} = 5.3 \text{ yr} = 1.67 \times 10^8 \text{ s}$. A ^{60}Co nucleus has mass $60 m_p$. 1 Ci = 3.70×10^{10} decays/s.

Solve: (a) $\lambda = \dfrac{\ln 2}{T_{1/2}} = \dfrac{\ln 2}{1.67 \times 10^8 \text{ s}} = 4.15 \times 10^{-9} \text{ s}^{-1}$

(b) $N = \dfrac{0.0400 \times 10^{-3} \text{ kg}}{60 m_p} = \dfrac{4.00 \times 10^{-5} \text{ kg}}{60(1.67 \times 10^{-27} \text{ kg})} = 3.99 \times 10^{20}$

(c) $\left|\dfrac{\Delta N}{\Delta t}\right| = \lambda N = (4.15 \times 10^{-9} \text{ s}^{-1})(3.99 \times 10^{20}) = 1.66 \times 10^{12}$ decays/s

(d) $(1.66 \times 10^{12}) \left(\dfrac{1 \text{ Ci}}{3.70 \times 10^{10} \text{ decays/s}}\right) = 44.9$ Ci

*30.61. Set Up: The number of radioactive nuclei left after time t is given by $N = N_0 e^{-\lambda t}$. The decay constant for ^{14}C can be calculated from its known half-life of 5730 years (Example 30.4): $\lambda = \ln 2 / T_{1/2}$. The problem says $N/N_0 = 0.21$; solve for t.

Solve: $0.21 = e^{-\lambda t}$ so $\ln(0.21) = -\lambda t$ and $t = -\ln(0.21)/\lambda$. We can calculate the decay constant for ^{14}C from its known half-life: $\lambda = \dfrac{\ln 2}{5730 \text{ y}} = 1.21 \times 10^{-4}$ y^{-1}. Thus $t = \dfrac{-\ln(0.21)}{1.21 \times 10^{-4} \text{ y}^{-1}} = 1.3 \times 10^4$ y.

Reflect: The half-life of ^{14}C is 5730 y, so our calculated t is more than two half-lives, so the fraction remaining is less than $\left(\frac{1}{2}\right)^2 = \frac{1}{4}$.

*30.65. Set Up: Table 30.5 gives $m(\text{K}^+) = 493.7$ MeV/c^2, $m(\pi^0) = 135.0$ MeV/c^2, and

$$m(\pi^{\pm}) = 139.6 \text{ MeV/}c^2.$$

Solve: (a) Charge must be conserved, so $\text{K}^+ \rightarrow \pi^0 + \pi^+$ is the only possible decay.
(b) The mass decrease is
$$m(\text{K}^+) - m(\pi^0) - m(\pi^+) = 493.7 \text{ MeV/}c^2 - 135.0 \text{ MeV/}c^2 - 139.6 \text{ MeV/}c^2 = 219.1 \text{ MeV/}c^2.$$
The energy released is 219.1 MeV.

*30.67. Set Up: Table 30.6 lists the properties of quarks. The values of Q, B, and S for an antiquark are opposite those given for the corresponding quark.
Solve: (a) $S = 1$ indicates the presence of one $\bar{s}$ antiquark and no s quark. To have baryon number 0 there can be only one other quark, and to have net charge $+e$ that quark must be a u, and the quark content is $u\bar{s}$.
(b) The particle has an $\bar{s}$ antiquark, and for a baryon number of -1 the particle must consist of three antiquarks. For a net charge of $-e$, the quark content must be $\overline{dd}\,\bar{s}$.
(c) Since $S = -2$ we know that there are two s quarks, and for baryon number 1 there must be one more quark. For a charge of 0 the third quark must be a u quark and the quark content is uss.
Reflect: From Table 30.5 we discover that the particle described in part (a) is a $\text{K}^+(u\bar{s})$. The other combinations are not listed.

Solutions to Passage Problems

*30.69. Set Up: We know that the energy of a photon is given by $E = hf = \dfrac{hc}{\lambda}$, thus $\lambda = \dfrac{hc}{E}$.

Solve: Using the result of the previous exercise we know that the energy of each photon is $E_{\text{photon}} = m_e c^2$. Thus we have $\lambda = \dfrac{hc}{E} = \dfrac{hc}{m_e c^2} = \dfrac{h}{m_e c}$. The correct answer is C.